From Internet of Things to Smart Cities

Enabling Technologies

CHAPMAN & HALL/CRC
COMPUTER and INFORMATION SCIENCE SERIES

Series Editor: Sartaj Sahni

PUBLISHED TITLES

ADVERSARIAL REASONING: COMPUTATIONAL APPROACHES TO READING THE OPPONENT'S MIND
Alexander Kott and William M. McEneaney

COMPUTER-AIDED GRAPHING AND SIMULATION TOOLS FOR AUTOCAD USERS
P. A. Simionescu

DELAUNAY MESH GENERATION
Siu-Wing Cheng, Tamal Krishna Dey, and Jonathan Richard Shewchuk

DISTRIBUTED SENSOR NETWORKS, SECOND EDITION
S. Sitharama Iyengar and Richard R. Brooks

DISTRIBUTED SYSTEMS: AN ALGORITHMIC APPROACH, SECOND EDITION
Sukumar Ghosh

ENERGY-AWARE MEMORY MANAGEMENT FOR EMBEDDED MULTIMEDIA SYSTEMS:
A COMPUTER-AIDED DESIGN APPROACH
Florin Balasa and Dhiraj K. Pradhan

ENERGY EFFICIENT HARDWARE-SOFTWARE CO-SYNTHESIS USING RECONFIGURABLE HARDWARE
Jingzhao Ou and Viktor K. Prasanna

FROM ACTION SYSTEMS TO DISTRIBUTED SYSTEMS: THE REFINEMENT APPROACH
Luigia Petre and Emil Sekerinski

FROM INTERNET OF THINGS TO SMART CITIES: ENABLING TECHNOLOGIES
Hongjian Sun, Chao Wang, and Bashar I. Ahmad

FUNDAMENTALS OF NATURAL COMPUTING: BASIC CONCEPTS, ALGORITHMS, AND APPLICATIONS
Leandro Nunes de Castro

HANDBOOK OF ALGORITHMS FOR WIRELESS NETWORKING AND MOBILE COMPUTING
Azzedine Boukerche

HANDBOOK OF APPROXIMATION ALGORITHMS AND METAHEURISTICS
Teofilo F. Gonzalez

HANDBOOK OF BIOINSPIRED ALGORITHMS AND APPLICATIONS
Stephan Olariu and Albert Y. Zomaya

HANDBOOK OF COMPUTATIONAL MOLECULAR BIOLOGY
Srinivas Aluru

HANDBOOK OF DATA STRUCTURES AND APPLICATIONS
Dinesh P. Mehta and Sartaj Sahni

PUBLISHED TITLES CONTINUED

PUBLISHED TITLES CONTINUED

From Internet of Things to Smart Cities
Enabling Technologies

Edited by
Hongjian Sun
Durham University, UK

Chao Wang
Tongji University, Shanghai, China

Bashar I. Ahmad
University of Cambridge, UK

CRC Press
Taylor & Francis Group
Boca Raton London New York

CRC Press is an imprint of the
Taylor & Francis Group, an **informa** business
A CHAPMAN & HALL BOOK

CRC Press
Taylor & Francis Group
6000 Broken Sound Parkway NW, Suite 300
Boca Raton, FL 33487-2742

© 2018 by Taylor & Francis Group, LLC
CRC Press is an imprint of Taylor & Francis Group, an Informa business

No claim to original U.S. Government works

Printed in Great Britain by Ashford Colour Press Ltd.
Version Date: 20170720

International Standard Book Number-13: 978-1-4987-7378-2 (Hardback)

Visit the Taylor & Francis Web site at
http://www.taylorandfrancis.com

and the CRC Press Web site at
http://www.crcpress.com

This book is dedicated to my parents Mr Shan Sun and Ms Lijun Chen, with love (actually they cannot read English).

This book is dedicated to my wife Jing, and my children Jessica Sun and William Sun, with love, too (actually my children cannot read books yet, just sometimes pretend to be reading stories).

But let's see how fast they can learn English and when they are able to understand the technologies in this book.

Hongjian Sun
Durham, United Kingdom
27 October 2016

This book is dedicated to my family and especially my baby daughter, Xiaoyun.

I would also like to acknowledge the support of the National Natural Science Foundation of China (Grants 61401314 and 61331009), and the EU FP7 QUICK project (PIRSES-GA-2013-612652).

Chao Wang

Contents

YUE CAO, DE MI, TONG WANG, and LEI ZHANG

CHAPTER   4 ▪ Routing Protocol for Low Power and Lossy IoT Networks   89

XIYUAN LIU, ZHENGGUO SHENG, and CHANGCHUAN YIN

SECTION III Towards Smart World from Interfaces to Homes to Cities

SALIM HANIFF, MARKKU TURUNEN, and ROOPE RAISAMO

CHAPTER 13 ■ Resources and Practical Factors in Smart Home and City   379

BO TAN, LILI TAO, and NI ZHU

Preface

Nowadays we are facing numerous environmental, technological, societal and economic challenges, including climate change, limited energy resources, aging populations and economic restructuring. To address these challenges, it becomes of vital importance to deploy information and communication technologies (ICT) throughout our homes and cities for enabling real-time responses to these challenges, for example, to reduce carbon emission, to improve resource utilization efficiency and to promote active engagement of citizens. However, such an ambitious target will require sustained efforts from the Societies of Communications, Signal Processing and Computing, etc., over the years to come.

This book aims to facilitate this sustained effort for introducing the latest ICT enabling technologies and for promoting international collaborations across Societies and sectors, and eventually demonstrating them to the general public. As such, this book consists of three tightly coupled parts:

- Part I: From Machine-to-Machine Communications to Internet of Things. We will introduce the evolvement of enabling technologies from basic machine-to-machine communications to Internet of Things technologies;

- Part II: Data Era: Data Analytic and Security. We will focus on the state-of-the-art data analytic and security techniques;

- Part III: Towards Smart World from Interfaces to Homes to Cities. We will discuss the design of human-machine interface that facilitates the integration of humans to smart homes and cities, either as decision–makers or as knowledge feeders to networks.

DISCLAIMER

We sincerely thank all contributors, who are world-leading experts in their research fields, for contributing chapters to this book. Due to the time limit, there might be errors or typos in this book, but as editors we tried our best to correct them. If you found any errors or typos, please feel free to contact one of our editors.

Editors:
Hongjian Sun, Email: hongjian.sun@durham.ac.uk;
Chao Wang, Email: chaowang@tongji.edu.cn;
Bashar Ahmad, Email: bia23@cam.ac.uk.

I

From Machine-to-Machine Communications to Internet of Things

From Machine-to-Machine Communications to Internet of Things: Enabling Communication Technologies

Hamidreza Shariatmadari

Aalto University

Sassan Iraji

Aalto University, Klick Technologies Ltd.

Riku Jäntti

Aalto University

CONTENTS

1.1 INTRODUCTION

The term Internet of Things (IoT) was first coined by Kevin Ashton in the late 90s, but in reality, IoT goes back a lot further than that. At the beginning it was only a vision but today with the development of communication technologies, and in particular wireless communications, it is becoming rapidly a reality in different domains and sectors [1]. IoT and its enabling services are revolutionizing the way we live and creating huge growth in our economy. According to IoT paradigm, everything and everyone can be part of the Internet. This vision redefines the way people interact with each other and objects they are surrounded by. With the development of communication technologies, billions of IoT devices are currently connected and it is expected there will be a few tens of billions connected devices within the next five years [2]. At the core of IoT, machine-to-machine (M2M) communications or machine-type communications (MTC) plays the fundamental role by providing connectivity between devices and servers. In that regard, it is concluded that IoT is a broader term, which contains devices, M2M communications, and data processing [3]. Thus, IoT vision cannot be materialized without providing efficient M2M communications.

IoT is constantly evolving by envisioning new services and applications in various domains. The envisioned services are associated with a diverse range of communication requirements. This fact has resulted in the emergence of different wireless communication solutions; generally, each solution satisfies a set of requirements. Although the plethora of communication solutions gives the opportunity to choose a proper solution according to the target application, it hinders the wide and fast deployment of the application. In order to overcome such limitations, extensive efforts are being undertaken to enhance the current solutions for supporting a wider range of applications. Meanwhile, the cellular standardization forums are also trying to consider the IoT vision in the development of the next generation of cellular systems.

The goal of this chapter is to make the readers familiar with the most important connectivity solutions for enabling M2M communications and supporting IoT applications. The remainder of the chapter is organized as follows. Section 1.2 discusses some of the IoT services and applications, along with their requirements. Section 1.3 describes some of the existing M2M connectivity technologies and their potentials for supporting IoT applications. Section 1.4 presents the future directions that enable a better support of M2M communications in the cellular systems. Finally, the chapter is concluded in Section 1.5.

1.2 IoT APPLICATIONS AND THEIR REQUIREMENTS

IoT has a wide range of applications in various sectors and domains [4], [5]. The applications can improve the ways people live, provide better interactions with the environment and facilitate the growth in the economy. Some of the viable IoT applications can be categorized as follows.

- **Transportation:** There are applications for transportation systems that help in monitoring the traffic, improving the road safety, and facilitating the assisted driving. Indeed, public safety is becoming an important concern for the increasing number of vehicles on the roads.

- **Monitoring:** Monitoring applications provide the real-time environmental observations enabled by deployed sensors.

- **Wearables:** The wearable devices can be utilized in game and leisure industries, also for the purposes of fitness, wellness, and health monitoring, which are increasingly emerging domains.

- **Smart environment:** The considered applications improve the efficiency of the environmental utilization from different perspectives, such as energy, resources, and carbon footprint. Some of the applications are applicable for smart homes, building automation, and smart cities.

- **Security:** The security applications are applicable to private residential, commercial, and public locations. They can offer remote surveillance, remote alarm, personal tracking, and public infrastructure protection.

- **Utilities:** There are applications that enable remote monitoring and controlling of user consumptions, such as water, gas, and electricity utilities. Other important applications are related to smart grids, which can efficiently balance between the production and consumption of electricity in large systems.

- **Industrial Internet:** Connecting industrial plants to the Internet facilitates the process management and improves the production. These can be achieved by support of logistics, business analytics, predictive maintenance, and factory automation.

The IoT applications, including those mentioned above, have different characteristics. The application characteristics imply constraints on the devices and underlying communications, and also determine how the data should be gathered and processed [3], [6], [7]. Some concerns related to the end devices are power consumption, computational abilities, size, and cost. The parameters that entail constraints on the communications include: the minimum transmission rate, transmission reliability, the maximum tolerable latency, security, mobility, and the number of supported devices. With increasing the number of IoT devices, the procedures of data gathering and processing are becoming more complex. Relevant concerns are related to the level of data processing locally, message routing, and information extraction. The system design for IoT applications requires the joint considerations of different elements, i.e., device, communication technology, and data processing. For instance, the low power consumption for the end device cannot be achieved unless the device consumes very low energy for sensing, processing, and transmitting data.

The M2M communications, as the underlying infrastructure for IoT connectivity, faces different data traffic types and deployment scenarios compared to other human-centric communications. General applications that directly interact with human users require transmitting high amounts of data only during the active periods. In addition, a limited number of users are in active mode at a time. On the contrary, many IoT applications are involved with short amounts of data, periodically generated by a massive number of devices. Some other applications, such as mission-critical applications, require reliable data transmissions with very low-latency. Excessive coverage is essential for those devices that are deployed in remote areas or in places with limited accessibility.

The diverse range of application requirements has hindered the development of a single solution for providing M2M communications. Thus, various wireless technologies have emerged, each one addressing a set of requirements, to support IoT applications. The next section describes some of the existing connectivity solutions, their features, and the supported applications.

1.3   IoT CONNECTIVITY LANDSCAPE

IoT systems are generally comprised of devices, or objects that need to exchange data with each other, or with a central server. As shown in Figure 1.1, the interconnection of the devices and the server can be provided by a local-based network or a cloud-based network. In the local-based network, the server runs the applications and interacts with the devices through the access points (APs). The server can provide Internet access for the devices in the network to make them remotely accessible.

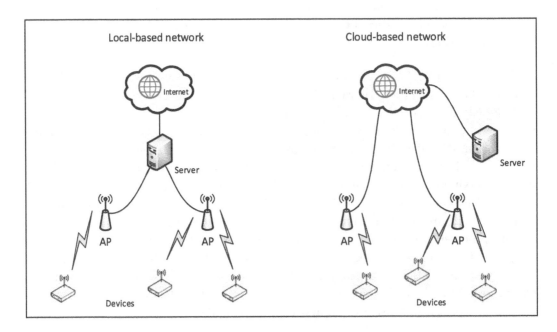

Figure 1.1   Typical network topologies for IoT.

The advantage of this topology is the high level of reliability as the applications can run even if the Internet access fails. In the cloud-based network, the devices and the server perform data exchange through the Internet, entailing all data flow is directed to the Internet. Hence, the stable Internet connection is essential for running the applications. This topology facilitates the scalability of application deployment as the devices can be deployed anywhere as long as they have the Internet access. In addition, the application can be run in a cloud server which eliminates the need of a dedicated server.

Wireless technologies are preferred to provide connectivity for IoT devices due to the ease of deployment, elimination of wiring cost, and mobility support. Initially, many manufacturers developed IoT applications using their own proprietary wireless systems. This approach has brought different technical issues, such as complexity of system design, and lack of compatibility. For instance, products from different manufacturers could not be easily integrated in a system. Those limitations could not be eliminated without employing standardized wireless technologies. In part, some standardization forums put effort to modify the existing wireless systems to support IoT applications. In addition, several wireless solutions have emerged, optimized for IoT applications. The rest of this section describes some of these technologies.

1.3.1 IEEE 802.15.4

IEEE 802.15.4 defines the physical (PHY) layer and medium access control (MAC) layer specifications for low-rate wireless personal area networks (LR-WPANs) [8]. It is designed to provide local connectivity for devices with low power consumption and over relatively short distances. The IEEE 802.15 working group released the first edition of the standard in 2003 and maintained it by providing additional amendments. The standard is based on the open system interconnection (OSI) model, without specifying the higher layers. The specifications of PHY and MAC layers enable interoperability between devices developed by different manufacturers. The upper layers can be designed and optimized according to the specific application. Indeed, various network protocols have been designed on top of IEEE 802.15.4 to support a wide range of applications, such as IPv6 over low power wireless personal area networks (6LoWPAN), ZigBee, WirelessHART, and ISA100.11a [9], [10], [11].

IEEE 802.15.4-2003 specifies the PHY layer over three different unlicensed industrial, scientific, and medical (ISM) frequency bands. The offered data rates depend on the operating frequency band, corresponding to 20 kbps in 868 MHz, 40 kbps in 915 MHz, and 250 kbps in 2.4 GHz. The higher data rates are offered in the later revisions of the standard. In most of the countries, the 2.4 GHz radio is utilized dominantly due to the higher data rate and the availability of the frequency band. In this frequency band, the total of 16 channels is available for network operation, each with 2 MHz bandwidth. The IEEE 802.15.4 transmits data with low power, i.e., less than 1 dBm. The low transmission power limits the communication range, typically less than 100 m, and also makes the data transmission susceptible to radio interference from other competing technologies operating in this frequency band, such as WiFi and Bluetooth [12].

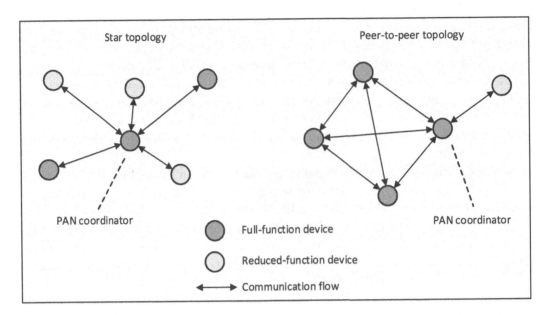

Figure 1.2 Network topologies in IEEE 802.15.4.

The MAC layer in IEEE 802.15.4 handles the access to the radio channel and provides an interface between the service specific convergence sublayer (SSCS) and PHY layer. It is also responsible for network synchronization, supporting device security, and providing a reliable link between two peer MAC entities. Two different device types are defined for the IEEE 802.15.4 network: a full-function device (FFD) that supports all network functionalities, and reduced function device (RFD) with reduced network functionalities [8], [13]. The network can utilize star or peer-to-peer topologies, as illustrated in Figure 1.2. In the star topology, devices can only exchange information with a single central controller unit in the network, known as the personal area network (PAN) coordinator. The PAN coordinator is responsible for managing the entire network. In the peer-to-peer topology, FFD devices have additional ability to communicate directly with each other.

The IEEE 802.15.4-2003 standard is complemented with amendments to further enhance the performance and offer additional functionalities that are required for some specific IoT applications. Some of these amendments are as follows.

- **IEEE 802.15.4a:** This amendment specifies two additional PHY layers, using ultra-wide band (UWB) and chirp spread spectrum (CSS) [13]. The UWB PHY enables precision ranging that can be utilized for positioning purpose. The CSS PHY provides robust data transmission over a large distance for devices moving at high speed, applicable for vehicular communications.

- **IEEE 802.15.4b:** This amendment offers the higher data rates, 200 kbps in 868 MHz and 500 kbps in 915 MHz bands.

- **IEEE 802.15.4e:** This amendment features functional improvements for the MAC layer to support industrial applications more efficiently [14]. It includes three new MAC schemes: time slotted channel hopping (TSCH), deterministic and synchronous multi-channel extension (DSME), and low latency deterministic network (LLDN). The new MAC options facilitate more reliable communications with low-latency.

- **IEEE 802.15.4q:** This amendment specifies two alternate PHY layers, based on amplitude shift keying with ternary amplitude sequence spreading (TASK) and rate switch-Gaussian frequency shift keying (RS-GFSK), to achieve ultra-low power consumption.

- **IEEE 802.15.4r:** This amendment provides PHY and MAC extensions to enable radio-based distance measurements.

- **IEEE 802.15.4s:** This amendment defines MAC functionalities that enable spectrum resource management.

- **IEEE 802.15.4t:** This amendment provides a higher data rate of 2 Mbps in 2.4 GHz band.

IEEE 802.15.4 enables a flexible integration of network stacks and protocols for supporting different applications. 6LoWPAN is a network protocol from the Internet Engineering Task Force (IETF) 6LoWPAN Working Group, which provides an efficient use of IPv6 over IEEE 802.15.4 networks [15]. The IP-based network not only allows remote access to the devices, but also facilitates the interconnection between different networks, required for building large-scale networks [10]. The normal IPv6 header is relatively large compared to the maximum IEEE 802.15.4 packet size. The header compression and fragmentation are utilized to reduce the header size, resulting in lower overhead for data transmissions, and consequently saving the energy. 6LoW-PAN supports unicast, multicast, and broadcast messaging. Hence, the transmitter can choose a specific device, or a group of devices, or all the devices in a network for message delivery. All these features have made the 6LoWPAN popular for developing IP-based networks; it has been implemented in other network stacks, such as ZigBee and ISA100.11a.

ZigBee is a network protocol built on top of IEEE 802.15.4, aimed at providing connectivity for control and monitoring applications. ZigBee alliance is an association of companies working together to develop and maintain the ZigBee standards [16]. ZigBee defines the network, security, and application framework profile layers that enable interoperability between products from different manufacturers. It supports star, mesh, and cluster-tree topologies. Empowered by these network topologies, a single network can cover a large area and accommodate over 65,000 devices [9]. The data transmissions between devices are performed with low-latency without the need for initial network synchronization. However, the reliable communication is not guaranteed as the network operates over a single channel, making the data transmissions susceptible to interference. Hence, a careful channel selection is essential for network planning [17], [18], [19]. The ZigBee alliance ensures the interoperability between devices by providing application profiles, defining message formats, and processing actions. The running applications can interact with devices in the ZigBee networks by the aid of application profiles. The specified application profiles cover building automation, health care, home automation, input device, network devices, remote control, retail services, smart energy, telecom services, and 3D sync. The plethora of profiles have made the ZigBee standards accessible for a wide range of applications.

WirelessHART and ISA100.11a are two major network standards for providing secure, reliable, and low-latency communications using IEEE 802.15.4 [11]. The reliable communications with low-latency is essential for the safe operation of mission-critical applications, such as industrial automation applications. WirelessHART was initially released in 2007 by HART Communication Foundation (HCF) as an extension to the HART Communication protocol. It is a centralized network, in which the whole network is managed by a single network manager. The interconnections of devices and the network manager can be provided by a single AP, known as a gateway, or multiple gateways in case of requiring extended coverage. WirelessHART supports mesh topology, so all the devices can cooperate for delivering data by forwarding packets to other devices in the network. This can reduce the transmission latency and improve the reliability of communications. Data transmissions are performed utilizing time slots, formed by combining time division multiple access (TDMA) with channel hopping. The duration of time slots is fixed at 10 milliseconds (ms). Time slots form a superframe that is repeated over time. The network manager allocates time slots to the devices and gateways in the network according to their traffics. In addition, the channel hopping technique is exploited that alleviates the effects of interference from other networks by switching between channels. Furthermore, the channel blacking technique can be applied. This is an optional feature in which the network administrator can exclude some channels manually from the hopping sequence. WirelessHART ensures the interoperability of the standard with previous and future releases of the HART protocol. So, the new devices can be added to the existing networks without any issue.

ISA100.11a is another standard for industrial automation applications, initially released in 2009 by the International Society of Automation (ISA). ISA100.11a is a centralized network and supports star, star-mesh, and mesh topologies. The routing capability is an optional feature for the devices in the network, enabling utilization of RFD for the end devices. The network employs the combination of TDMA and channel-hopping, with the configurable time slot duration. The channel blacking technique can be performed adaptively, enabling each device to blacklist its desired channels. The ISA100.11a provides a flexible network implementation by allowing to optimize the stack parameters. This flexibility can result in achieving a better performance compared to the WirelessHART. However, it causes interoperability issues between devices from different manufacturers. In addition, the network design is more complex and requires more configurations.

1.3.2 WiFi

IEEE 802.11 standard, known as WiFi, was initially designed for wireless local area networks (WLAN), aiming at providing high data transmission rates for a limited number of connected devices, known as stations, over short distances. The standard has undergone extensive modifications in order to boost the transmission rates. As a result, several IEEE 802.11 amendments have emerged, exceeding the transmission rates to more than 1 Gbps in the latest versions, e.g., 802.11ad and 802.11ac. The amendments define PHY layers over various frequency bands, including 2.4 GHz, 5 GHz, and 60 GHz, with different bandwidths, ranging from 20 MHz to 160 MHz. The high transmission rates have made the standard a prevailing indoor broadband wireless technology in most the countries. The recent IEEE 802.11 amendments are as follows.

- **IEEE 802.11n:** This amendment defines PHY layers operating in 2.4 and 5 GHz bands. The maximum transmission rate varies between 54 Mbps to 600 Mbps.

- **IEEE 802.11s:** This amendment supports mesh networking that can be utilized for extending the network coverage with a limited number of APs [20].

- **IEEE 802.11ac:** This amendment utilizes wider channels in 5 GHz band, which yields the transmission rate up to 1300 Mbps.

- **IEEE 802.11ad:** This amendment defines a new PHY layer in 60 GHz band, which significantly increases the transmission rate up to 7 Gbps. However, the coverage is limited as the frequency band has different propagation characteristics compared to the 2.4 and 5 GHz bands.

- **IEEE 802.11af:** This amendment defines the PHY layer operating in the white space spectrum in the frequency bands between 54 and 790 MHz. The propagation loss in this band is low which improves the communication range, reaching up to 1 km. The maximum achievable data rate is 426 Mbps.

- **IEEE 802.11ah:** This amendment defines the PHY layer operating in the sub 1 GHz (S1G) band. The modifications for PHY and MAC layers address some of the important IoT requirements.

IEEE 802.11 achieves high data rates by utilizing wide channels that are available in high frequency bands, i.e., 2.4, 5, and 60 GHz. The wide-band operation results in making the transceiver expensive and increasing the power consumption, while the coverage is limited due to the high penetration loss. In order to make the standard suitable for IoT applications, IEEE 802.11ah Task Group (TGah), also called low-power WiFi, was formed in 2010. They provide a new amendment by considering the IoT requirements, aiming at improving the transmission coverage, reducing the power consumption, and supporting a large number of stations connected to an AP [21]. The IEEE 802.11ah PHY layer is designed by down-clocking ten times the IEEE 802.11ac PHY layer. In order to achieve an extended coverage range, the IEEE 802.11ah utilizes the S1G frequency band. This band has better penetration properties that increase the coverage up to 1 km without boosting the transmission power. The PHY layer supports different channel bandwidths: 1, 2, 4, 8, and 16 MHz, offering data rates from 150 kbps up to 78 Mbps. This enables balancing the power consumption and transmission rate according to the application demand. The MAC layer adopts various enhancements, including: new compact frame formats, enhanced channel access, improved power management mechanisms, and throughput enhancements. The compact frame formats permit reducing the protocol overheads, consequently resulting in a higher throughput. The traffic indications map (TIM) and page segmentation (PS) are the channel access schemes, which reduce the time that stations need to compete for accessing the channel. The employed hierarchical association identifier (AID) supports up to 8191 stations connected to a single AP, much higher compared to the legacy IEEE 802.11 standard. The new power management modes allow the station to turn off the radio for a long period, while the AP buffers the downlink (transmission from the AP to the device) packets until the station wakes up again. This reduces the power consumption significantly in the station. By employing the mentioned features, the IEEE 802.11ah has become a suitable technology for supporting a wide range of IoT applications for monitoring, smart environment, and industrial automation.

1.3.3 Bluetooth

Bluetooth is a low cost wireless technology for establishing personal area networks (PANs) over short distances, typically less than 100 m. Since 1998, the Special Interest Group (SIG) is responsible for developing and maintaining the Bluetooth open standard. Bluetooth operates in 2.4 GHz, and employs frequency hopping spread spectrum (FHSS) to alleviate the effects of interference. Upto now, five generations of Bluetooth standard have been defined, with the following specifications.

- **Bluetooth V1.0:** The initial version that operates in basic rate (BR) mode with data rate of 1 Mbps.

- **Bluetooth V2.0:** It introduces the enhanced data rate (EDR) mode that provides data rate of 3 Mbps, in addition to support of BR mode.

- **Bluetooth V3.0:** It supports BR mode, with the optional support of EDR mode and high speed (HS) mode that provides data rate of 24 Mbps.

- **Bluetooth V4.0:** It supports BR mode, with the optional support of EDR, HS, and low energy (LE) modes.

- **Bluetooth V5.0:** SIG announced the Bluetooth 5 in June 2016. It will improve significantly the coverage range, speed, and broadcast messaging capacity.

The Bluetooth low energy (BLE), also known as Bluetooth smart, was introduced in 2010 and featured very low power consumption with enhanced transmission range, compared to previous generations [22]. These features have made the BLE suitable for

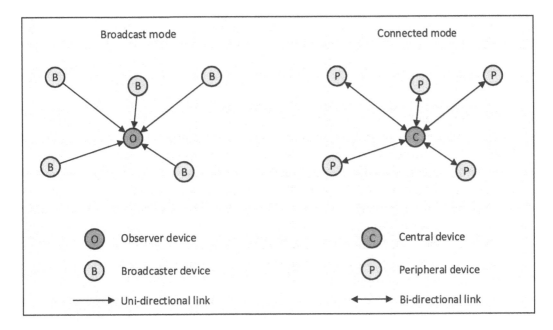

Figure 1.3 BLE communication topologies.

low-power control and monitoring applications. BLE defines 3 advertising channels and 37 data channels, each channel with 2 MHz bandwidth. It employs Gaussian frequency shift keying (GFSK) modulation with increased modulation index compared to the classic Bluetooth, to achieve robust transmission with high coverage and low power consumption [23]. The LE mode achieves data rate up to 1 Mbps. A BLE device can communicate with other devices in one of broadcast or connected modes, as shown in Figure 1.3. The broadcast mode provides unidirectional data transmissions from a broadcaster device to surrounding devices that listen to advertising channels. The data transmissions in the broadcast mode are inherently unreliable due to the lack of acknowledgment feedback. In addition, a limited amount of information can be transferred in this mode. The standard advertising packet contains 31-byte payload. In case the payload is not large enough to fit all the information, there is an optional secondary advertising payload. A scanner device can request the second advertisement frame upon receiving the initial part. Hence, the maximum of 62 bytes can be transferred. The connected mode provides bidirectional data transmissions for a group of devices, forming a piconet. A piconet consists of a central device and one or more peripheral devices. The central device is the master in the piconet and can support multiple connections. The peripheral devices are considered as slaves and are able to communicate only with the master device. In order to join the piconet, a peripheral device broadcasts connectable advertisement messages over the advertising channels. When the central device receives the advertisement message, it can initiate a connection with the peripheral device by sending the connection request. Once the connection is established, the master device and the peripheral device can exchange data over the data channels. Within a piconet, peripheral devices can only communicate with the central device, and not with each other. To reduce the power consumption, the peripheral devices stay in sleep mode by default and wake up periodically to listen for possible packet receptions from the central device. The central device provides the common clock and hopping pattern for the peripheral devices in the piconet. It also manages the medium access by using TDMA scheme and determines the instances that each peripheral device needs to wake up.

BLE includes features that enable achieving very low power consumption, particularly in peripheral devices. This is an important feature for battery-powered devices, such as sensors in health monitoring applications. Studies show that a peripheral device can operate for years while powered with a battery [23]. In addition, a peripheral device can have reduced functionalities. This reduces the design complexity of hardware, while a device can be integrated in a small system-on-chip (SOC). Currently, BLE can be utilized in applications requiring a short-range communication due to the support of single-hop topology. In 2015, Bluetooth SIG formed Bluetooth Smart Mesh Working Group to provide mesh capability for BLE that extends the network coverage [24]. The IoT vision is also considered in the development of Bluetooth v5.0. It is expected that this new release will increase the communication range, the speed of low energy connections, and the capacity of connectionless data broadcast. Employing these features will make the Bluetooth technology applicable for a wider range of applications.

1.3.4   RFID and Ambient Backscattering

Radio-frequency identification (RFID) technologies were initially designed to provide short-range connectivity for the purpose of identification, utilized in various domains such as logistic, manufacturing, health care, security, and access control. Later, a specialized subset of RFID, known as near field communication (NFC), was developed that facilitated the secure data exchange. NFC has been widely deployed in smart mobile devices for performing contactless payment, ticketing, and device pairing. Ambient backscattering is another form of technology that utilizes the ambient radio waves for data transmissions [28]. It is considered as an appealing method of communications for the future smart sensing systems.

RFID enables the unique identification of objects. The objects that are equipped with small tags can communicate with an RFID reader. A reader transmits a query signal to the tags and receives the reflected signals from them. The RFID tags are categorized as passive, semi-passive, and active. The passive tag does not have a source of energy and harvests the electromagnetic energy radiated from the reader for sending response messages. Due to the limited energy, the reading range is short and the tag can only perform very simple computation processes. The semi-passive has a limited access to the power source, mainly for powering the chip while the reader signal is absent. The chip might be connected to sensors for sensing the environment. In this way, the tag can transmit identification information along with the data collected from the sensors. The active tag has access to the source of energy, allowing the support of peripheral sensors and data transmissions over longer distances [2].

Three different frequency bands are allocated for the RFID systems, identified as low frequency (LF), high frequency (HF), and ultra high frequency (UHF) bands. The most common carriers used for LF are the 125 and 134.2 kHz. The LF operation provides low data transmissions over very short range, typically from a couple of centimeters to a couple of meters. The 13.56 MHz band is allocated for HF operation worldwide. It offers higher transmission rates and ability to read several objects at the same time. The UHF systems utilize frequency bands in the range of 300 MHz and 3 GHz, while the 860–960 MHz band is predominantly used. The UHF operation offers the higher data rates, extended range, and ability of reading a larger number of tags simultaneously. The range is limited to several meters for the passive tags [26].

NFC is a bidirectional communication technology based on ISO/IEC 14443 and ISO/IEC 18000-3 specifications [9], [27]. The former specification defines the smart cards utilized for storing information, while the latter specification determines the communication for NFC devices. The NFC devices are categorized as active and passive. A passive device, e.g., an NFC tag, can only provide information for other devices, without the ability to obtain information from others. An active device, can read a passive device and alter the stored information if it is authorized. Additionally, it can exchange data with other active devices. NFC operates in 13.56 MHz band and provides transmission rates ranging from 106 to 424 kbps. The communication range is limited to a few centimeters in order to avoid eavesdropping the data transmissions. An additional level of security is achievable by establishing a secure channel for sending sensitive information. The technology development ensures the interoperability among all NFC products, as well as with other wireless technologies. Nowadays, NFC is embedded in many smart phones to facilitate data sharing and performing payments.

Ambient backscattering takes advantage of existing RF signals, for instance from TV, WiFi, and cellular systems, to harvest energy and perform the data transmissions [25], [28]. It provides communication flexibility, as a device can communicate with all other devices, not exclusively with the readers. The transmitter reuses the ambient signals for conveying the information by changing its antenna between reflecting and non-reflecting states. The reflecting state provides an additional path for the receiver. The receiver distinguishes the transmitter state by assessing the received signals. As the transmitter does not need to have a dedicated source of energy, its size can be reduced significantly. This technology is still in the early development stage; however, it is foreseen that it can be utilized widely in future monitoring systems. For instance, a new type of sensors can be deployed in a home to monitor the environment, while they obtain the energy from the WiFi APs.

1.3.5 Dedicated Short Range Communications

Dedicated short range communications (DSRC) was developed to provide two-way communications, mainly for intelligent transportation systems (ITS) with a wide breadth of applications based on vehicular communications. The envisioned applications have the potential to improve the road safety and utilization [29]. DSRC can support various communication types for vehicular communications, including vehicle-to-vehicle (V2I) and vehicle-to-infrastructure (V2I) communications. DSRC was designed considering the stringent requirements imposed by safety applications, such as low-latency for communications, high reliability, and strict security.

DSRC benefits from other existing technologies [30]. The PHY and MAC layers are based on IEEE 802.11p wireless access for vehicular environments (WAVE), providing data transmissions with a rate of 6–27 Mbps and and a single-hop range of 300–1000 m [31]. The middle layers employ a suite of standards defined by IEEE 1609 Working Group: IEEE 1609.4 for channel switching, IEEE 1609.3 for network services, and IEEE 1609.2 for security services. DSRC supports Internet protocols for the network and transport layers.

A dedicated frequency band in 5.9 GHz is allocated for DSRC operation. Channels have 10 MHz bandwidth and are divided into control and service channels. The control channels are used for the broadcast transmissions and link establishment, while the service channels are utilized for bidirectional communications. Message broadcast is a means for disseminating safety information, including cooperative awareness message (CAM) and decentralized environmental notification message (DENM). The broadcast message is sent without establishing a basic service set (BSS), which eliminates the link establishment latency. However, there are chances for transmission collisions as the channel access is based on the carrier sense multiple access with collision avoidance (CSMA/CA) mechanism.

DSRC still suffers from technical challenges to be widely implemented in ITS. One issue is related to the performance of CSMA/CA in dense networks. The performance of the data transmissions, in terms of reliability and latency, is degraded under high traffic loads. Another concern is related to interoperability of devices. Although the communication protocols are defined, it is not clear how the system can efficiently operate with different applications and in challenging circumstances [31].

1.3.6 Low Power Wide Area Network

Low power wide area network (LPWAN) technologies were developed to provide connectivity for IoT applications that require low-cost device, wide-area coverage, low-power consumption, and exchanging small amounts of data. Such requirements were not efficiently fulfilled with the legacy cellular systems, such as Long-Term Evolution (LTE) Rel-8 and its predecessors. Hence, several preparatory technologies have emerged for enabling LPWAN, including Amber Qireless, Coronis, Huawei's CIoT, Ingenu, LoRa, M2M Spectrum Networks, Nwave, Senaptic, Sigfox, and Weightless [24].

Most of the LPWAN technologies operate in unlicensed bands while utilizing narrow bandwidths. The offered data rates vary across the technologies, ranging from several bps to several hundred kbps. Their coverage ranges can be up to several kilometers, which enable covering very large areas with a limited number of APs. These solutions can achieve low-power consumption while keeping the hardware cost low by featuring simplified functionalities. For instance, an end device may not require performing link establishment and handover in a network. It can send data without establishing a link. Data might be received by multiple APs and are delivered to the cloud-based server. The server then filters the redundant received messages. Another example is the support of single-hop communication that is simple to be implemented, without requiring precise synchronization. In some cases, the transmission latency in download is relaxed, allowing a device to remain in the sleep mode most of the time. The simplified functionalities enable a battery-powered device to operate for several years.

Despite the general similarities in the LPWAN technologies, there are differences in terms of network deployment, operational model, and device categories. The main features for some of the well known LPWAN technologies are as follows.

- **Ingenu:** The On-Ramp Wireless has changed its name to Ingenu. It aims at building a nationwide network. This can accelerate the application deployment as the IoT devices can utilize the network with subscription fee. However, currently only a limited number of countries have the network coverage. The aggregated traffic rates for an AP are limited to 624 kbps and 156 kbps in uplink and downlink, respectively.

- **LoRa and LoRaWAN:** Lora Alliance comprises of different companies participating in the development of LoRa and LoRaWAN. LoRa defines the PHY layer while LoRaWAN specifies the communication protocol and system architecture. The data transmission rates vary according to the communication range and the regional spectrum allocation, ranging from 0.3 kbps to 50 kbps. There are three device classes: one that allows downlink transmissions only during a window period after uplink data delivery; one that permits periodic downlink transmissions; and one that allows downlink transmissions at any time.

- **NWave:** The NWave supports only the uplink data transmissions with the maximum rate of 100 bps.

- **Platanus:** The protocol was designed to support ultra-dense device deployments over modest ranges. The transmission rate can be up to 500 kbps.

- **Sigfox:** This company also aims at deploying a managed worldwide network, currently covering a limited number of countries. The transmission rate is limited to 1 kpbs, with the maximum message size of 12 bytes.

- **Weightless:** It is comprised of three protocols: Weightless-W that is a bidirectional communication protocol operating over licensed TV spectrum; and Weightless-N and Weightless-P, which are narrow-band protocols utilizing unlicensed bands. The Weightless-N supports only uplink transmissions, while Weightless-P provides bidirectional data transmissions. The transmission rates are limited to 100 kbps.

Despite the appealing features of LPWAN technologies, they are faced with some technical challenges, mainly raised from the use of unlicensed bands for long range transmissions. Many countries enforce heavy regulations on utilizing the unlicensed bands in terms of effective radiated power (ERP), duty cycle, and access mechanism [1], [24]. These might limit the transmission rate, the message size, and the number of messages that can be sent over a period of time. Another challenge is related to the asymmetric link performance in uplink and downlink directions, imposed by ERP limitation that is applied at the output of the antenna. The antenna gains for the APs are significantly higher than for the devices, resulting in the better performance in the uplink direction. Consequently, the network cannot effectively control the devices as control information is delivered in the downlink with poor performance. In order to overcome these barriers, some of these technologies, such as Sigfox and Weigthless, are involved with standardization activities to obtain licensed spectrum for the network implementations. The current solutions are able to offer inexpensive connectivity for applications that tolerate the unreliable data transmissions with high latency.

1.3.7 Cellular Systems

The traditional cellular systems, including LTE Rel-8 and its predecessors, were mainly designed to serve human-to-human communications. However, their inherent features, such as wide coverage, easy deployment, access to the dedicated spectrum, and high security level, have attracted many IoT applications to exploit the cellular systems for their connectivity. The fast growth of IoT applications has encouraged the standardization forums to consider enhancements for the cellular systems in order to support IoT applications more efficiently. Consequently, the new releases of LTE encounter new features that facilitate the M2M communications.

The first release of LTE specifications, i.e., LTE Rel-8, was introduced in 2008. The LTE network architecture is fully IP-based that enables accommodating a large set of devices. As shown in Figure 1.4, IoT devices can be connected to the cellular network directly or through gateways [32]. In the direct connection, a device directly interacts with a serving base station, called eNodeB in LTE. This entails that the device be compatible with the cellular air interface. In the indirect connection, a device can benefit from other wired or wireless technologies to be connected to a gateway that acts as a mediator between the cellular network and the device. The indirect connection facilitates the migration of the existing IoT devices to the cellular networks.

LTE Rel-8 defines different user equipment (UE) categories with various performances and capabilities [33]. UE Cat-1 has the basic capabilities and provides data rates up to 5 Mbps in uplink and 10 Mbps in downlink. The achieved transmission rates in basic UE categories are satisfactory for many IoT applications. However, there are several issues that prevent the wide usage of these UE categories for IoT applications. For instance, the price of transceivers is not suitable for low-cost IoT applications. In addition, the transceivers consume a high amount of energy and cannot operate for a long period of time when powered with batteries. Another challenge is related to the network coverage. The transceivers may not operate in locations with high penetration loss, e.g., in a meter closet in which smart meters are generally deployed. In order to eliminate such shortcomings for the wide deployment of IoT applications, The Third Generation Partnership Project (3GPP) has introduced new UE categories, optimized for M2M communications. The first IoT-specific UE

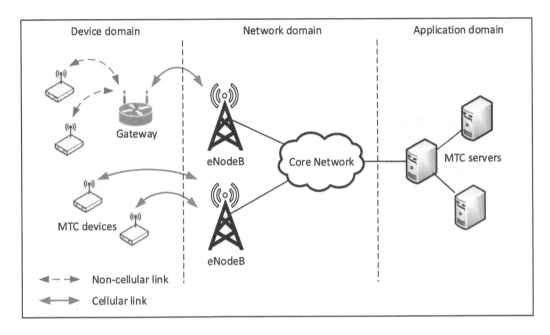

Figure 1.4 M2M communications in a cellular network.

was introduced in LTE Rel-12, known as LTE Cat-0 or LTE-M. This category was enhanced in Rel-13 and appeared as enhanced MTC (eMTC). LTE Rel-13 also introduced another category with narrow-band operation that is called Narrow-band IoT (NB-IoT). Some of the important features for these new categories are as follows.

- **LTE Cat-0:** This is a low-cost MTC UE that operates in 20 MHz bandwidth with the maximum throughput of 1 Mbps in both uplink and downlink. Rel-12 has defined a set of reduced requirements in order to scale down the chip cost and power consumption compared to the basic LTE Rel-8 category, i.e., Cat-1. For instance, Cat-0 incorporates a single antenna, while previous UE categories needed to have at least two receive antennas. This inevitably resulted in reduced coverage and transmission rate due to the loss of receiver combining gain and channel diversity. In addition, Cat-0 has an optional half-duplex operation in frequency division duplex (FDD) mode. This brings further cost reduction by removing the duplexer, as the UE needs to only transmit or receive at a time. With the mentioned reduced capabilities, Cat-0 can achieve approximately 50% cost saving over Cat-1 [34]. In order to reduce the energy consumption, Rel-12 added a power saving mode (PSM) feature to minimize the energy consumption. An UE in PSM mode is basically registered in the network but cannot be reached. It does not monitor the control channels for possible incoming data. So, the data should be buffered in the network side until the UE becomes available, for instance, when it wants to transmit something or when the PSM timer expires. Therefore, data transmission in the downlink may face huge delay, making the PSM mode more applicable for delay tolerant or opportunistic systems, which are not sensitive to delays.

- **LTE eMTC:** This category is based on the LTE Cat-0 featuring additional reduced requirements and coverage enhancements. For instance, the RF bandwidth is limited to 1.4 MHz for both uplink and downlink. The maximum of 1 Mbps data rate can be achieved in uplink and downlink. In addition, the maximum transmit power is reduced to 20 dBm compared to 23 dBm in LTE Cat-1, allowing in integrating the power amplifier (PA) and the radio transceiver in a single chip. The employed features enable to achieve approximately 75% cost saving over Cat-0, making the eMTC more cost-effective. For reducing the power consumption, more advanced discontinues reception (DRX), called enhanced DRX (eDRX), was introduced. A device in DRX mode can avoid monitoring the incoming control information, allowing to save energy by entering the idle mode. LTE Rel-13 specifies coverage enhancements to alleviate the coverage loss due to the reduced capabilities. The enhanced coverage allows to support devices located in places with high penetration loss. The coverage enhancement corresponds to 15 dB improvement for the maximum coupling loss (MCL) compared to the FDD MCL in Cat-0, i.e., 140.7 dB.

- **LTE NB-IoT:** This category was designed to enable: low-cost device, long battery life, high coverage, and deployment of a large number of devices. It inherits basic functionalities from the LTE, while it operates in a narrow band. As shown in Figure 1.5, the NB-IoT has flexible network deployment options: stand-alone, in-band, and guard-band modes. The stand-alone mode occupies a single GSM channel, i.e., 200 kHz. The in-band mode operates over a wideband LTE carrier, while guard-band mode operates out-band of the existing LTE carrier. The occupied bandwidths in both in-band and guard-band modes are equal to the bandwidth of a single LTE physical resource block (PRB), i.e., 180 kHz. The transmission rates are in the range of 100 - 200 kbps. The narrowband operation reduces the complexity of transceiver elements, such as analog-to-digital (A/D) and digital-to-analog (A/D) conversion, buffering, and channel estimation. NB-IoT has the reduced transmission power of 20 dBm. All these features provide cost reduction for the radio chip.

  Another important feature of NB-IoT is the coverage enhancement. The target for MCS is 164 dB, which is almost 23 dB more than the MCL for FDD Cat-0. This provides a good coverage, even for devices located in places with high penetration loss. Additionally, NB-IoT benefits from eDRX for reducing the power consumption, the battery for a device reporting short packets in long intervals can last for up to ten years. It is apparent that NB-IoT covers the important features of LPWAN technologies, while it provides a better performance with access to dedicated radio spectrum.

The new LTE UE categories that are optimized for M2M communications can provide connectivity for a wide range of IoT applications. The LTE Cat-0 and eMTC are suitable for devices requiring high transmission rates, while the NB-IoT is efficient for devices that require long time operation. However, the application deployments depend on the support of new LTE releases by network operators.

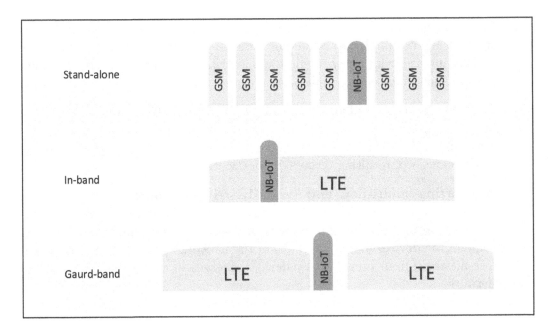

Figure 1.5 Deployment modes for NB-IoT.

1.4 CHALLENGES AND SOLUTIONS FOR CONNECTIVITY IN 5G ERA

Cellular systems are gaining more attention for accommodating IoT applications. The enhancements in the new releases of LTE provide better support of M2M communications. However, these enhancement cannot satisfy all the identified requirements for M2M communications, imposed by the envisioned IoT applications. Thus, further advancements are essential for the future cellular systems, including the fifth generation (5G) of wireless systems, to support M2M communications more efficiently. The rest of this section describes some of the existing challenges that should be addressed, along with some solutions that have been proposed to overcome them.

1.4.1 Low-power Consumption

Power consumption is an important issue for devices that are powered with batteries. In some applications, such as environmental monitoring, the devices are located in areas with limited access. To accommodate such devices in the cellular networks, the low-power consumption mode is essential, permitting a device operates for several years without needing to change the batteries. Some enhancements have been considered in LTE Rel-12 and Rel-13 to reduce the power consumption. Some of the considered enhancements and other possible solutions are as follows.

- **Narrow-band operation:** The narrow-band operation reduces the power consumption and complexity of radio transceiver. The transceiver needs to scan a narrow spectrum, which directly results in lower power consumption.

- **Power saving mode:** This mode was initially introduced in LTE Rel-12 and permits an UE device enters the dormant mode while it remains registered in the network. A device in PSM mode is not reachable immediately by the network and downlink data should be buffered until the device exits this mode. This is applicable for delay-tolerant applications as the delay in downlink data transmissions can be very high.

- **Discontinuous reception:** This feature enables an UE device stays in the sleep mode without requiring to decode incoming information form the cellular network. LTE Rel-12 introduced eDRX by supporting longer periods that the device can sleep, resulting in saving more energy [35].

- **Supporting multi-hop and group-based communications:** Multi-hop communications can bring power saving for devices located far from the serving base station [36]. Group-based scheme takes the same approach by selecting some of the devices as the gateways in the network. The selected devices collect messages from other nearby devices, aggregate, and deliver them to the network.

1.4.2 Enhanced Coverage

As mentioned earlier, some of the simplified functionalities that are defined for LTE UE categories for reducing the chip cost result in lower signal energy at the receiver. In addition, some IoT devices might be deployed in locations with high penetration loss. In order to overcome these challenges, coverage enhancement techniques can be applied. LTE Rel-12 and Rel-13 introduced some techniques to achieved this goal. Some of employed enhancements in these releases and other possible enhancements are as follows.

- **Retransmission:** Data retransmission, using automatic repeat request (ARQ) or hybrid ARQ (HARQ), can be utilized to ensure the receiver can decode the message correctly. In the ARQ scheme, the receiver tries to decode the message by utilizing received information in the last transmission round, while in the HARQ scheme, the receiver utilizes all the received information to retrieve the message.

- **Transmission time interval bundling:** Transmission time interval (TTI) is the time unit for scheduling uplink and downlink transmissions. In the TTI bundling, several consecutive TTIs are combined to transmit data over a longer period. Hence, data transmissions can be performed with a lower rate, improving the success rate of decoding data.

- **Frequency hopping:** Through the frequency hopping, data can be transmitted over different frequency bands. This scheme alleviates the effects of frequency fading and provides more robust data transmissions.

- **Power boosting and power spectral density boosting:** In the downlink, the base station can increase the transmission power for devices with poor channel conditions. In the uplink, a device can employ power spectral density (PSD) boosting by concentrating the transmission power on a decreased bandwidth, which results in higher power density over the bandwidth.

- **Relaxed requirements:** Some control channel performance requirements can be relaxed for IoT devices, such as the minimum probability of decoding the random access response.

- **Increasing reference signal density:** The number of resources allocated for reference signal (RS) can be increased to provide better channel estimations.

1.4.3 Ultra-reliable Low-latency Communications

Ultra-reliable low-latency communications (URLLC) refers to provision of a certain level of communication service almost all the time. It is essential for supporting time-critical applications, including: industrial automation, autonomous driving, vehicular safety, and tactile Internet [37], [38]. URLLC implies requirements on the availability, reliability, and latency, according to the applications. For instance, some factory automation use cases need communications with end-to-end latency less than 1 ms with reliability of $1 - 10^{-9}$ [39]. The current wireless solutions, including cellular and non-cellular technologies, cannot meet such stringent requirements. However, cellular systems have a better chance to support URLLC, due to the access to dedicated spectrum. In order to realize URLLC in the future cellular systems, extensive enhancements are required in different parts of the networks, such as device, radio access network (RAN), and core network (CN). The enhancements target at reducing the transmission latency, improving the link reliability, and enhancing the resource utilization. Some of the possible enhancements are as follows.

- **Employing shorter TTI:** LTE defines a frame structure with 10 ms duration. Each frame consist of 10 subframes, which results in having TTI of 1 ms. It is required to employ shorter TTI in order to meet 1 ms end-to-end latency. It is agreed that in 5G, an integer division of 1 ms, e.g., 0.25 or 0.125 ms, would be considered as TTI, at least for delay-sensitive communications.

- **Employing more mode changes in time-division multiplexing:** LTE have two operational mode: time-division multiplexing (TDD) and FDD. In the former mode, downlink and uplink transmissions are occurred in the same frequency band while they are separated in time domain. In the latter mode, uplink and downlink transmissions are occurred over different frequency bands. For TDD mode, different configurations are defined for uplink and downlink subframe allocations. For better support of URLLC, new configurations can be defined to reduce the time gap between uplink and downlink transmissions, e.g., changing the uplink and downlink directions every subframe.

- **Flexible frame structure:** In LTE system, the physical downlink control channel (PDCCH) is located at the beginning of each downlink subframe and it carries control information, including scheduling assignments in downlink and uplink. PDCCH can be spanned over up to three orthogonal frequency-division multiplexing (OFDM) symbols. This design enables the base stations to deliver control information to a limited number of users through each subframe. On flip side, it causes excessive delay under high demand for the data transmissions. To address this problem, the enhanced PDCCH (ePDCCH) was introduced in LTE Rel-11 to increase the signaling capacity by utilizing physical downlink shared channel (PDSCH) resources. For more dynamic resource allocation, some schemes based on the flexible frame structure have been proposed [40]. In this way, the amount of the allocated radio resources to PDCCH can be adaptively changed according to service requirements and the number of users.

- **Network slicing:** The available resources, in the RAN and CN, can be sliced in order to efficiently support different services [41]. This approach gives the opportunity to optimize each slice according to the dedicated service.

- **Device-to-device communications:** The initial design of the LTE system entails directing all data traffic from the UE devices to the cellular network. For the communication between two UE devices, the base station receives a message from the transmitter and then forwards it to the receiver. The delay in this scenario can be reduced by supporting device-to-device (D2D) communications for devices located in a close vicinity. In D2D mode, two devices can communicate directly without needing to transmit data to the cellular network. This feature is already defined in LTE Release 12, called as proximity services(ProSe), specified for public safety services. D2D is considered as a promising solution for reducing the communication delay, as well as improving the spectral efficiency.

- **Mobile edge computing:** In the scenario that a device interacts with a server located in a different place, data are passed through the CN and possibly other networks, all causing delays for exchanging information. The edge computing aims at eliminating these network delays, by running the application on a server close to the device. For instance, the server can be run in the serving base station. The scalable solution can be provided by operators allowing running the applications in the base stations.

- **Small cells:** Deploying small cells, including microcell, femtocell, and picocell, is an effective solution to improve the link quality and reduce the signal outage probability. Indeed, the outage probability, i.e., the probability that the received signal energy is less than a threshold, is an important concern as the data transmissions cannot be performed in the outage state [42]. In addition, small cells can benefit from distribute antenna systems that offer more spatial gain. The deployment of small cells and distributed antenna systems is expected to be facilitated in 5G with exploiting millimeter wave spectrum. This is due to the fact that the antenna size and cost are reduced in high frequencies.

- **Massive multiple-input and multiple-output antenna systems:** The link quality can be improved by employing massive multiple-input and multiple-output (MIMO) antenna systems, providing a large number of antennas at the base stations. Massive MIMO provides a high degree of freedom which eliminates the channel frequency dependency. This results in obtaining quite stable link quality. In addition, the resource allocation and per-coding procedures are simplified, which can further reduce the processing latency [43].

- **Multi-connectivity:** The reliability of communications can be increased by employing multi-connectivity. For instance, a device can be connected to several base stations or operates in multiple modes, e.g., cellular and D2D modes, at a time [44]. In this way, the handover disruption is also eliminated. The multi-connectivity can be also realized by using different communication technologies, e.g., connecting to the cellular and WiFi networks. In this case, the device can send the data through both networks, or switch to the backup network when the primary network fails.

- **Applying robust data transmissions:** In LTE system, data transmissions are performed using a combination of modulation and coding schemes (MCSs) that are selected from the predefined MCS set. Generally, an MCS with the lower rate has a better performance in terms of block error rate (BLER). According to the link quality, an MCS is selected that offers the highest rate while ensuring a BLER not exceeding 10%. In case of failure in decoding the block correctly, the retransmissions is performed with the same or different MCS. Data retransmission continues until the receiver can decode the block correctly or the maximum number of retransmission rounds is reached. Although the data retransmission improves the reliability and efficiency of transmission, it introduces additional delay. For URLLC, it is proposed to limit the data retransmission to only one round. This limitation would affect the overall transmission reliability. To compensate this effect, a tighter BLER target, e.g., 1%, can be utilized to choose a more robust MCS for data transmission. Another solution is using the adaptive retransmission. In this method, the initial transmission is performed with high rate, while retransmission is performed with lower rate [45].

- **Semi-persistent scheduling:** In LTE, a device sends a scheduling request (SR) message when it intends to transmit data in uplink. The base station replies with the scheduling grant (SG), indicating the allocated radio resources for the transmissions. The device can send SR at specific subframes. This process causes additional delay before the device can perform uplink transmission. This delay can be eliminated by employing semi-persistent scheduling (SPS). The base station reserves some uplink radio resources for the uplink transmissions. The device can start transmitting data without sending the SR. However, this scheme results in low resource utilization when the device has no uplink traffic.

- **Resource pooling:** This scheme also eliminates the latency for sending SR prior to uplink data transmissions. Each device is allocated with the dedicated resources periodically for uplink transmissions. In case a device needs additional radio resources, it informs the base station and will be assigned from the shared pool of resources [46].

- **Adaptive transmission and resource allocations:** Conventionally, the delay budget is equally divided between uplink and downlink when two devices communicate through the cellular network. For instance, if 1 ms is considered for the end-to-end latency, the time budget for passing the message in each link, i.e., uplink and downlink, is set to 0.5 ms. However, it is more efficient to divide the time budget adaptively, for instance according to the link qualities. In this way, a link with a better quality is given less time compared to the other link with poor quality [47].

- **Enhanced control channels:** Control channels carry important information, such as channel side information (CSI), RG, ACK/NACK, that is essential for establishing a reliable link for data transmissions. The accuracy and reliability of control information is more important for URLLC. Various enhancements are considered for control channels to improve their performances. For instance, a device reports the estimated channel quality in downlink by the means of channel quality indicator (CQI), which triggers the employed modulation and coding schemes for data transmissions. In the fading channels, there is an impairment between the reported channel quality and the actual one that results in transmission disruption. To alleviate this problem, a back off value can be utilized, to achieve a robust link adaptation [48]. Another example is related to the process of data retransmission in case of failure in decoding the data. It is observed that the accuracy of NACK signal is more important than the accuracy of the ACK signal [45]. So, some techniques, such as asymmetric signal detection, can be employed to protect the NACK signal.

1.4.4 Massive Number of Devices

It is expected that the number of IoT devices utilizing cellular systems become in order of magnitude more than the mobile devices. Surge of a massive number of devices in the cellular networks can cause different issues, mainly related to random access (RA) procedure. In addition, it is essential that the existence of IoT devices does not sacrifice the performance of the normal mobile users.

LTE defines two operation modes for devices: idle and connected. In the idle mode, a device is not connected to any base station and not granted radio resources for data transmissions. The device consumes very low energy as the radio transceiver is mostly off. The device needs to transit to the connected mode before it can communicate with the base station. This transition is initiated by performing the RA procedure, sending a randomly selected preamble over the shared physical random access channel (PRACH). The performance of PRACH is degraded significantly when a massive number of devices perform the RA procedure simultaneously. This would result in undesirable delays and waste of radio resources, in addition to increasing the power consumption in the devices [49], [50]. Some solutions that can improve the performance of RA procedure are as follows [5].

- **Dynamic PRACH resource allocation:** The performance of PRACH depends on the amount of allocated resources and the number of devices that try to perform RA procedures. The dynamic resource allocation changes the amount of resources allocated for PRACH according to the traffic load. This approach can marginally improve the performance of PRACH under very high demand for link establishment, as there is a limit on the amount of resources that can be allocated.

- **Separation of PRACH resources:** The allocated resources for PRACH can be separated for different services. This approach guarantees that the high demand for link establishment from a service would not degrade the performance of other services. For instance, mobile users or URLLC service can have dedicated radio resources for performing RA procedures.

- **Backoff scheme:** In case a device performs the RA procedure and fails to establish a link, it should wait before it can perform RA again. The backoff indicator determines how long the device should wait. Services can be assigned with different backoff indicator values in order to control their priorities in accessing the PRACH. This scheme is effective under a moderate traffic load and cannot solve the congestion problem under heavy overload situations.

- **Access class barring scheme:** This scheme enables the base station to control the access of the devices to the PRACH, while supporting different service classes. The base station broadcasts an access probability and access class barring (ACB) time. When a device intends to perform the link establishment, it picks up a random value and compares it with the access probability. If the value is less than the access probability, it performs the RA procedure; otherwise it waits for the AC barring duration. The base station can assign a small value for the access probability under excessive PRACH overload.

- **Pull-based scheme:** In this scheme, the base station indicates which devices can perform the RA procedure at a time. In this way, the base station can control the number of devices performing the RA procedure considering the traffic load and resource availability. The drawback is that additional radio resources are required for paging the devices.

1.4.5   Handling Small Bursts of Data

There are IoT applications that mainly need to exchange very short amounts of data. The current design of cellular systems, entailing establishing a link before performing data transmissions, is not efficient for such traffic type. For instance, in LTE for transmitting 100 bytes in the uplink, approximately 59 and 139 bytes of signaling are transmitted in the uplink and downlink, respectively [32]. In this case, the amount of signaling information exceeds the message size, reducing the transmission efficiency. Some solutions for handling the small bursts of data are as follows.

- **Data aggregation:** A device can aggregate messages and send all together [51]. This approach improves the efficiency of data transmissions and reduces the demand for link establishment [52]. However, it cannot be utilized for delay-sensitive applications as the data aggregation introduces the delay. Data aggregation is also applicable for gateways that connect external networks to the cellular networks. The gateway aggregates messages from several devices and sends all together [53].

- **Contention-based scheme:** Some part of radio resources can be assigned for devices needed to transmit small bursts of data in a contention-based manner. These devices can send their data without performing the link establishment [54]. This scheme is promising for a massive number of devices with sporadic traffic types.

- **Connection-less communication:** In this approach, a device attaches the message to the PRACH preamble, trying to deliver the message without establishing a link. This scheme can be utilized for transmitting very short amounts of data, as preambles can carry very short amounts of information [55].

1.5 CONCLUSIONS

This chapter summarized the main wireless technologies that enable the realization of IoT applications. As presented in Table 1.1, the connectivity solutions are currently fragmented with multiple competing standards and proprietary technologies. Four different connectivity strategies can be identified: short-range radio access to local server, short-range radio access to the Internet, low power wide area access over a license-free band, and direct cellular access, e.g., NB-IOT.

Short-range solutions can be seen as complementary technologies that provide Internet connectivity through gateways connected to cellular or fixed networks. These technologies can achieve low cost and low power consumption connectivity, having different performances in terms of communication range, the number of supported nodes, and the communication protocol. In addition, many deployed IoT applications utilize these technologies for their connectivity and are expected to operate for many years. Thus, most of these technologies will be utilized for IoT applications. It is noteworthy that interoperability is a prevalent challenge for these systems; even if two devices utilize the same radio technology, they still could use different protocol stacks that are not compatible with each other. This problem can be solved by employing gateways or protocol conversions.

Table 1.1 Comparison of connectivity technologies

Technology	Frequency band	Bandwidth (MHz)	data rate (Mbps)	Coverage	Peak Power consumption
IEEE 802.15.4-2003	868 MHz 915 MHz 2.4 GHz	0.3 0.6 2	0.02 0.04 0.25	Low	Low
IEEE 802.11 ah	Sub-1 GHz	1, 2, 4, 8, 16	0.15–78	High	Low
BLE	2.4 GHz	2	1	Low	Very low
RFID	LF, HF, UHF	Variable	Typically less than 1	Very low	Very low
DSRC	5.9 GHz	10	6–27	Very low	High
LPWAN	ISM bands	Variable	Typically less than 0.1	Very high	Very low
LTE Cat-1	LTE bands	20	Downlink: 10 Uplink: 5	High	High
LTE Cat-0	LTE bands	20	1	High	Low
LTE eMTC	LTE bands	1.4	1	High	Low
LTE NB-IoT	GSM bands LTE bands	0.18 0.2	0.1–0.2	Very high	Very low

Meanwhile, there is a serious competition between the cellular access and LPWAN solutions. NB-IoT has the advantage over the LPWAN because it is a global standard, likely to be deployed ubiquitously by the operators. The LPWAN market is more fragmented by different proprietary technologies, typically utilizes unlicensed bands and can be deployed independently by application developers. However, providing city or nationwide coverage can be expensive, exposed by site rental costs. In addition, the unlicensed spectrum regulations entail constraints on the duty cycle, channel access, and ERP. Hence, LPWAN technologies are limited with the transmission rate, maximum message size, and link performance. In these regards, the NB-IoT might eliminate LPWAN technologies when it becomes available widely.

The enhancements in the new releases of LTE have addressed some of the IoT requirements, including low-cost device, low-power consumption, and extended coverage. However, there are still some challenges which should be considered in the future cellular systems. These challenges are related to enabling URLLC, supporting massive number of devices, and handling small bursts of data. Considering these challenges, the future cellular systems can accommodate a wider range of IoT applications.

Bibliography

[1] S. Andreev, O. Galinina, A. Pyattaev, M. Gerasimenko, T. Tirronen, J. Torsner, J. Sachs, M. Dohler, and Y. Koucheryavy. Understanding the IoT connectivity landscape: A contemporary M2M radio technology roadmap. *IEEE Communications Magazine*, 53(9):32–40, September 2015.

[2] A. Al-Fuqaha, M. Guizani, M. Mohammadi, M. Aledhari, and M. Ayyash. Internet of things: A survey on enabling technologies, protocols, and applications. *IEEE Communications Surveys Tutorials*, 17(4):2347–2376, 2015.

[3] J. Gubbi, R. Buyya, S. Marusic, and M. Palaniswami. Internet of things (IoT): A vision, architectural elements, and future directions. *Future Generation Computer Systems*, 29(7):1645 – 1660, 2013.

[4] 3GPP TR 22.891. Technical Specification Group Services and System Aspects; Feasibility Study on New Services and Markets Technology Enablers; Stage 1 (Release 14), June 2016.

[5] F. Ghavimi and H. H. Chen. M2M Communications in 3GPP LTE/LTE-A networks: Architectures, service requirements, challenges, and applications. *IEEE Communications Surveys Tutorials*, 17(2):525–549, 2015.

[6] J. Kim, J. Lee, J. Kim, and J. Yun. M2M service platforms: Survey, issues, and enabling technologies. *IEEE Communications Surveys Tutorials*, 16(1):61–76, 2014.

[7] M. A. Razzaque, M. Milojevic-Jevric, A. Palade, and S. Clarke. Middleware for internet of things: A survey. *IEEE Internet of Things Journal*, 3(1):70–95, February 2016.

[8] Part 15.4: Wireless Medium Access Control (MAC) and Physical Layer (PHY) Specifications for Low-Rate Wireless Personal Area Networks (LR-WPANs), October 2003.

[9] D. Minoli. *Building the Internet of Things with IPv6 and MIPv6: The Evolving World of M2M Communications*. Wiley, 2013.

[10] G. Mulligan. The 6LoWPAN Architecture. In *Proceedings of Workshop on Embedded Networked Sensors*, pages 78–82, 2007.

[11] S. Petersen and S. Carlsen. WirelessHART Versus ISA100.11a: The format war hits the factory floor. *IEEE Industrial Electronics Magazine*, 5(4):23–34, December 2011.

[12] L. Angrisani, M. Bertocco, D. Fortin, and A. Sona. Experimental study of coexistence issues between IEEE 802.11b and IEEE 802.15.4 wireless networks. *IEEE Transactions on Instrumentation and Measurement*, 57(8):1514–1523, August 2008.

[13] E. Karapistoli, F. N. Pavlidou, I. Gragopoulos, and I. Tsetsinas. An overview of the IEEE 802.15.4a Standard. *IEEE Communications Magazine*, 48(1):47–53, January 2010.

[14] IEEE Standard for Local and Metropolitan Area Networks–Part 15.4: Low-Rate Wireless Personal Area Networks (LR-WPANs) Amendment 1: MAC Sublayer, 2012.

[15] G. Montenegro, J. Hui, D. Culler, and N. Kushalnagar. Transmission of IPv6 Packets over IEEE 802.15.4 Networks. RFC 4944, October 2015.

[16] P. Baronti, P. Pillai, V. Chook, S. Chessa, A. Gotta, and Y. Hu. Wireless sensor networks: A survey on the state of the art and the 802.15.4 and ZigBee standards. *Computer Communications*, 30(7):1655–1695, 2007.

[17] M. M. A. Hossian, A. Mahmood, and R. Jäntti. Channel ranking algorithms for cognitive coexistence of IEEE 802.15.4. In *Proceedings of IEEE Symposium on Personal, Indoor and Mobile Radio Communications*, pages 112–116, September 2009.

[18] A. Mahmood and R. Jäntti. A decision theoretic approach for channel ranking in crowded unlicensed bands. *Wireless Networks*, 17(4):907–919, 2011.

[19] H. Shariatmadari, A. Mahmood, and R. Jäntti. Channel ranking based on packet delivery ratio estimation in wireless sensor networks. In *Proceedings of IEEE Wireless Communications and Networking Conference (WCNC)*, pages 59–64, April 2013.

[20] G. R. Hiertz, D. Denteneer, S. Max, R. Taori, J. Cardona, L. Berlemann, and B. Walke. IEEE 802.11s: The WLAN mesh standard. *IEEE Wireless Communications*, 17(1):104–111, February 2010.

[21] T. Adame, A. Bel, B. Bellalta, J. Barcelo, and M. Oliver. IEEE 802.11AH: The WiFi approach for M2M communications. *IEEE Wireless Communications*, 21(6):144–152, December 2014.

[22] The Bluetooth Special Interest Group. Specification of the Bluetooth system, covered core package version: 4.2, December 2014.

[23] C. Gomez, J. Oller, and J. Paradells. Overview and evaluation of Bluetooth low energy: An emerging low-power wireless technology. *Sensors*, 12(9):11734, 2012.

[24] M. R. Palattella, M. Dohler, A. Grieco, G. Rizzo, J. Torsner, T. Engel, and L. Ladid. Internet of things in the 5G era: Enablers, architecture, and business models. *IEEE Journal on Selected Areas in Communications*, 34(3):510–527, March 2016.

[25] D. Bharadia, K. Joshi, Raj, M. Kotaru, and S. Katti. BackFi: High throughput WiFi backscatter. *SIGCOMM Comput. Commun. Rev.*, 45(4):283–296, August 2015.

[26] B. Fennani, H. Hamam, and A. O. Dahmane. RFID overview. In *Proceedings of ICM*, pages 1–5, December 2011.

[27] G. Madlmayr, J. Langer, C. Kantner, and J. Scharinger. NFC devices: Security and privacy. In *Proceedings of Conference on Availability, Reliability and Security*, pages 642–647, March 2008.

[28] V. Liu, A. Parks, V. Talla, S. Gollakota, D. Wetherall, and J. Smith. Ambient backscatter: Wireless communication out of thin air. *SIGCOMM Comput. Commun. Rev.*, 43(4):39–50, August 2013.

[29] Intelligent Transport Systems (ITS); Vehicular Communications; Basic Set of Applications; Definitions, June 2009.

[30] J. B. Kenney. Dedicated short-range communications (DSRC) standards in the United States. *Proceedings of the IEEE*, 99(7):1162–1182, July 2011.

[31] X. Wu, S. Subramanian, R. Guha, R. G. White, J. Li, K. W. Lu, A. Bucceri, and T. Zhang. Vehicular communications using DSRC: Challenges, enhancements, and evolution. *IEEE Journal on Selected Areas in Communications*, 31(9):399–408, September 2013.

[32] H. Shariatmadari, R. Ratasuk, S. Iraji, A. Laya, T. Taleb, R. Jäntti, and A. Ghosh. Machine-type communications: Current status and future perspectives toward 5G systems. *IEEE Communications Magazine*, 53(9):10–17, September 2015.

[33] S. Sesia, I. Toufik, and M. Baker. *LTE, The UMTS Long Term Evolution: From Theory to Practice*. Wiley Publishing, 2009.

[34] R. Ratasuk, A. Prasad, Z. Li, A. Ghosh, and M. A. Uusitalo. Recent advancements in M2M communications in 4G networks and evolution towards 5G. In *Proceedings of Conference on Intelligence in Next Generation Networks (ICIN)*, pages 52–57, February 2015.

[35] C. S. Bontu and E. Illidge. DRX mechanism for power saving in LTE. *IEEE Communications Magazine*, 47(6):48–55, June 2009.

[36] C. Xie, K. Chen, and X. Wang. To hop or not to hop in massive machine-to-machine communications. In *Proceedings of IEEE Wireless Communications and Networking Conference (WCNC)*, pages 1021–1026, April 2013.

[37] P. Popovski. Ultra-reliable communication in 5G wireless systems. In *Proceedings of Conference on 5G for Ubiquitous Connectivity (5GU)*, pages 146–151, November 2014.

[38] M. Simsek, A. Aijaz, M. Dohler, J. Sachs, and G. Fettweis. 5G-Enabled tactile internet. *IEEE Journal on Selected Areas in Communications*, 34(3):460–473, March 2016.

[39] B. Holfeld, D. Wieruch, T. Wirth, L. Thiele, S. A. Ashraf, J. Huschke, I. Aktas, and J. Ansari. Wireless communication for factory automation: An opportunity for LTE and 5G systems. *IEEE Communications Magazine*, 54(6):36–43, June 2016.

[40] K. I. Pedersen, G. Berardinelli, F. Frederiksen, P. Mogensen, and A. Szufarska. A flexible 5G frame structure design for frequency-division duplex cases. *IEEE Communications Magazine*, 54(3):53–59, March 2016.

[41] M. Jiang, M. Condoluci, and T. Mahmoodi. Network slicing management amp; prioritization in 5G mobile systems. In *Proceedings of European Wireless Conference European Wireless*, pages 1–6, May 2016.

[42] H. Shariatmadari, S. Iraji, and R. Jäntti. Analysis of transmission methods for ultra-reliable communications. In *Proceedings of IEEE Symposium on Personal, Indoor, and Mobile Radio Communications (PIMRC)*, pages 2303–2308, August 2015.

[43] I. Akyildiz, S. Nie, S. Lin, and M. Chandrasekaran. 5G roadmap: 10 key enabling technologies. *Computer Networks*, 106:17–48, 2016.

[44] H. Shariatmadari, R. Duan, S. Iraji, Z. Li, M. A. Uusitalo, and R. Jäntti. Analysis of transmission modes for ultra-reliable communications. In *Proceedings of IEEE Symposium on Personal, Indoor, and Mobile Radio Communications (PIMRC)*, September 2016.

[45] H. Shariatmadari, Z. Li, M. A. Uusitalo, S. Iraji, and R. Jäntti. Link adaptation design for ultra-reliable communications. In *Proceedings of IEEE International Conference on Communications (ICC)*, pages 1–5, May 2016.

[46] G. Corrales Madueño, Č. Stefanović, and P. Popovski. Reliable Reporting for Massive M2M Communications with Periodic Resource Pooling. *IEEE Wireless Communications Letters*, 3(4):429–432, August 2014.

[47] H. Shariatmadari, S. Iraji, Z. Li, M. A. Uusitalo, and R. Jäntti. Optimized transmission and resource allocation strategies for ultra-reliable communications. In *Proceedings of IEEE Symposium on Personal, Indoor, and Mobile Radio Communications (PIMRC)*, September 2016.

[48] U. Oruthota, F. Ahmed, and O. Tirkkonen. Ultra-reliable link adaptation for downlink MISO transmission in 5G cellular networks. *Information*, 7(1):14, 2016.

[49] A. Laya, L. Alonso, and J. Alonso-Zarate. Is the random access channel of LTE and LTE-A suitable for M2M communications? A Survey of Alternatives. *IEEE Communications Surveys Tutorials*, 16(1):4–16, 2014.

[50] P. Osti, P. Lassila, S. Aalto, A. Larmo, and T. Tirronen. Analysis of PDCCH performance for M2M traffic in LTE. *IEEE Transactions on Vehicular Technology*, 63(9):4357–4371, November 2014.

[51] K. Zhou and N. Nikaein. Packet aggregation for machine type communications in LTE with random access channel. In *Proceedings of IEEE Wireless Communications and Networking Conference (WCNC)*, pages 262–267, April 2013.

[52] K. Zhou and N. Nikaein. Random access with adaptive packet aggregation in LTE/LTE-A. *Journal on Wireless Communications and Networking*, 2016(1):1–15, 2016.

[53] H. Shariatmadari, P. Osti, S. Iraji, and R. Jäntti. Data aggregation in capillary networks for machine-to-machine communications. In *Proceedings of IEEE Symposium on Personal, Indoor, and Mobile Radio Communications (PIMRC)*, pages 2277–2282, August 2015.

[54] Y. Beyene, N. Malm, J. Kerttula, L. Zhou, K. Ruttik, R. Jäntti, O. Tirkkonen, and C. Bockelmann. Spectrum sharing for MTC devices in LTE. In *Proceedings of IEEE International Symposium on Dynamic Spectrum Access Networks (DySPAN)*, pages 269–270, September 2015.

[55] R. P. Jover and I. Murynets. Connection-less communication of IoT devices over LTE mobile networks. In *Proceedings of IEEE Conference on Sensing, Communication, and Networking (SECON)*, pages 247–255, June 2015.

CHAPTER 2

Power Control for Reliable M2M Communication

Ling Wang

Wayne State University

Hongwei Zhang

Wayne State University

CONTENTS

FROM industrial automation to connected and automated vehicles, machine to machine (M2M) applications pose stringent requirements for reliability and timeliness in wireless communication. For example, networked control systems for industrial automation are required to guarantee control information delivery before a preset deadline. Active vehicle-safety standards suggest message exchange intervals of 100 ms or less. Wireless communication, however, is subject to complex cyber-physical dynamics and uncertainties due to harsh environments and /or mobility. Among all wireless techniques, including MIMO, MAC scheduling, routing, congestion control, etc., which can be jointly designed to support reliability and low latency, power control is one of the most direct ways of responding to channel dynamics and guaranteeing link reliability. In this chapter, we examine M2M channel characteristics and power control approaches, with a focus on fundamental principles and representative methods. We aim to investigate the possibility of power control in applications of M2M communication systems. We also summarize the literature to illustrate research trends and challenges in the area of power control. Throughout this chapter, we emphasize channel dynamics and narrow down our discussion on enabling reliability in M2M communication systems.

2.1 INTRODUCTION

Power control has been widely used in cellular networks ranging from GSM to LTE. By adjusting the transmission power of individual links in an independent or cooperative manner, power control can be used to improve system throughput and reliability. M2M communication systems are emerging concepts and usually refer to a broad range of application systems that depend on machine to machine communication. They differ from cellular systems in terms of the network architecture and application requirements, but they have a lot in common, for instance, in channel characteristics and the co-channel interference model. As in cellular networks, power control will play an important role in M2M communication systems. Thus, we first explore the history of power control in cellular networks.

2.1.1 History of Power Control in Cellular Networks

Power control has been playing important roles in cellular networks, ranging from the 2G GSM or CDMA systems, to the 3G networks based on WCDMA or CDMA2000, and to the 4G networks based on LTE or LTE-Advanced. The cellular systems have experienced great changes ranging from user requirements to techniques. Despite those changes, power control has remained a critical mechanism for cellular networks, and power control is a technique that cannot be ignored.

The research on power control in cellular networks dates back to the 1990s when GSM systems started to be commercially developed. In order to maintain fixed voice data rate, power control was introduced in GSM systems to compensate channel changes and support overall acceptable voice quality. Around that time, power control was drawing broad attention in the research community. Of all algorithms, Foschini–Miljanic distributed power control [12] (usually denoted as DPC) is taken as a canonical power control algorithm. This work first proposed a simple and autonomous method to track average channel variation and regulate interference among users in different cells to meet certain required signal-to-interference-plus-noise-ratios (SINRs). With interference regulated, channel reuse is maximized. Many extensions [30] [23] [19] have discussed this algorithm's characteristics and generalized it to a class of algorithms. There are also many variants with special requirements in performance or settings such as base station assignment [6]. The GSM standard [31] implemented a discrete version of DPC, where each user's transmission power is altered by a fixed step-size update of 2 dB or 5 dB in extreme situations. The update frequency of transmission power is once every 480 ms, which corresponds to one update every 104 frames. Compared to cellular systems to be discussed shortly, this update rate is very low.

Power control is a mandatory component in CDMA systems. We can even say that without power control there would not have been the success of CDMA systems. In the early IS-95 system (correspondingly 2G CDMA), the received signals of all links must be equal in order to decode successfully since they are not perfectly orthogonal. Power control was introduced in all IS-95 systems to solve the well-known near-far problem and ensure insignificant intra-cell interference. The actual power control scheme in IS-95 systems has an open-loop and closed-loop component. The open-loop power control scheme (OLPC) [28] estimated the uplink power required by measuring downlink channel strength via a pilot signal. The OLPC scheme was augmented by the closed-loop power control (CLPC) scheme [13] by adding a 1-bit or 2-bits feedback considering that the uplink and downlink channel typically differ in carrier frequency and are not identical. The update rate of power control in IS-95 systems is set as 800 Hz, and the step-size is 1 dB.

In addition to voice, 3G and 4G systems support data of varying rates and aim to extend system capacity. Rather than enabling power control to support fixed SINR, power control and rate control are jointly designed to maximize system capacity. In CDMA2000 systems, on the downlink, the transmit power is fixed and the uplink, however, is not scheduled and relies on power control to achieve a required rate. As described in [3], two independent control mechanisms together determine the power control scheme of CDMA2000 systems. The first component is the basic power control scheme like CLPC, whose update rate is 600 Hz with step-size 1 dB. The second control mechanism determines the data rate of transmission. All base stations measure the interference level and set a control bit referred to as "Reverse Activity Bit." Each user adjusts their transmission rate by these control bits. The RAB-bits are fed back at the rate of 37.5 Hz. Similarly, LTE systems adjust coding and modulation schemes with the channel strength. In the meantime, fractional power control [20] is adopted in the 4G LTE system to increase the overall system throughput. It has been proved in [20] when each link only compensates a part of channel attenuation, the overall system throughput can be maximized.

From GSM to LTE, the philosophy of all power control schemes is similar to the classical DPC scheme. They try to compensate channel attenuation and mitigate co-channel interference. However, their objectives are a bit different. 3G and 4G systems aim to improve system capacity and support QoS while GSM and IS-95 would like to maintain fixed SINR. Moreover, they differ in both update rates and step-sizes. These differences not only depend on specific system architectures but also consider the overall system requirements with a tradeoff between Doppler tolerance, robustness, and spectral efficiency.

2.1.2 Objectives

Although we can borrow ideas and experiences in cellular networks to design M2M communication systems, M2M communication systems are different from cellular networks in a few respects. Firstly, most M2M communication systems are ad hoc networks. Without the support of central controllers, distributed protocol design is challenging. Secondly, M2M communication systems such as wireless sensing and control networks and vehicular networks may face much harsher network and environmental uncertainties as compared with traditional cellular networks. In supporting safety-critical, real-time applications, in the meantime, they have more stringent requirements for communication reliability and timeliness. Thirdly, different from wireless cellular networks, where system throughput is the main performance metric, packet delivery reliability in M2M networks tends to be critical. For example, industrial wireless networks [39] need to support mission-critical tasks such as industrial process control, and packet delivery is required to be reliable. At the early development stage of wireless ad hoc networks, reliable packet delivery may be able to be guaranteed due to the fact that the traffic load is low and co-channel interference can be controlled by limiting concurrent users. As wireless ad hoc networks develop with dense users, however, the co-channel interference will dominantly affect the packet delivery reliability. The emergence of vehicular networks makes the issue even more urgent [12]. The main application of vehicular networks is to support vehicle active safety. The reliable delivery of warning information between vehicles is crucial. Moreover, the broadcast of safety messages makes the traffic load high. For most wireless networks, there is a tradeoff between reliability, delay, and throughput. Reliability guarantee of high-load traffic is challenging, especially when the channel is dynamic.

For wireless communication systems, one basic task of the link layer is to address channel variation or channel fading [32]. In addition, an efficient media access control mechanism is required to support as many concurrent links as possible since high system capacity is always desirable and will finally affect the timeliness and decide if the system can work well in a dense network. Rate control, scheduling, and power control are all link-layer mechanisms. Rate control is finally reflected in coding and modulation schemes. Scheduling controls all links' media access so as to control co-channel interference. Power control is implemented to respond to channel variations by directly adjusting transmission power. However, the optimum transmission power is not simply proportional to individual link's channel attenuation due to co-channel interference. When all links adjust transmission power by their own channel attenuation, they cannot necessarily transmit successfully. The optimum transmission power is a basis of all power control related topics. Feasibility is another issue. That is, there may not be a transmission power assignment for ensuring the success of the transmissions along all the links. In M2M communication systems, power control schemes tend to be implemented in a distributed way. Thus the timescale of channel variations becomes a critical factor in power control design. Theoretically, distributed power control should converge much faster than the speed of channel variations. Otherwise, failure in tracking instantaneous channel change would result in channel outage [15].

In this chapter, we will focus on power control theory as well as representative methods. We will analyze the basic mathematical theory behind power control schemes to investigate how power control can affect and support M2M communication systems. We will also briefly discuss rate control and scheduling, but, due to the limitation of space, we will not dive into specific algorithms. In this chapter, we assume TDMA-based scheduling and constant transmission rates unless mentioned otherwise.

2.1.3 Organization

The remaining parts of the chapter will be presented as follows. First, we will describe the system architecture and channel characteristics of M2M communication systems. Then, we will examine the theoretical fundamentals of power control in terms of optimal power control and infeasibility of power control. Next, we will introduce typical power control approaches for constant and fading channels, followed by the discussion on adaptive power control approach for prospective applications in M2M communication systems. We review literature and summarize research topics and challenges in the area of power control. Finally, we will conclude this chapter with open challenges and emerging trends.

2.2 M2M COMMUNICATION SYSTEMS

Machine to machine (M2M) communication distinguishes itself from human-oriented communication, and, unlike traditional cellular networks with specific network architectures, it represents a wide range of networks. In this section, we introduce the co-channel interference model and discuss the general network architecture. We assume the ad hoc network architecture for all M2M communication system unless mentioned otherwise. Following this, we discuss the SINR model and the metrics of channel reliability. Then we analyze the origin of channel dynamics and present the statistical models. Lastly, we discuss the timescale of channel variation and the instantaneous characteristics since these metrics are so important for power control design and implementation. This section aims to demonstrate the relationships among channel dynamics, network reliability, and timeliness requirements.

2.2.1 Co-channel Interference and Network Architecture

There is no unified network architecture for M2M communication systems. Different M2M application systems may have different network architectures. For example, wireless sensor networks' architecture tends to be hierarchical, where the whole network is divided into multiple levels and all nodes in lower levels converge to higher levels and ultimately to a sink. Vehicular networks are currently designed as vehicle-to-vehicle communication networks and there are no central control nodes. But it is very likely in the future that vehicular networks will evolve into a mixed and more complicated network architecture with vehicle-to-vehicle and vehicle-to-infrastructure (or vehicle-to-cell) networks coexisting. The vehicle-to-infrastructure networks are more like cluster-based networks just as cellular networks while vehicle-to-vehicle networks are real ad hoc networks. Whatever network architecture, however, we can model the whole network or a part of the whole network as an ad hoc network if we only consider co-channel interference. Indeed, power control is originally used to manage co-channel interference. It is quite reasonable to model all M2M communication systems as ad hoc networks as far as power control is concerned.

Co-channel interference refers to interference from links operating at the same frequency. Due to the scarcity of wireless spectrum, it is impossible that all links transmit at orthogonal frequency bands. Since power control started from cellular networks and there are extensive studies in cellular networks, let's take cellular networks as an example. In cellular networks, all transmitters in a cell may be designed to ensure orthogonal transmissions. That is, there is no intra-cell interference. However, channel frequency is reused among all cells and the neighboring cells are assigned the same frequency resources. This is indeed the case for CDMA and LTE networks, where any desired downlink signal in a cell receives interference from other base stations and any desired uplink signal receives interference from other cell phones in the neighboring cells. If we only consider download links or upload links, all links can form an ad hoc network. Different from the general ad hoc network, the network nodes and links of this ad hoc network will change over time due to the burst of users entering or leaving. Compared to cellular networks, most M2M communication systems have more limited frequency resources and all links interfere with each other. Therefore, similar to cellular networks we can model all M2M communication systems as ad hoc networks.

In Figure 2.1, we show co-channel interference among links and the ad hoc network architecture. For simplicity, we only show partial interfering links in Figure 2.1. For example, link i will receive interference from all links, but we only show the interference from the nearby links such as link 3, 5, and 7. The co-channel interference is the main limiting factor in general wireless systems. Commonly, we call cellular networks as interference-limited systems. That is because modern cellular networks' performance, especially capacity, is limited by co-channel interference. The co-channel interference in the M2M communication system may be more severe than cellular networks since few M2M communication systems have powerful base stations like cellular networks to assign all frequency resources and temporal resources orthogonally. We investigate the possibility of power control in M2M communication systems for managing interference. We expect that power control can bring a bunch of benefits in terms of link reliability, energy consumption, system throughput, and end-to-end delay.

2.2.2 SINR Model and Link Reliability

Despite decades of research on interference-oriented channel access control, most existing literature are either based on the physical interference model or the protocol interference model [37]. In the protocol model, a transmission from a node S to its

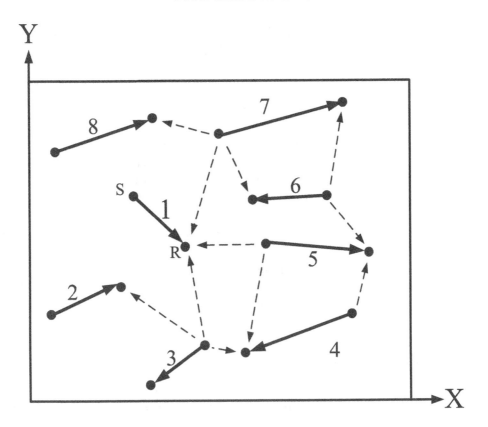

Figure 2.1 Co-channel fading model and ad hoc network architecture.

receiver R is regarded as not being interfered by a concurrent transmitter C if

$$D(C, R) \geq K \times D(S, R) \tag{2.1}$$

where $D(C, R)$ is the geographic distance between C and R, $D(S, R)$ is the geographic distance between S and R, and K is a constant number. In the physical model, a transmitter can send a packet successfully if and only if its receiver's signal-to-interference-plus-noise rate (SINR) is over a certain threshold. The SINR can be written as

$$SINR = \frac{S}{I + N} \tag{2.2}$$

where S is received signal, I is the interference, and N is thermal noise.

According to the SINR model, a set of concurrent transmissions is regarded as not interfering with one another if the SINR requirements hold for all links. The physical model is commonly known as the SINR model. The SINR model is a high-fidelity interference model in general, but interference relations defined by the physical model are non-local and combinatorial; that is because as we can see from (2.3), whether one transmission interferes with another explicitly depends on all other transmissions in the network. For the consideration of reliability, a SINR physical model is a preferred model. Throughout the whole chapter, we use the SINR model as a reliability reference model unless mentioned otherwise.

Due to the broadcast nature of the electromagnetic wave, a transmission signal decays over distance and the received signal is related to transmission power and channel attenuation. Thus we write the SINR model as

$$\frac{P_i G_{ii}}{\sum_{j \neq i} P_j G_{ij} + n_i} \geq \beta_i \qquad (2.3)$$

where P_i is the transmission power of link i; β_i is link i' required SINR threshold; n_i is the noise received by link i. G_{ii} is the path gain between link i's sender and receiver; G_{ij} is the path gain between link i's receiver and link j's sender.

In (2.3), the SINR threshold depends on the modulation scheme, bit error rate (BER) requirement, and packet size. Generally, the SINR threshold increases when any one of the transmission rate, BER requirement, and packet size goes up. The channel gain changes over time in a real system, but we can assume it as a constant or a random variable, which depends on network environment and node mobility. From (2.3), we see once the channel gains change, the SINR requirements are possibly no longer satisfied and packet loss can happen. So we introduce power control to respond to channel variation and guarantee channel reliability.

2.2.3 Channel Dynamics and Statistical Models

From the last part, we have known that channel gain variation is directly related to packet delivery reliability. In this part, we discuss in detail the origin of channel dynamics and obtain a deep understanding of channel dynamics.

A radio link in a network may suffer from signal reflection, diffraction, and scattering from surrounding objects when the signal propagates from the transmitter to its receiver. The multipath propagation and aggregation of the original wave is the main factor that results in instantaneous channel variation, usually called multipath fading [32]. Mutipath fading is generally called fading for short. When signal propagates along multiple paths, the differences in delays among different paths will cause distortion of the original sinusoidal signal in terms of amplitude and phase, and most importantly, any tiny change in these path delays can result in significant channel variation. This is why we mention fast channel variation when we mention fading. But whether the fading is fast or not depends on actual node mobility; that is, fast fading is only a relative concept compared with the system requirements.

Let us explain multipath fading with the well-known example where a receiver is moving. If the receiver moves with velocity v, there may exist two waves along two different directions, one with a frequency of $f(1 - v/c)$ and experiencing a Doppler shift $D_{min} := -fv/c$, and the other with a frequency of $f(1 + v/c)$ and experiencing a Doppler shift $D_{max} := +fv/c$. The frequency shift

$$f_m = fv/c \qquad (2.4)$$

is called the *Doppler shift*. Here, f is the carrier frequency, and $c = 3 \times 10^8$ m/s is the speed of light. *Doppler spread* is the biggest difference between the Doppler shifts. We can write

$$D_s = D_{max} - D_{min} \qquad (2.5)$$

where D_{max} is the maximum Doppler shift, and D_{min} is the minimum Doppler shift. The frequency of channel variation depends on Doppler spread. The coherence time T_c of a wireless channel is defined as the interval over which the magnitude of signal changes significantly. In [32], $Tc = \frac{1}{4D_s}$. This relation is imprecise and many people instead replace the factor of 4 by 1. Whatever, the important thing is to realize that the coherence time depends on Doppler spread and the larger the Doppler spread, the smaller the time coherence. Assume $Tc = \frac{1}{4D_s}$; if a mobile is moving at 60 km/h and the carrier frequency $f = 1800$ MHz, the Doppler shift is 100 Hz, and the coherence time is 1.25 ms.

Most of the time, we may mistakenly think that multipath fading results from transmitter or receiver's mobility. Actually, the movement of surrounding objects or other changes in propagation path can also result in fading if the propagation path delay or propagation path itself is experiencing time-varying change. That is, a stationary network can have multipath fading. The truth is just because the example of receiver mobility is easier for us to explain and to analyze multipath fading, and they also represent the characteristics of multipath fading.

Shadowing is slowly varying fading. The randomness of scatters in the environment makes channel change slowly. This is called shadowing because it is similar to the effect of clouds partly blocking sunlight [32]. The duration of shadowing lasts for multiple seconds or minutes and occurs at a much slower timescale compared to multipath fading. For convenience, we usually refer to multipath fading as fading and shadow fading as shadowing. Whether fading or shadowing, the spatial change of scatters or transmitters finally manifests itself as time diversity, and this is why a wireless channel changes over time.

Path loss is due to natural radio energy attenuation. In free space, the path loss is inversely proportional to power 2 of link length. We call the number 2 as *path loss index*. The path loss index depends on the environments. In the urban or suburban areas, path loss indexes are different. Generally, the path index of wireless networks ranges from 2.5 to 6. For analysis and by experimental results, cellular networks usually use 3.5 as the path loss index. Some experimental results can be found in [22].

There are statistical models to represent shadowing and fading. Although statistical models cannot accurately represent actual systems, thanks to these models we have the opportunities to obtain a clearer perspective and understanding of wireless communication systems. In the channel statistical models, we take each link's fading at any time t as an independent and identically distributed (i. i. d.) random variable. Shadowing is usually modeled as a random variable with log-normal distribution. Typical fading distributions are Rician fading, Rayleigh fading, and Nakagami fading [29]. When there is a line-of-sight path between transmitter and receiver, or there is a specular path between transmitter and receiver, the channel is represented by a Rician fading model. When there is not a main path component, we can think of the channel consisting of many small paths. Rayleigh fading model is the most widely used model. The Nakagami model is known to provide a closer match to some measurement data than either Rayleigh or Rician distributions [4]. The Nakagami model can be used to model the channel which is more or less severe than Rayleigh fading. The Nakagami model defines a Nakagami shape factor m. When $m = 1$, the

Nakagami distribution becomes the Rayleigh distribution, and when $m \to \infty$ the distribution approaches an impulse (no fading). The Nakagami model has been recently used in vehicular networks.

The magnitude of the received complex envelop with a Rayleigh distribution can be written as

$$p_\alpha(x) = \frac{x}{b_0}\exp\{-\frac{x^2}{2b_0}\} \tag{2.6}$$

where b_0 is variation value. The corresponding squared envelope α^2 is

$$p_{\alpha^2}(x) = \frac{1}{\Omega_p}\exp\{-\frac{x}{\Omega_p}\} \tag{2.7}$$

where $\Omega_p = 2b_0$. We can see that $p_{\alpha^2}(x)$ is an exponential distribution. This distribution is very important. We will discuss it later.

Nakagami fading describes the magnitude of the received complex envelope as

$$p_\alpha(x) = 2(\frac{m}{\Omega_p})^m \frac{x^{2m-1}}{\Gamma(m)}\exp\{-\frac{mx^2}{\Omega_p}\}, \quad m \geq 1/2. \tag{2.8}$$

where $\Gamma(m)$ is Gamma distribution. With Nakagami fading, the squared envelope has the Gamma distribution

$$p_{\alpha^2}(x) = (\frac{m}{\Omega_p})^m \frac{x^{m-1}}{\Gamma(m)}\exp\{-\frac{mx}{\Omega_p}\} \tag{2.9}$$

We plot the Nakagami pdf for comparison and analysis as in [29]. From Figure 2.2, we see that the Rayleigh distribution (i.e., when $m = 1$) covers a wide range of values while the value of Nakagami distribution is mostly around the mean value. The physical meaning here is Rayleigh channels generally have more frequent fluctuation with larger variation compared to Nakagami fading.

It is easy to confuse the envelope distribution and squared envelope distribution. The squared envelope is more important for the performance analysis of M2M communication systems because it is proportional to the received signal power and, hence, the received signal-to-interference-plus-noise ratio.

In a M2M communication system with fading, fading changes much faster than shadowing and path loss. Thus we assume that shadowing and path loss represent large-scale path gain G_{ii} and can be denoted with a constant, and h_{ii} is an i. i. d. random variable. Compared to (2.3), we add random variable h for fading. The SINR model in fading can be rewritten as

$$\frac{P_i h_{ii} G_{ii}}{\sum_{j \neq i} P_j h_{ij} G_{ij} + n_i} \geq \beta_i \tag{2.10}$$

Without fading, we may be able to find a transmission power for each link to satisfy the SINR requirements and guarantee 100% reliability. In the case of fading, it is impossible to guarantee 100% reliability since h is a random variable and can be of a very large value. Thus reliability in this case refers to outage probability or package delivery rate.

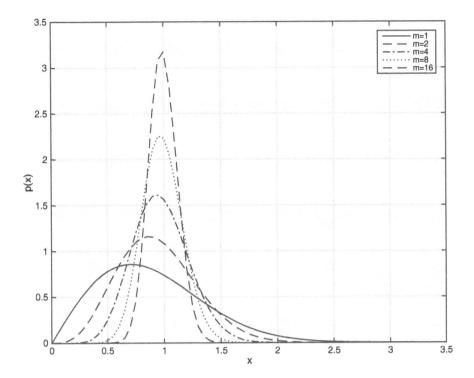

Figure 2.2 The Nakagami pdf with $\Omega_p = 1$.

2.2.4 Multiscale and Instantaneous Characteristics

Many concerns on instantaneous or short-term reliability and delay have risen in M2M communication systems. While channel variations result in the requirements for power control, the timescale of channel variations determines or limits the design and implementation of power control. In this part, we discuss the timescale and instantaneous characteristics of channel dynamics.

The SINR model demonstrates that communication reliability depends on whether wireless channels are constant or dynamic. In real-world networks, however, there are no completely constant channels. Lin et al. in [24] did extensive empirical studies to confirm that the quality of radio communication for low power sensor devices in static wireless sensor networks varies significantly over time and environment. The relative timescales of channel variation and application delay requirement determine the final channel model and power control design. Thus, we discuss the multiscale and instantaneous characteristics of wireless channels in M2M communications systems.

Two-level timescale exists in many wireless channels. The short timescale is related to fading, and the longer timescale comes from shadowing or path loss change. The multipath fading results in fast channel variation at shorter timescales while shadowing or path loss brings average channel change at longer timescales. The timescale of channel variation from shadowing or path loss is generally in the order of seconds or minutes, which is much longer than the timescale of fast variation from fading.

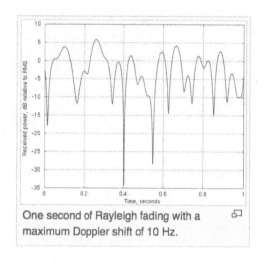
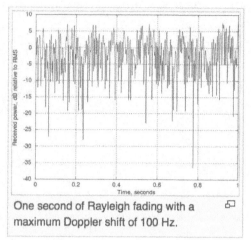

One second of Rayleigh fading with a maximum Doppler shift of 10 Hz.

One second of Rayleigh fading with a maximum Doppler shift of 100 Hz.

Figure 2.3 Instantaneous channel characteristics with fading.

Figure 2.3 shows instantaneous channel variation and demonstrates the difference between fading with different Doppler shift. From the figure, we can see that the received power with 100 Hz Doppler shift has much faster channel change than that with 10 Hz Doppler shift. These Doppler shifts correspond to velocities of about 60 km/h (40 mph) and 6 km/h (4 mph), respectively, at 1800 MHz, one of the operating frequencies for GSM mobile phones [36]. This is the classic shape of Rayleigh fading.

How do these signal variations affect the design of protocols and power control? As it is well known, modern wireless communication systems are discrete systems. Let's first transform the continuous system into discrete format. We use the block fading model to represent the continuous fading channel. As shown in Figure 2.4, we assume that the channel gain during the coherence time is constant and any two channel gains are independent although the actual channel gain $h(t)$ is correlated and changes over time. Based on the inherent multiscale channel characteristics, modern communication systems adopt multiple-level timescale design. The multiple-level timescales include symbol time, time slot duration, and frame length. The symbol time is determined by the carrier bandwidth; the selection of time slot duration and frame length depends on channel variation and system delay requirements. Take LTE system as an example. The symbol time of the LTE system is 0.0667 ms with the sub-carrier bandwidth 15 kHz; the time slot is 1 ms and the frame length is 10 ms [10]. These parameters are appropriate to meet current LTE requirements. Considering the more stringent delay requirement, however, a scheme on shorter time slot duration has been proposed in future 5G cellular systems. For vehicular networks, the shorter time slot is also required since the coherence time can be in the order of 5 ms in the case of high vehicle velocity.

In general, if the wireless channels change over a frame (or a few time slots), it is reasonable or feasible to obtain a desirable transmission power; otherwise, it may be difficult to track channel change and guarantee packet delivery rate. We will explain this in the next section by introducing power control theory. In this situation, we can

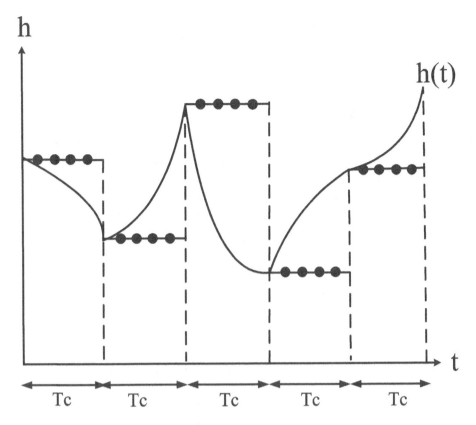

Figure 2.4 Block fading model.

only draw support from other techniques such as interleaved coding or transmission repetition to guarantee reliability. This is the limitation of power control and this fact also tells us the philosophy of wireless communication system design that only when all techniques work together can we obtain a desirable system.

2.3 POWER CONTROL THEORY

The application requirements of cellular networks drive the development of power control approaches. There exist extensive work about power control in the research community and industrial community. Power control is essentially an optimization issue. A minor change in objectives or constraints can generate different problems. However, all power control topics cannot leave the basic SINR model we mentioned in the previous section. Based on the SINR model, power control approaches are not confined in a specific type of network. In fact, many literature don't specify the network type in their power control schemes. Therefore, throughout the whole chapter, we will not specify the network type of given power control methods, and we assume that all power control approaches discussed in this chapter can be used in both cellular networks and M2M communication systems unless mentioned otherwise. In this section, we mainly discuss the feasible and optimal power control and the infeasibility of power control, combining with the mathematical models Linear Programming [8] and Mixed Integer Programming [2]. But one thing should be kept in mind: base stations in cellular networks can centrally do channel measurement and control. Thus power control in M2M communication systems with the ad hoc network architecture tends to be more challenging.

2.3.1 Feasible and optimal power control

Given a set of transmitter-receiver pairs, we would like to find a transmission power for each link to satisfy their SINR requirements. In the SINR requirement model (2.3), each link's transmission power depends on all other links' transmission power. To obtain a transmission power for each link, we can transform the SINR model in (2.3) into a matrix form and we have the transformed form

$$P \geq FP + \eta \tag{2.11}$$

and

$$F_{ij} = \begin{cases} \beta_i G_{ij}/G_{ii}, & \text{if } i \neq j \\ 0, & \text{if } i = j \end{cases} \tag{2.12}$$

and

$$\eta_i = \beta_i n_i / G_{ii} \tag{2.13}$$

where P is a vector of each link's transmission power. Each entry of F represents the normalized interference multiplied by the SINR target. The normalized interference is obtained by dividing each link's interference by its channel gain. The inequality (2.11) meets the form of Linear Programming. Therefore, we can utilize the theory of Linear Programming to get the solution of all transmission powers. According to Linear Programing, if there exist solutions for the inequality (2.11), all solutions form a cone and the vertex of the cone is the point that lets the equation condition hold. All those solutions are called feasible solutions and the vertex of the cone is usually called *fixed point* [8] by optimization convention. By solving the linear equation, we have the fixed point

$$P^* = (I - F)^{-1}\eta \tag{2.14}$$

P^* is the minimum one among all solutions, so it is the optimal solution in the perspective of power consumption. This characteristic is usually utilized to calculate the minimum power consumption in a given network.

It is theoretically easy to obtain the feasible and optimal transmission power for all links. However, it is challenging to obtain the fixed point in a distributed way. Foschini and Miljanic [12] first proposed the simple and autonomous algorithm to obtain the fixed point. The algorithm is as (2.15)

$$P_{t+1}^i = \beta_i P_t^i / r_t^i \tag{2.15}$$

where P_{t+1}^i is the transmission power of link i at time $t + 1$; P_t^i is the transmission power of link i at time t; β_i is the SINR threshold of link i; r_t^i is the actual received SINR at time t for link i. Because each link updates its current transmission power only by its previous SINR, this method is easy to implement. Foschini and Miljanic in [12] proved this algorithm can synchronously converge to the fixed point. Most of the following distributed power control are based on this algorithm.

2.3.2 Infeasibility of power control

In contrast, the Linear Programming constraint in (2.11) may have no solutions. That is, we cannot find a transmission power for each link to make sure they can transmit concurrently. We can introduce the Perron–Frobenius Theory [26] to explain the feasibility of the Linear Programming problems.

Theorem 2.1 *[26] if A is a square non-negative matrix, there exists an eigenvalue λ such that*

- *λ is real and non-negative;*

- *λ is larger or equal to any eigenvalue of A;*

- *there exists an eigenvector $x > 0$ such that $Ax = \lambda x$*

Here, λ is the largest eigenvalue of A. We take it as the *spectral radius* of A and we also call it the *Perron root* of A. Applying the Perron–Frobenius theory with the SINR model, we can find if $\lambda(F) < 1$ when $\eta \neq 0$ or $\lambda(F) \leq 1$ when $\eta = 0$, there exists feasible power assignments. The proof can be found in [26].

Once a set of links are infeasible, we need to introduce scheduling to remove a subset of links to ensure remaining links are feasible. Joint scheduling and power control is an important topic in wireless systems since a real system almost needs scheduling to remove strong interference. The objective of joint scheduling and power control is to find the active links and their feasible transmission power. We can introduce an indicator variable X_i to represent scheduling, with $X_i = 1$ meaning active and $X_i = 0$ meaning inactive. The mathematical form of joint scheduling and power control can be

$$\text{Maximize} \sum_{i=1}^{N} X_i \tag{2.16}$$

Subject to

$$\frac{P_i G_{ii} X_i}{\sum_{j \neq i} P_j G_{ij} X_j + n_i} \geq \beta_i X_i \tag{2.17}$$

This problem is a mixed integer linear programming (MILP) problem, which is known to be NP-hard. Any NP-hard problems cannot find the solution in a reasonable computation time. Thus most real-world joint scheduling and power control algorithms are approximation methods. One type of heuristic method uses the approaches of adding links one by one and testing its feasibility. These heuristic methods may be helpful for a centralized system, but it is difficult to implement them in a distributed way.

2.4 POWER CONTROL APPROACHES FOR CONSTANT AND FADING CHANNELS

Power control schemes depend on channel variations of wireless networks. In short, power control schemes are totally different in the case of fading or not. Although it is impossible in real-world wireless systems to have constant channels, static networks are usually modeled as constant channels for the purpose of analysis and it is also reasonable since the coherence time is relatively large compared to the communication time. Assuming static networks have constant channel and mobile networks such as vehicular networks have fading channels, we discuss power control approaches for constant channels and fading channels. These two approaches are applicable to static and mobile M2M communication systems, respectively.

2.4.1 Conflict Graph-based Power Control for Constant Channels

In a static network, we assume wireless channels don't change over the communication time. Given a large network with fixed channel gain, power control mainly cares about finding the maximum feasible set and their corresponding optimal transmission power. Just as we discussed in the previous section, such an issue is modeled as joint scheduling and power control, and it is theoretically NP-hard. Therefore, all power control approaches in constant channels are approximate methods. Most approximate algorithms utilize the fact that the total interference from links beyond a certain distance can be upper bounded [16]. Once the interference is upper bounded, the SINR can be guaranteed and all packets can be delivered successfully.

Given a set of links, if we use a simple path loss model, we can calculate the accumulated interference as [16]

$$I = \sum_{d_{ij} > \rho} c/d_{ij}^{\alpha} \tag{2.18}$$

where c is a constant related to path loss and transmission power. Assuming all nodes are uniformly distributed in a given area, we can obtain an upper bound of the accumulated interference from nodes within distance ρ away. Furthermore, we can calculate exact ρ by ensuring any specific SINR requirement in (2.3). Given a link, the value of ρ means all links within the distance ρ interfere with the given link. Therefore, each link has a corresponding distance beyond which other links can transmit simultaneously, and all links within which should be disabled as conflict links. We can build a conflict graph to represent the conflict relationship between any two links and then utilize this conflict graph to obtain a maximum independent set.

In a conflict graph, a circle represents a link, and all links are vertices of the graph. If two links can transmit at the same time, they are not connected in the conflict graph; otherwise, they are connected. In Figure 2.5, link 5 is in conflict with the link 1, 2, 6, and 7. When Link 5 is transmitting a packet, Link 1, 2, 6, and 7 cannot transmit at the same time with link 5.

There is a disadvantage for the conflict graph-based approach. When we build the conflict graph, we mean that a link is conflicting with all links within the distance ρ.

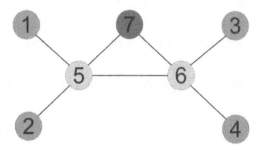

Figure 2.5 Link conflict graph [25].

This implicitly indicates that the conflict links cannot transmit concurrently because all links beyond the distance ρ are interfering. But the fact is that there is very small probability that all those links have traffic requirements at the same time in a real-world system and thus the accumulated interference is far less than the upper bounded interference. In this case, two conflict links may be allowed to transmit concurrently. The direct result is big sacrifice in concurrency. Zhang et al. [37] proposed the Physical-Ratio-K (PRK) interference model, which is similar to graph theory but defines a ratio value K to more accurately build the conflict relationship between any two links.

For static networks with constant channels, scheduling and power control is used as a means to guarantee reliability. The performance of conflict graph-based approaches depends on the accuracy of conflict links. A large guard area can bring significant degradation in concurrency while a small guard area may be unable to guarantee reliability. Whatever, all existing algorithms are time consuming. There are few applications of these algorithms in real M2M communication systems.

2.4.2 Geometric Programming-based Power Control for Fading Channels

For a real-world system, we cannot ignore fading, especially when the system is mobile. Fading is the most important factor that affects the instantaneous packet delivery rate (PDR). If fading is considered, the SINR model will be a bit different. Due to the fact that shadowing changes slowly, most mathematical models don't consider shadowing. The SINR requirement is as (2.10).

For the random channel, it is impossible to guarantee 100% packet delivery. We use packet delivery rate or outage rate to measure link reliability. For Rayleigh fading, the distribution is an exponential function as we discussed in the previous section and is easy to analyze. Most analytical models assume that channel fading follows the Rayleigh fading model. We can obtain a closed form of outage probability [21]

$$O_i = 1 - \prod_{j \neq i} \frac{1}{1 + \frac{\beta_i G_{ij} P_j}{G_{ii} P_i}} \tag{2.19}$$

where O_i is the outage probability. We can use some mathematical methods such as Laplace Transform to obtain the results, but Kandukuri and Boyd first gave the conclusion in [21]. We note that the term $\beta_i G_{ij}/G_{ii}$ is exactly the entry of channel gain matrix F. This indicates that the outage probability must be related to static or average channel characteristics. Byod et al. in [1] has proved the relationship between O_i and channel gain matrix F. Here, we define

$$O^* = \min_P \max_i O_i, \quad P_i > 0. \tag{2.20}$$

We can obtain that

$$\frac{\lambda}{1+\lambda} \le O^* \le 1 - \exp^{-\lambda} \tag{2.21}$$

where λ is the *Perron root* of F, and F is the same as defined in 2.3. There is a very interesting quantitative result here. If SINR is fixed and $\lambda(F)$ approaches 1, the maximum outage probability can be larger than 50% if we assume G_{ii} is constant. The physical meaning is that if we don't respond to fading and use fixed transmission power during the process of fading, in the worst case, some feasible links can obtain at most 50% package delivery rate. Obviously, this result is unacceptable. This is the disadvantage of power control with fixed transmission power in a fading network. Therefore, this model is usually combined with rate control. Only by adjusting transmission rate, that is, SINR threshold, can the packet delivery rate be guaranteed. The mathematical model can be

$$\text{Maximize } R_i \tag{2.22}$$

$$\text{Subject to}$$

$$O_i \ge O_{i,min} \tag{2.23}$$

$$R_i \ge R_{i,min} \tag{2.24}$$

where $O_{i,min}$ is the minimum outage requirement and $R_{i,min}$ is the minimum transmission rate requirement. We have $R_i = \log(1 + \beta_i)$ by the well-known Shannon theory [27]. This issue is difficult to solve due to the nonlinear relationship between SINR threshold and transmission rate. If β_i is much larger than 1, however, we can get $R_i = \log\beta_i$. We can use geometric programming to solve it. Due to the limitation of space, we will not introduce geometric programming. But we would like to mention that almost all joint power control and rate control issues are based on this geometric programming model. This model is complex, and it is non-convex especially in the low SINR region. Therefore current effort is focusing on efficiently converting a non-convex issue into a convex issue.

2.5 DISCUSSION ON ADAPTIVE POWER CONTROL FOR M2M COMMUNICATION SYSTEMS

In general, the conflict graph-based power control approaches and geometric programming-based power control approach are designed for two different systems. The conflict graph-based power control approach is for static networks with constant channels, and the geometric programming-based power control approach is for mobile networks with fading channels. These two approaches assume that the large-scale channel gain is constant over a long time. There are obvious drawbacks for the above two approaches: time consuming and low concurrency. Thus we suggest adaptive power control in M2M communication systems. Especially in a real-world system, the large-scale channel gain changes over time, and we have to adjust to this change even though it may be small. In this section, we will discuss the possibility of adaptive power control and its limitation in M2M communication systems.

Zhang et al. [37] proposed the PRK interference model and the corresponding adaptive scheduling algorithms [38]. The PRK model leverages some inherent characteristics of wireless networks like bounded interference and removes some unreasonable assumptions such as constant channel over time. Although power control has not been completely implemented in the PRK model, Zhang et al.'s method presents the potential application of power control in a real M2M communication system. The PRK model defined a loose conflict graph. Different from conflict graph in static networks, where conflict graph is based on a simple path loss model and is related to the transmitter-receiver pair's position, this graph does not assume the simple path loss model and the PRK model defined the conflict graph based on the ratio K of the link' instantaneous received signal to the instantaneous interference signal.

In the PRK model, a node C' is regarded as not interfering and thus can transmit concurrently with the transmission from another node S to its receiver R if and only if the following holds

$$P(C', R) < \frac{P(S, R)}{K_{S,R,Ts,R}} \tag{2.25}$$

where $P(C', R)$ and $P(S, R)$ is the average strength of signals reaching R from C' and S, respectively, and K is the minimum real number chosen such that, in the presence of cumulative interference from all concurrent transmitters, the probability for R to successfully receive packets is satisfied. Therefore, K defined the conflict graph between any links. However, PRK-based scheduling can achieve only average or long-term packet delivery rate.

We may be able to consider other potential power control methods. Noncooperative power control is another type of power control. For these power control schemes, each link's transmission power depends on their own channel gain. These algorithms have proved that they can obtain an increase in throughput for a random network. It is a potential direction for power control to use simple transmission power that is related to channel gain or received SINR to obtain an increase in throughput and reliability. Here we would like to introduce a few adaptive power control schemes including channel inversion [35], fractional power control [20], and step-by-step power control. [18].

Channel inversion sets transmission power inversely proportional to channel gain. If the channel of link i can be represented by $h_i G_{ii}$, transmission power by channel inversion is

$$P_i = \frac{1}{h_i G_{ii}} \tag{2.26}$$

Therefore, the received power equals to 1 for all links. The main purpose of this approach is to completely compensate the channel attenuation. On the other hand, the transmission power by fraction power control is

$$P_i = \frac{1}{(h_i G_{ii})^\alpha} \tag{2.27}$$

where α is the fractional number between 0 and 1. Jindal et al. in [20] proved that if h meets Rayleigh fading distribution and the network can be modeled as a Poisson network, any link can obtain the maximum package delivery rate when $\alpha = 0.5$. Fractional power control has been adopted in LTE to improve system capacity. These approaches are common in quickly responding to channel variations. They may be able to obtain long-term packet delivery rate. But obviously there is not any proof that they can guarantee short-time or instantaneous packet delivery rate since most of them only care about their own channel variation.

Another adaptive power control is to use the canonical distributed power control, which is itself an iterative power control method. Tim Holliday et al. [18] have directly applied DPC into fading networks. Applied in a fading channel rather than a constant channel, the algorithm will not converge any more. The authors also consider adjusting transmission power by a fixed step size or adaptive step size and have obtained some experimental results. Those experimental results showed that the DPC algorithm can bring great SINR variation and average SINR overshot. The fixed step-size algorithm can perform better. But the issue is how to select the appropriate step size. Moreover, no theoretical analyses have demonstrated the instantaneous SINR characteristics.

So far no power control has achieved short-term reliability in a dynamic system. One main reason is the convergence rate of power control algorithms. To achieve short-term reliability, the convergence rate of power control should be much faster than the channel variation rate. It is challenging in a dynamic system. The scheme which combines fractional power control or its variants with PRK model is under study. We believe these adaptive power control algorithms will build a basis for reliability guarantee and timeless requirement in dynamical M2M communication systems.

2.6 EXTENSIVE STUDIES ON POWER CONTROL

There is extensive research on power control. They mostly focus on performance metrics other than reliability. In this section, we give a summary of studies on power control. We expect to convey a high level idea about the research topics and challenges in the area of power control.

All power control related research started from Foschini and Miljanic's work in [12]. The simple, autonomous, and distributed power control first demonstrated how power control can work to satisfy SINR requirements. Foschini and Miljanic proved that their proposed algorithm can converge to the fixed point. Huang et al. [19] proposed the discrete version where each link updates their transmission by a fixed step and discussed its application in admission control. The convergence property of distributed power control is very important. Thus Leung et al. [23] proposed a general class of power control algorithms and proposed the conditions of convergence. They claimed that any functions that satisfy these conditions can converge. Compared with the convergence property, it is equally important that a power control algorithm can converge quickly to a fixed point or quickly detect the case of infeasibility. Huang and Yates [19] showed that Foschini–Miljanic's algorithm converges to a unique fixed point at a geometric rate. Other than power control alone, there are lots of variants regarding to combine Beamforming and BS assignment. The interested readers can see more algorithms in [6].

The study of power control in wireless ad hoc networks started in the early 2000s. Gupta et al. [14] discussed the system capacity limitation due to co-channel interference and proved that when identical randomly located nodes, each capable of transmitting at bits per second, form a wireless network, the throughput for each node can approach 0. This work told us that it is crucial to mitigate co-channel interference by optimally utilizing power control and scheduling. Ebatt and Ephremides [11] in 2004 introduced power control as a solution to the multiple access problem in contention-based wireless adhoc networks. The authors showed that the classical Foschini–Miljanic algorithm [12] in cellular networks is directly applicable to wireless ad hoc networks. Other than this, the general framework of joint scheduling and power control was first proposed. Wan et al. [33] further mathematically formulated the scheduling issue as selecting a maximum set of independent links given a set of links. The authors proved that the cumulative interference beyond a certain distance can be upper bounded. That is, we can guarantee link reliability by removing all links within a distance from the receiver. Leveraging this finding, heuristic methods are mostly used in finding the maximum independent set. These algorithms go through links in a certain order, and all the links are added to form an independent set as in [5] and [33].

Graph theory is used for solving scheduling and power control. Leveraging the finding that the cumulative interference can be bounded, a conflict graph is built to obtain the maximum independent set and any independent set. In the conflict graph, all links are the vertices of the graph. A link can connect to another link if they are far away or satisfy a certain relationship. Magnús M. Halldórsson focuses on the research of joint scheduling and power control, especially their asymptotical properties. In [16], Magnús M. Halldórsson divided all links into subsets with equal link length. Each subset is then scheduled separately through graph coloring. Halldorsson and Tonoyan [17] presented the first-approximation algorithm, which is claimed as the best among oblivious power schemes. Although all these approximation algorithms have an good asymptotical bound, their practical concurrency is very low.

In mobile networks, fading is an inevitable characteristic. Once fading is considered, the theoretical basis of power control is different, and studies on power control extend to joint power control and rate control. Kandukuri and Boyd [21] proposed optimal power control in interference-limited fading wireless channels with outage-probability specifications, where power control is updated in the timescale of shadowing rather than by fading. Chiang et al. [7] extended Kandukuri and Boyd's work and applied the method into joint power control and rate control in random wireless networks. Of all applications, one is to maximize the overall system throughput while meeting each user's minimum transmission rate constraint and outage probability constraint. The authors concluded that at the high SINR regimes the issue can be solved by geometric programming (GP) [1] and efficiently solvable for global optimality. The variants of the problem, e.g. a total power consumption constraint or objective function, can be also solved by GP. In the median or low SINR area, the issue is intractable since the Shannon equation cannot be approximated as a linear function between transmission power and transmission rate. However, the successive convex approximation method, which converges to a point satisfying the Kaurush–Kuhn–Tucker (KKT) conditions, can be a good approach as in [7]. Cruz and Santhanam [9] studied joint power control, rate control, and scheduling to minimize total average transmission power with the minimum average data rate constraints per link in a long term. Cruz and Santhanam formulated the issue as a duality problem via the Langrage Multiplier method and decomposed the whole issue into a single-slot optimization issue. Cruz and Santhanam concluded that for the optimal policy each node is either not transmitting at all or transmitting at the maximum possible peak power. As for scheduling, the authors recommended a pseudo-random number generator to select which link is activated. The author also mentioned hierarchical link scheduling and power control, where all links are partitioned into clusters. Links in one cluster are scheduled somewhat independently of links in other clusters. Each cluster is constrained to accommodate a limited number of links. The inter-cluster interference is modeled as static ambient noise. If the desired data rates on links are sufficiently low, the optimal policy activates a large number of clusters. All analyses and conclusions are based on the assumption that the achieved data rate is a linear function of SIR. In fact, this assumption hints that the SIR is high; otherwise, it is unreasonable.

Recent studies mainly care about QoS requirements, especially delay. The system is also toward M2M communication with the coexistence between cellular networks and wireless ad hoc networks. However, when these studies attempt to obtain optimum system design including scheduling, power control, and rate control, they face the curse of dimensionality. So current work mainly focus on turning intractable issues into tractable ones. The computation complexity and the ease of implementation are not the point. Wang et al. [34] considered dynamic power control in Device-to-Device Communications with delay constrained. The D2D networks are similar to wireless ad hoc networks, but the cellular networks can assist D2D networks to make centralized resource allocation. The delay is measured by the ratio of queue length to packet arrival rate. The paper simplified the scheduling process. The scheduling is controlled by a CSMA-like policy, where any two links' distance must be larger than a constant. The objective of the paper was to minimize the weighted average delay and average power consumption in the long term. The authors formulated the issue as a Markov Decision Process (MDP). In the formulation, the admitted links, the instantaneous channel gain, and the queue's size a ternary state. The action is the transmission power of each link. The transmission probability depends on each link's queue size, traffic arrival rate, and channel gain. The problem is an infinite horizon average cost MDP, which is known as a very difficult problem. The authors gave a sufficient condition for optimality by solving the equivalent Bellman equation. The authors explained that at each stage (time slot), the optimal power has to strike a balance between the current costs and the future cost because the action taken will affect the future evolution of queue size. Similarly, the authors used an approximation method and decomposed the issue into a per-stage (one time slot) power control problem. The per-stage issue is similar to the weighted sum-rate optimization subject to the power constraint. From the design, we see that the calculation of the transmit power is very complex.

2.7 OPEN CHALLENGES AND EMERGING TRENDS

There are different perspectives for Internet of Things and M2M communications. It is difficult to reach consensus on the system model and network architecture of M2M communication systems. But based on the co-channel interference model, all M2M communication systems (including cellular networks) can be modeled in ways similar to ad hoc networks. In this sense, we can potentially extend power control schemes from a cellular system to a M2M communication system. But there is a very important difference: base stations in cellular networks can centrally do channel measurement and control. Thus power control in M2M communication systems in the ad hoc network architecture tends to be more challenging.

There are extensive studies on power control in the research community. For static M2M communication systems, the joint scheduling and power control can be used to guarantee reliability. However, most scheduling-related issues are NP-hard, and there are still many open problems. For instance, how to enable distributed scheduling, power control, and rate control in the presence of non-local co-channel interference remains a major challenge. Recent work on high-fidelity and local interference models such as the PRK interference model and related scheduling methods may be leveraged in developing field-deployable solutions. For mobile M2M communication systems, although geometric programming is theoretically feasible, we can see the big degradation in concurrency and its inability in ensuring short-timescale reliability. Adaptive power control is seen as a prospective scheme for future M2M communication systems.

Although there are extensive studies in research community, few power control algorithms have been tested or used in the real-world M2M communication systems. There are reasons from technical aspects and application requirements. For most static wireless sensor networks, the network density is low and the traffic has not reached the system capacity. Without power control, the system can function well. The benefits of power control in energy efficiency and concurrency may be not enough to outweigh the communication overhead power control introduces. Technically, distributed implementation of power control is still challenging. For vehicular networks, we have seen the scenario where a large number of vehicles gather together due to traffic congestion such that they pose stringent requirements on communication reliability. Unfortunately, due to inherent challenges of reliability and channel dynamics, power control schemes for vehicular networks are still open challenges.

Power control is an important tool for optimizing network performance. However, the adoption of power control faces the tradeoff between optimization performance and overhead. Moreover, power control alone cannot guarantee communication reliability. Other mechanisms such as packet retransmission and interleaving-coding can be used to further improve the reliability and predictability of wireless communication.

Bibliography

[1] S. Boyd, S.-J. Kim, L. Vandenberghe, and A. Hassibi. A tutorial on geometric programming. *Optimization and Engineering*, 8(1):67–127, 2007.

[2] S. Boyd and L. Vandenberghe. *Convex optimization*. Cambridge University Press, 2004.

[3] S. Chakravarty, R. Pankaj, and E. Esteves. An algorithm for reverse traffic channel rate control for cdma2000 high rate packet data systems. In *Global Telecommunications Conference, 2001. GLOBECOM'01. IEEE*, volume 6, pages 3733–3737. IEEE, 2001.

[4] U. Charash. Reception through Nakagami fading multipath channels with random delays. *IEEE Transactions on Communications*, 27(4):657–670, 1979.

[5] X. Che, H. Zhang, and Xi Ju. The case for addressing the ordering effect in interference-limited wireless scheduling. *IEEE Transactions on Wireless Communications*, 13(9):5028–5042, 2014.

[6] M. Chiang, P. Hande, T. Lan, and C. W. Tan. Power control in wireless cellular networks. *Foundations and Trends in Networking*, 2(4):381–533, 2008.

[7] M. Chiang, C.-W. Tan, D. P. Palomar, D. O'Neill, and D. Julian. Power control by geometric programming. *IEEE Transactions on Wireless Communications*, 6(7):2640–2651, 2007.

[8] E.K.P. Chong and S.H. Zak. *An Introduction to Optimization*. Wiley Series in Discrete Mathematics and Optimization. Wiley, 2011.

[9] R. L. Cruz and A. V Santhanam. Optimal routing, link scheduling and power control in multihop wireless networks. In *INFOCOM 2003. Twenty-second annual joint conference of the IEEE computer and communications. IEEE Societies*, volume 1, pages 702–711. IEEE, 2003.

[10] E. Dahlman, S. Parkvall, and J. Skold. *4G: LTE/LTE-advanced for mobile broadband*. Academic Press, 2013.

[11] T. ElBatt and A. Ephremides. Joint scheduling and power control for wireless ad hoc networks. *IEEE Transactions on Wireless Communications*, 3(1):74–85, 2004.

[12] G. J. Foschini and Z. Miljanic. A simple distributed autonomous power control algorithm and its convergence. *IEEE Transactions on Vehicular Technology*, 42(4):641–646, 1993.

[13] K. S. Gilhousen, R. Padovani, and C.E. Wheatley. Method and apparatus for controlling transmission power in a CDMA cellular mobile telephone system, October 1991. US Patent 5,056,109.

[14] P. Gupta and P. R. Kumar. The capacity of wireless networks. *IEEE Transactions on Information Theory*, 46(2):388–404, 2000.

[15] M. Haenggi and R. K. Ganti. *Interference in large wireless networks*. Now Publishers Inc., 2009.

[16] M. M. Halldórsson. Wireless scheduling with power control. *ACM Transactions on Algorithms (TALG)*, 9(1):7, 2012.

[17] M. M. Halldórsson and T. Tonoyan. The price of local power control in wireless scheduling. *arXiv preprint arXiv:1502.05279*, 2015.

[18] T. Holliday, A. Goldsmith, N. Bambos, and P. Glynn. Distributed power and admission control for time-varying wireless networks. In *Proc. 2004 IEEE Global Telecommun. Conf.*, volume 2, pages 768–774, 2004.

[19] C.-Y. Huang and R. D. Yates. Rate of convergence for minimum power assignment algorithms in cellular radio systems. *Wireless Networks*, 4(3):223–231, 1998.

[20] N. Jindal, S. Weber, and J. G. Andrews. Fractional power control for decentralized wireless networks. *IEEE Transactions on Wireless Communications*, 7(12):5482–5492, 2008.

[21] S. Kandukuri and S. Boyd. Optimal power control in interference-limited fading wireless channels with outage-probability specifications. *IEEE Transactions on Wireless Communications*, 1(1):46–55, 2002.

[22] W. C.Y. Lee. *Mobile communications design fundamentals*, volume 25. John Wiley & Sons, 2010.

[23] K. K. Leung, C. W. Sung, W. S. Wong, and T.-M. Lok. Convergence theorem for a general class of power-control algorithms. *IEEE Transactions on Communications*, 52(9):1566–1574, 2004.

[24] S. Lin, J. Zhang, G. Zhou, L. Gu, J. A. Stankovic, and T. He. Atpc: Adaptive transmission power control for wireless sensor networks. In *Proceedings of the 4th International Conference on Embedded Networked Sensor Systems*, pages 223–236. ACM, 2006.

[25] X. Liu, Y. Chen, and H. Zhang. A maximal concurrency and low latency distributed scheduling protocol for wireless sensor networks. *International Journal of Distributed Sensor Networks*, 2015:153, 2015.

[26] C. D. Meyer. *Matrix analysis and applied linear algebra*, volume 2. Siam, 2000.

[27] C. E. Shannon. A mathematical theory of communication. *ACM SIGMOBILE Mobile Computing and Communications Review*, 5(1):3–55, 2001.

[28] S. Soliman, C. Wheatley, and R. Padovani. Cdma reverse link open loop power control. In *Global Telecommunications Conference, 1992. Conference Record., GLOBECOM'92. Communication for Global Users, IEEE*, pages 69–73. IEEE, 1992.

[29] G. L. Stüber. *Principles of mobile communication*. Springer Science & Business Media, 2011.

[30] C. W. Sung and W. S. Wong. A distributed fixed-step power control algorithm with quantization and active link quality protection. *IEEE Transactions on Vehicular Technology*, 48(2):553–562, 1999.

[31] ETSI TC-SMG. Radio subsystem link control, July 1996.

[32] D. Tse and P. Viswanath. *Fundamentals of wireless communication*. Cambridge University Press, 2005.

[33] P.-J. Wan, X. Jia, and F. Yao. Maximum independent set of links under physical interference model. In *International Conference on Wireless Algorithms, Systems, and Applications*, pages 169–178. Springer, 2009.

[34] W. Wang, F. Zhang, and V. K.N. Lau. Dynamic power control for delay-aware device-to-device communications. *IEEE Journal on Selected Areas in Communications*, 33(1):14–27, 2015.

[35] S. Weber, J. G. Andrews, and N. Jindal. The effect of fading, channel inversion, and threshold scheduling on ad hoc networks. *IEEE Transactions on Information Theory*, 53(11):4127–4149, 2007.

[36] Wikipedia. Rayleigh fading — Wikipedia, the free encyclopedia, 2016. [Online; accessed 13-July-2016].

[37] H. Zhang, X. Che, X. Liu, and X. Ju. Adaptive instantiation of the protocol interference model in wireless networked sensing and control. *ACM Transactions on Sensor Networks (TOSN)*, 10(2):28, 2014.

[38] H. Zhang, X. Liu, C. Li, Y. Chen, X. Che, F. Lin, L. Y. Wang, and G. Yin. Scheduling with predictable link reliability for wireless networked control. In *2015 IEEE 23rd International Symposium on Quality of Service (IWQoS)*, pages 339–348. IEEE, 2015.

[39] R. Zurawski. *The Industrial Communication Technology Handbook*. Industrial Information Technology. CRC Press, 2014.

CHAPTER 3

Enabling Geo-centric Communication Technologies in Opportunistic Networks

Yue Cao

University of Surrey, UK

De Mi

University of Surrey, UK

Tong Wang

Harbin Engineering University, PRC

Lei Zhang

University of Surrey, UK

CONTENTS

OPPORTUNISTIC NETWORKS (ONs) have been attracting a great interest from the research community. The nature of ONs that data communication naturally does not require contemporaneous end-to-end connectivity has been enabling a range of applications in Smart Cities. Although suffering from a large variation of network topology due to nodal mobility, numerous previous routing protocols proposed in ONs still make an effort on qualifying delivery potential via network topology information only. Geographic routing is an alternative, conceptually, by relying on the geographic information instead of topological information. However, this approach has not been adequately investigated in ONs. In this chapter, the research motivation and challenges for bringing geographic routing protocols in ONs are introduced. An up-to-date review on well known geographic routing protocols in ONs is provided. Finally, potential future directions leading ongoing research in this explicit topic are given.

3.1 INTRODUCTION

Internet of Things (IoT) has been receiving attention increasingly to make Smart Cities greener, safer and more efficient [44]. By connecting a large number of entities such as mobile devices, vehicles and infrastructure everywhere in Smart Cities, governments and their partners can reduce energy and water consumption, keep people moving efficiently and improve economy, safety and quality of human life.

Mobility in Smart Cities represents a fascinating complex system that involves social relationship, daily constraints and random explorations. Collection and analysis of data that capture human mobility not only help to understand their underlying patterns but also to design intelligent systems to facilitate their convenience. Such an effort would facilitate Smart Cities to reduce traffic and to develop other applications to provide human convenience. Some recent studies demonstrated that the majority of human travels occur between a limited number of places, with less frequent trips to new places outside an individual radius.

The continuously increasing number of mobile entities, e.g., mobile phones, vehicles, Unmanned Aerial Vehicles (UAVs) in Smart Cities, not only for personal communication purposes but also as a distributed network of sensors, inevitably generates massive data that challenges the traditional wireless communication infrastructure. The opportunistic communication paradigm is deemed as a suitable solution, by exploiting resources that are temporarily available when two mobile entities are encountered, thereby providing cheaper and more energy efficient alternatives using the cellular network communication and actively contributing to its offloading.

3.2 BACKGROUND

3.2.1 Opportunistic Networks (ONs)

In Opportunistic Networks (ONs) [32], there is no contemporaneous end-to-end path towards destination most of the time, due to the large variation of network topology and sparse network density. Mobile nodes are capable of communicating with each other even if the contemporaneous end-to-end connectivity is unavailable. Here, the connectivity is maintained when pairwise nodes come into the transmission ranges of each other. Each node receives a message among its current neighbors, stores this message and waits for the future encounter opportunities with other nodes to relay the message, which is known as the Store-Carry-Forward (SCF) mechanism.

Apart from ONs, there are also a few terms to characterize such types of networks, e.g., Intermittently Connected Networks (ICNs) [20] and Delay/Disruption Tolerant Networks (DTNs) [4]. In [32], the authors provide the concept of Opportunistic Networks (ONs) and interpret it is as a more flexible environment than Delay/Disruption Tolerant Networks (ONs). Thanks to the most recent tutorial [20], providing a rigorous definition of the difference between Delay/Disruption Networking (DTN) and ICNs. Indeed, these three terms are interchangeable, while ONs are considered closer to mobility driven applications in Smart Cities.

Given the examples illustrated in Figure 3.1(a) and Figure 3.1(b) where message M is relayed from node A to node C via node B, the difference between routing in Mobile Ad Hoc NETworks (MANETs) and ONs is that the former relies more on symmetric relaying of the message with a multi-hop routing behavior, thanks to the contemporaneous end-to-end connectivity (high network density and large nodal communication range), whereas the latter relies more on the mobility of mobile nodes to bring an encounter opportunity for an asymmetric routing behavior, under the assumption of intermittent connectivity (sparse network density and short communication range).

3.2.2 Applications of ONs in Smart Cities

In ONs, the communication between mobile nodes brings potential capacity to exchange data messages, and have been envisioned for a range of applications in Smart Cities, e.g., Mobile Social Networks (MSNs) [41], UnderWater Sensor Networks (UWSNs) [19], Vehicular Ad hoc NETworks (VANETs) [33] and Airborne Networks (ANs) [14].

3.2.2.1 *Vehicular Ad Hoc NETworks (VANETs)*

Connected and autonomous vehicles have been identified as enabling technologies for the next generation of driving. Such technologies propose the use of a great deal

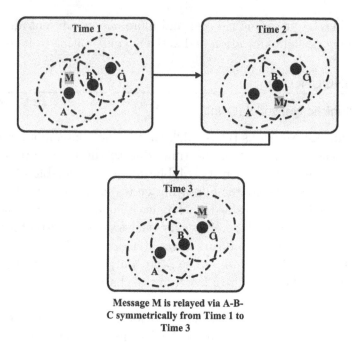

(a) Routing in MANETs

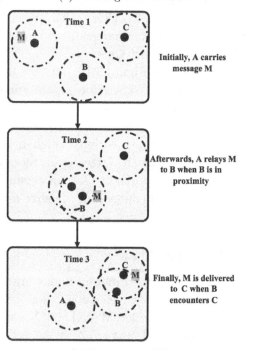

(b) Routing in ONs

Figure 3.1 Illustration of routing in MANETs and ONs

of information from vehicular sensors. A notable advantage of this approach is that it allows vehicles to effectively sense their local environment and global position, thus obtaining the information of roads and obstacles and the position of surrounding vehicles, in order to achieve safer and more efficient driving. In addition, those on-board sensors provide the connectivity over mobile networks, which are capable of keeping vehicle occupants connected to the Internet.

3.2.2.2 *Airborne Networks (ANs)*

In recent years, automotive sectors, e.g., Unmanned Aerial Vehicles (UAVs), have seen unprecedented growth. Unlike the previous major applications in military work, UAVs now attract numerous interest from the civil sectors, e.g., agriculture, geology, meteorology, etc. Due to the fact that UAV can achieve image acquisition with high imaging resolution, they are gradually replacing traditional satellite sensing systems that can be significantly affected by harsh environments. This technology enables city management and monitoring in the development of Smart Cities, in an effective and cost-efficient way. UAVs normally communicate with each other cooperatively and meet the tasks, also with the assistance of other physical infrastructures for data delivery towards destination.

3.2.2.3 *Mobile Social Networks (MSNs)*

With continued and dramatically increased interests in the civil applications alongside collaboration with adjacent sectors, MSNs have been extensively studied and identified among promising wireless mobile ad hoc networking architectures for the next generation of wireless communications. MSNs make use of both human mobility and its connectivity to relay the message between mobile users' devices. However, the applications of MSNs in the Smart Cities is still an open issue and a timely research topic. For example, considering the services from MSNs based on the GPS-enabled mobile devices, users can get access to the location information of themselves in both virtual and physical ways, which enables users to discover the local environment, nearby recommendations, and even locate their friends or families, although the users can be unaware of potential hazards in their surroundings. This leads to security and privacy problems for the MSNs' users and providers.

3.2.2.4 *UnderWater Sensor Networks (UWSNs)*

UnderWater Acoustic (UWA) communication systems differ from terrestrial telemetry due to differences in system geometry and environmental conditions. With the autonomous underwater vehicles delivering data to sink node, UWSNs are envisioned for oceanographic applications such as pollution monitoring, offshore exploration, disaster prevention, assisted navigation and tactical surveillance applications.

3.3 MOTIVATION AND CHALLENGES FOR GEOGRAPHIC ROUTING IN ONS

3.3.1 Geo-centric Technologies in Smart Cities

Cities have long made use of Geographic Information Systems (GIS) for internal purposes. As citizens start to use mobile phones to access city information, GIS has become central to delivering accurate and relevant data to citizens on the move. Likewise, Location Based Services (LBS) would be extremely beneficial and appropriate in handling the current state of economic development, urbanization, population growth, level of infrastructure and rising expectations of the people.

Recent years have witnessed the wide proliferation of geo-information applications. Urban computing is a process of acquisition, integration and analysis of big GPS data generated by a diversity of sources in urban spaces [39, 46]. To tackle the major issues that cities face, e.g. increased energy consumption and traffic congestion, Taxi GPS trajectory data [30] are employed to help us efficiently serve real-time requests sent by taxi users and generate ridesharing schedules that reduce the total travel distance significantly. Numerous studies [31, 43] mine the time-dependent and practically quickest driving route for end users using GPS-equipped taxicabs traveling in a city. Xu et al. [42] proposes a Taxi-hunting Recommendation System (Taxi-RS) processing the large-scale taxi trajectory data in order to provide passengers with a waiting time to get a taxi ride in a particular location. In addition, geographic information can be applied in localization of sensor applications in Smart Cities [36]. How to exchange and deliver such geographic information is an important issue, particularly given the opportunistic communication nature.

3.3.2 Introduction on Geographic Routing

Geographic routing, also called position based routing, requires that each node can determine its own location and that the source is aware of the location of the destination. Different from topological routing concerning the network topology, geographic routing exploits the geographic information instead of topological connectivity information for message relay, to gradually approach and eventually reach the intended destination. According to Cao and Sun [4], it can be observed that previous routing protocols in ONs, mainly, have adopted historically topological information[1] to predict the future encounter opportunity. In contrast, the focus of this chapter is to highlight the research vision and potential for applying geographic routing in ONs.

3.3.3 Motivation for Geographic Routing in ONs

Under the mobile scenario, the variation of network topology arising from nodal mobility is mainly the challenge for routing in ONs. However, it is difficult to obtain the most recent network topological information given such conditions. Up to now, numerous previous works in the literature have mainly adopted the historical

---

[1]General knowledge such as how frequent and how long a pair of nodes will meet is utilized.

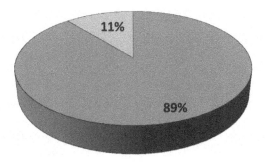

Figure 3.2 Proportion of topological and geographic routing protocols reviewed in [4]

network topology information, to predict how possible that pairwise nodes would be encountered in the the future.

In terms of research opportunities and based on data in Figure 3.2 obtained from [4], there have been 7 up-to-date geographic routing protocols reviewed compared to other numerous topological protocols in [4]. Following the guidance from previous surveys [20, 45], the reason for this lack of attention is that researchers only paid attention to the limitations of traditional topological routing protocols such as Dynamic Source Routing (DSR) [13] when it was utilized in ONs.

In terms of research feasibility [2], geographic routing inherently is without the requirement of contemporaneous end-to-end connectivity. This is because only the one-hop geographic information (e.g., distance and direction towards destination) is exploited. Therefore, geographic routing can adapt to topological variation by its geometric relaying behavior, which is more reliable for message delivery especially for the VANETs scenario. As already shown in Figure 3.3(a), it is observed that traditional geographic routing protocol, e.g., Greedy Perimeter Stateless Routing (GPSR) [18], outperforms topological protocol DSR, in terms of message delivery ratio in MANETs.

3.3.4 Challenges for Geographic Routing in ONs

Conventional geographic routing protocols designed for MANETs assume that the location of destination is always available for all nodes in networks, such that they would make individual routing decisions for message delivery towards destination. However, applying geographic routing in ONs further brings the following challenges as shown in Figure 3.4.

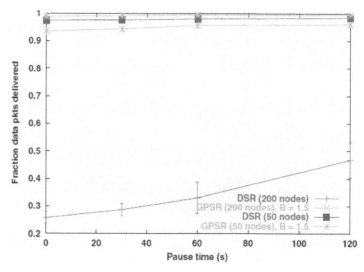

(a) Message delivery ratio in MANETs (1 node/9000 m^2 for dense networks)

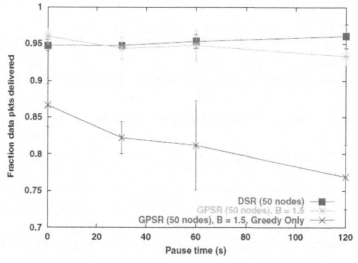

(b) Message delivery ratio in ONs (1 node/35912 m^2 for sparse networks)

Figure 3.3 Comparison results between DSR and GPSR, in [17]

- **Reliability for Message Relaying**: In MANETs, the message is greedily relayed towards destination via the continuously connected path within a short time. However in ONs, the network node which is currently closer to the destination may not be so in the future. This is so that it may not encounter others within a short time, particularly concerning the high mobility in a sparse network. It is worthy noting that in ONs (refer to Figure 3.3(b)), GPSR begins to perform worse than DSR in sparse networks.

Figure 3.4 Challenges for geographic routing in ONs

- **Locating Mobile Destination**: Concerning the mobility of destination, this challenge limits the feasibility of applying a centralized location server to track the real-time location of a mobile destination in sparse networks. This happens due to the fact that there is a long delay to request/reply the information from the location server in the ONs, while the obtained location information may be outdated and inaccurate for making a routing decision.

- **Difficulty for Handling the Local Maximum Problem**:[2] Conventional geographic routing protocols in MANETs rely on the high network density, which is infeasible in ONs because there are insufficient numbers of encountered nodes for handling this problem. Particularly, Figure 3.3(b) also shows its importance in sparse networks, where GPSR (without addressing this problem) suffers more from a performance degradation, as compared to the original GPSR.

The mobility prediction has been covered in literature, whereas the second and third challenges have not been adequately addressed. In addition to the above challenges

---

[2]This problem implies that if a better relay node is unavailable, the message carrier will keep on carrying its message. In light of this, the message delivery is delayed or even degraded if a better relay node is never met. Using distance metric as an example, any node closer to destination is generally qualified with a better delivery potential. However, a message cannot be relayed if any encountered node which is farther away from destination.

identified for geographic routing in ONs, other factors (e.g., limited buffer space, bandwidth and energy) also play important roles in general routing performance in ONs. Different from the work [4] that is a review for general routing protocols in ONs addressing "Unicasting," "Multicasting" and "Anycasting" issues,[3] this chapter focuses on exploiting geographic routing in ONs, with identified motivation, challenges, state of the art based on an original taxonomy and further potential directions.

3.4 TAXONOMY AND REVIEW OF GEOGRAPHIC ROUTING IN ONS

Existing geographic routing protocols in ONs are classified into three classes, depending on the awareness of destination. Following the taxonomy in Figure 3.5, the details of each class are discussed in this section.

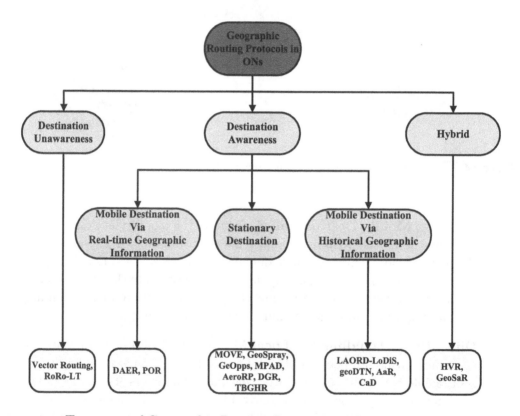

Figure 3.5 Taxonomy of Geographic Routing Protocols in ONs

---

[3]The term "Unicasting" describes communication where the information is sent from one sender to one receiver, both with unique addresses. "Multicasting" is the term used to describe communication where the information is sent to a group of receivers with unique addresses. Given "Anycasting," information is routed to a single member of a group of potential receivers, but are all identified by the same destination address.

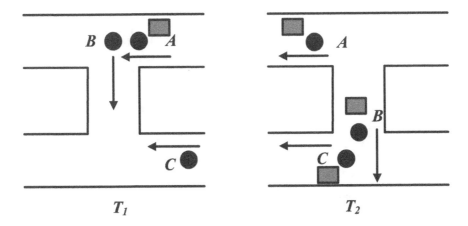

T_1 T_2

Figure 3.6 Drawback of vector routing

3.4.1 Destination Unawareness Class

Protocols in this branch aim to achieve efficient message replication[4] using geometric utility, without the requirement to track where the destination is.

[Vector Routing]

The key insight of Vector Routing [15] is to replicate the message according to an encounter angle $\omega \in [0, \pi]$ between pairwise encountered nodes. This protocol intends to replicate a smaller number of messages given a small value of ω, because pairwise encountered nodes moving with a similar direction would result in redundant replication. Even though these two nodes move with quite different direction, it is also redundant to replicate a large number of messages given that $\omega = \pi$, because the encountered node is currently moving with a previous trajectory of message carrier. Results show the low routing overhead of Vector Routing compared to blind flooding protocol [38].

However, Vector Routing is inefficient given the case shown in Figure 3.6. At the T_1 time slot, the message in node A is replicated to B according to the policy of Vector Routing, as their encounter angle is $\frac{\pi}{2}$. Even if the moving directions of nodes A and C are consistent, node B still replicates the message received from node A to C at the T_2 time slot. This procedure brings replication redundancy, as message copies are carried by both nodes A and C with the same moving direction. Since Vector Routing only handles each message identically, a huge overhead will occur if there are a large number of messages for replication.

[RoRo-LT]

In RoRo-LT [37], the LT stands for Location at corresponding Time slot, such that a long term observation (for several weeks) from such spatiotemporal history can predict future locations. Basically, the self-periodicity measures how similar that current routine of a node is, compared to its historical habit. If a high similarity is matched,

---

[4]The term "replication" means a copy of a message is generated through routing procedure.

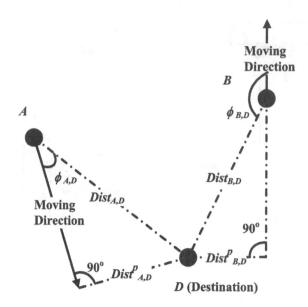

Figure 3.7 Illustration of geometric utilities for MOVE [24] and AeroRP [34]

the current routine is used to predict its mobility. With the estimated future trajectories of pairwise encountered nodes, the message carried by node A would be replicated to an encountered node B, only if they are predicted to be distant from each other in a near future. This is different from Vector Routing which determines how many messages are replicated depending on the mobility of pairwise encountered nodes.

However, results show that RoRo-LT does not perform well, although using the knowledge from a certain increased depth (or referred to number of weeks). This certainly indicates the limitation of routing protocols in "Destination Unawareness" Class. Driven by this limitation, the following protocols discuss the importance of tracking the message destination on the routing performance.

3.4.2 Destination Awareness Class

3.4.2.1 Stationary Destination

By knowing the location of stationary destination in advance, protocols under this branch focus more on exploring various geometric metrics to select relay.

[MOVE]

MOtion VEctor (MOVE) [24] forwards[5] the message towards its destination, mainly based on the nodal moving direction. Furthermore, based on the consistent moving direction between pairwise encountered nodes, the distance is adopted to identify the node which does not extensively contribute to message delivery (further away from the destination). As an example shown in Figure 3.7 where node A is the

---

[5]The term "forward" refers to the policy that a copy of a message will not be generated through a routing process.

message carrier, the condition $(Dist_{A,D} > Dist_{B,D})$ makes message forwarding decision to node B, if both nodes A and B are moving away from D. In contrast, the condition $(Dist^p_{A,D} > Dist^p_{B,D})$ is applied, if both nodes A and B are moving towards D.

One disadvantage of MOVE is that it does not consider the factor of nodal moving speed, since the node with a faster proximity to destination is able to reduce delivery delay. In addition, MOVE does not address the local maximum problem, where the relay node with a closer distance to the destination may not be always available. Therefore, additional delay or unsuccessful message delivery would occur, if greedily waiting for an appropriate relay node.

[GeOpps]

Assisted by the navigation system to calculate a suggested route towards the destination, the core of GeOpps [25] is to select the Nearest Point (NP) along the suggested route where the NP is the nearest point to destination. With message forwarding policy, the metric Minimum Estimated Time of Delivery (METD) as calculated in Equation (3.1) is used to qualify the relay node:

$$\text{METD} = \text{ETA} + \frac{Dist_{NP,D}}{\text{Average Moving Speed}} \tag{3.1}$$

Here, the ETA is the estimated time traveling from the location of current node to the NP, while $Dist_{NP,D}$ is denoted as the distance from the NP to destination. By considering the nodal moving speed, GeOpps achieves a lower delivery latency compared with MOVE. This protocol is further extended as GeoSpray [35], by limiting the number of copies that a message is controlled for replication. Besides, GeoSpray further tackles the limited network resources including bandwidth and buffer space, by prioritizing messages for transmission and storage.

It is observed that the selection of NP is an important factor for GeOpps, where running a weighted shortest path algorithm to find NP is complication, time-wise, by considering the actual road topology. This brings scalability concerns in a large scale road system, as it would need a longer time to find the NP, thus increasing the computational complexity. Besides, the local maximum problem that the node with a lower value of METD may be currently unavailable is not addressed in GeOpps. Particularly, due to nodal mobility, missing a communication opportunity brings additional calculation of the NP for the next encounter.

[MPAD]

Mobility Prediction based Adaptive Data gathering (MPAD) [47] is designed for mobile sensor networks. Under such a scenario, the message gathering is a major challenge, because traditional protocols mainly rely on a large number of densely deployed sensor nodes to construct a contemporaneous end-to-end path towards sink nodes. Considering mobility prediction, MPAD adopts a metric as an intersection between the moving path of a corresponding node and the transmission range of a sink node, as the moving direction of node A shown in Figrue 3.8. If the above condition is not met, the communication angle calculated by the two tangents between node A and sink node is adopted as a back-up metric. Here, a closer distance to the

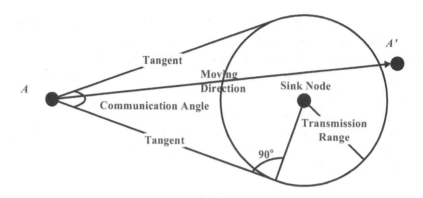

Figure 3.8 Illustration of geometric utility in MPAD

sink node indicates a larger communication angle, thus increasing the potential for message delivery.

MPAD does not consider the factor of message lifetime for making a routing decision, because it is important for message gathering before expiration especially in a sensor network. The message that is close to expiration deadline should be greedily replicated, regardless of the relay node selection policy. In other words, the message with a short lifetime is still suggested for replication, although its carrier is further away from a sink node.

[AeroRP]

Envisioning for ANs, AeroRP [34] selects the relay node to forward the message, by jointly considering distance and moving direction as well as speed. As the example shown in Figure 3.7, the metric Time-To-Intersect (TTI) for node A to approach D is calculated as $\text{TTI}_{A,D} = \frac{Dist_{A,D}}{Speed_A \times \cos\phi_{A,D}}$, where $Speed_A$ is the current moving speed of node A.

Since AeroRP only considers the case that pairwise encountered nodes are moving towards the destination, its routing policy is limited in case the message carrier is moving away from the destination even if its encountered node is moving towards the destination. Here, according to its routing policy, the message will not be forwarded. This is because the negative value of TTI is invalid for making a routing decision. Due to the very sparse network density and high speed in ANs, failing to relay a message due to such a limitation degrades routing performance.

[DGR]

Delegation Geographic Routing (DGR) [6] overcomes the limitation of the routing decision in AeroRP [34], by comparing the TTI of a historically encountered node with that of a currently encountered node, instead of comparing that between the current encountered node and message carrier. Therefore, any failure of routing decisions due to the inconsistent moving directions between pairwise encountered nodes is overcome. Such an operation is implemented via the Delegation Forwarding (DF) [11] optimization policy, by always recording the TTI of a selected relay node after successful message transmission. Thus, with the update of its level, the number of copies duplicated for a message is expected to be decreased.

DGR also addresses the local maximum problem, where the heuristic approach is presented as:

$$T_M^{ela} + T_M^{TTI} > T_M^{ini} \tag{3.2}$$

Here, T_M^{ela} and T_M^{ini} are the elapsed time since message generation and the initial message lifetime, respectively, while T_M^{TTI} is the recorded TTI in a message. This inequality implies that the message will expire, if the remaining message lifetime $T_M^{ini} - T_M^{ela}$ is shorter than T_M^{TTI}. In this context, rather than just keeping the message that is close to expiration, a message copy is replicated to increase the possibility to encounter other relay nodes.

[TBHGR]

The-Best-Heterogeneity-based-Geographic-Relay (TBHGR) [9] geographically relays L message copies to the destination, located in the area around which other nodes are heterogeneous. The heterogeneity refers to the behavior that mobile nodes in the same group would have a common preference, to move within a certain area consisting of some popular places, whereas this behavior is differentiated among those in diverse groups. In TBHGR, the heterogeneity (in terms of visiting preference) is concerned with moving direction and nodal temporarily stop status. The routing decision in TBHGR is decoupled into two phases, illustrated via an example in Figure 3.9. The first phase aims to replicate message copies to a relay node which has potential to move towards the area where the destination is located (e.g., the routing procedure from source to N_2, and that from N_1 to N_3). The second phase further forwards message copies to a relay node (e.g., N_4 or N_5) within this area, and contributes to delivery towards destination.

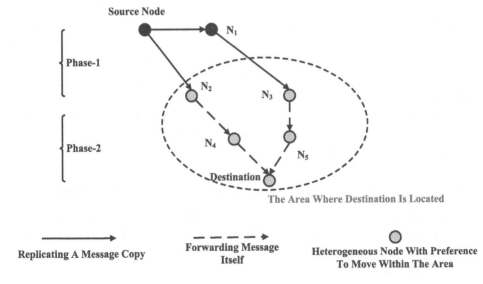

Figure 3.9 Message delivery in TBHGR

As TBHGR assumes the location of destination is fixed, it does not face the difficulty of tracking where the destination (concerning its mobility) is whereas it indeed provides a guidance to efficiently deliver a message among areas where nodal mobility is heterogeneous.

3.4.2.2 Considering Mobile Destination via Real-time Geographic Information

Protocols under this branch consider the mobility of destination. However, they ignore the feasibility to track mobile destination in sparse networks. Here, a centralized location server is still assumed to support real-time location request/reply.

[DAER]

Distance Aware Epidemic Routing (DAER) [12] assumes the real-time location of mobile destination can be collected from a centralized location service system, regardless of the delay to exchange such information in sparse networks. Any encountered node with a closer distance to destination than the message carrier itself is replicated with a message copy. Upon a successful message transmission, the original message carrier moving away from destination would discard its message in a local buffer. This operation aims to prevent additional message replication redundancy away from destination, which reduces the routing overhead.

The major concern is how to obtain the real-time location of destination given the identified challenges in section II. Along with that, DAER does not predict the nodal mobility, due to only relying on a distance metric. As such, a high speed may contribute to faster proximity to destination and vice versa. Considering the encountered node with a closer distance to destination may not be always available, such a local maximum problem needs to be addressed.

[POR]

As an extension based on DAER [12], the routing decision in Packet Oriented Routing (POR) [27] takes the balance between the distance and messages replicated along this distance into account. The idea behind this is to select a longer distance for replicating less messages, so as to enhance the transmission reliability. Results show the advantage of POR over DAER, particularly given limited transmission bandwidth.

However, POR still faces the limitation of design assumption in DAER, in terms of relying on the centralized location server and mobility prediction as well as local maximum problem handling.

3.4.2.3 Considering Mobile Destination via Historical Geographic Information

Protocols under this branch tackle the limitation in the former branch. Instead of assuming the real-time location of mobile destination can be always tracked, the historical geographic information is applied under this branch to estimate where the destination would be.

[LAROD-LoDiS]

The main contribution of Location Aware ROuting for Delay tolerant networks (LAROD)-Location Dissemination Service (LoDiS) [22] is the design of a location service system in sparse networks, by maintaining a local database of nodal locations. Such information is updated using broadcast gossip together with routing overhearing. Depending on an estimated replication area to limit routing redundancy, the message is replicated towards the location of destination using the distance metric.

One concern for LAROD-LoDiS is that the current location of destination may differ from the recorded information, which influences the reliability of routing decision. If using an obsolete location information, the routing decision making would be inaccurate and even degrade routing performance. Besides, the local maximum problem in terms of distance factor and the mobility prediction concerning direction and speed are not addressed.

[geoDTN]

As designed for MSN's scenario, geoDTN [29] records the historical nodal movement information and calculates an intersection area that any two nodes encountered in the past, as a metric to score the encounter possibility. This is common in reality that two people may move within a same area because of their social habits. In detail, the distance model of geoDTN adopts the minimum distance to the destination if the scores of pairwise nodes are below a predefined threshold value. Considering the local maximum problem that the node with a closer distance to destination is unavailable, geoDTN randomly forwards the message based on a predefined probability in the rescue model. If both pairwise nodes have higher scores than the predefined threshold value, the node with a higher score is selected as relay.

However, there are several predefined parameters defined in geoDTN, which affect the scalability. In particular, the predefined probability value in the rescue model for handling the local maximum problem should adapt to the network condition. It is important that this probability should become larger when the message is close to expiration, while an infrequent nodal encounter also enlarges such a value.

[AaR]

Referring to Location Aided Routing (LAR) [21], Approach and Roam (AaR) [5] adopts the historical geographic information of destination to estimate its movement range, via its moving speed and location recorded in the past. The key insight is to make faster message replication towards this estimated movement range via the Approach phase, and to guarantee message replication within this range for a longer duration via the Roam phase.

Since AaR consists of two routing phases, the local maximum problem in the Approach phase implies that the relay node which approaches the movement range of destination faster than the current message carrier is unavailable. Meanwhile, the unavailability of relay node which roams in this range with a longer time is considered as the local maximum problem in the Roam phase. These two problems are both handled based on the similar idea proposed in DGR [6], by considering the message lifetime and nodal mobility. Further effort is also paid in Converge-and-Diverge (CaD) [8], which reduces routing overhead via DF optimization policy [11].

3.4.3 Hybrid Class

Protocols under this branch combine the advantage of those in the "Destination Unawareness" class when the location of destination is unavailable in initial stage, and those with the historical geographic information in the "Destination Awareness" class when an approximate location of destination is found.

[HVR]

History based Vector Routing (HVR) [16] enables each mobile node to record the historical geographic information including location and moving speed of other encountered nodes. This information is used for relay node selection, by calculating the overlapped area between the circle estimated for destination and the transmission range of node. An example is shown in Figure 3.10, where t_{cur} is the current time and W is the estimation time window. Assuming a constant speed of node D is $Speed_D$, the radius R'_D for node D estimated at the $(t_{cur}+W)$ time slot is calculated as $Speed_D * (t_{cur} + W - t_{A,D})$; here $t_{A,D}$ is the historical encounter time for node D recorded by node A. In light of this, the one with a larger overlapped area with destination is selected as the relay node. In addition, HVR adopts Vector Routing [15] for message replication if the historical geographic information of destination is unavailable.

Although HVR is more reliable than other approaches considering the mobile destination, one concern is that there may not be an overlapped area for calculation,

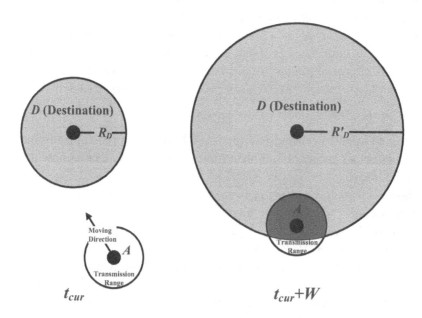

Figure 3.10 Illustration of geometric estimation in HVR.

because of a quite long distance between the corresponding node and destination. Therefore, the routing decision has limitations when both pairwise encountered nodes do not have the overlapping with the movement range estimated for destination, even if they have the knowledge about destination. In addition to this, HVR does not discuss the local maximum problem that the relay node with a larger overlapped area may be unavailable.

[GSaR]

Geographic-based Spray-and-Relay (GSaR) [7] delivers messages given a limited number of L replications, where the initial value of L is predefined based on scenario and distributed to the selected relay node. This protocol assumes that when network nodes are sufficiently mobile, only generating a small number of message copies is able to achieve successful delivery. Given the historical location, moving speed, as well as encounter time recorded in the past, the movement range of destination is estimated by referring to [5]. Here, with L copies allowed for each message, GSaR decouples routing decisions threeways; 1) to expedite message copies being replicated towards this range in order to reduce delivery delay; 2) to prevent message copies being replicated away from this range in order to reduce routing overhead; 3) to postpone message copies being replicated out of this range in order to increase delivery probability. As a back-up scheme, if the above historical geographic information of destination is unavailable, messages are replicated considering the relative moving direction between pairwise nodes as well as their moving speeds.

3.5 COMPARISON AND ANALYSIS

Table 3.1 illustrates the reviewed geographic routing protocols in ONs, where their years of publication and the application scenarios are highlighted. In addition, the concerns regarding limited bandwidth, buffer space and energy are discussed, because ONs are generally assumed with limited network resources. Details regarding how these factors affect routing performance have already been introduced in [4].

The term "Not Discussed" for the bandwidth concern is further defined, if a reviewed protocol either does not define the message priority for transmission, or the performance evaluation addresses the varied traffic load. A similar judgment is applied for the concern on buffer space and energy as well. Here, providing a quantified comparison among reviewed protocols is not feasible, since they are designed for different application scenarios. Also, the algorithm complexity is out of discussion due to subjectivity.

Apart from the knowledge required for making routing decisions, in which policy messages are relayed is discussed, given that reviewed protocols are either forwarding based via a single message copy or replication based via multiple message copies. Here, using forwarding policy is more efficient because there are no more additional message copies existing in the network. Even with more redundancy, using replication policy is more reliable in sparse networks. This is because generating additional message copies increases the possibility that at least any one of them could be delivered.

Table 3.1 Comparison Among Geographic Routing Protocols in ONs

Name of Protocol	Application Scenario	Year	Knowledge for Routing Decision	Relay Policy	Bandwidth	Buffer Space	Energy	Local Maximum Problem
★Destination Unawareness Class★								
Vector Routing [15]	VANETs	2008	Encounter Angle, Moving Speed	Replication	Yes	Not Discussed	Not Discussed	Not Applicable
RoRo-LT [37]	MSNs	2013	Distance	Replication	Not Discussed	Not Discussed	Not Discussed	Not Applicable
★Destination Awareness Class★								
——Stationary Destination——								
MOVE [24]	VANETs	2005	Distance, Moving Direction	Forwarding	Not Discussed	Yes	Not Discussed	Not Discussed
GeOpps [25]	VANETs	2007	Selection of the NP, Moving Speed, Distance	Forwarding	Not Discussed	Not Discussed	Not Discussed	Not Discussed
GeoSpray [35]	VANETs	2014	Selection of the NP, Moving Speed, Distance	Replication	Yes	Yes	Not Discussed	Not Discussed
MPAD [47]	UWSNs	2008	Moving Direction, Communication Angle	Replication	Yes	Yes	Not Discussed	Not Discussed
AeroRP [34]	ANs	2011	Moving Direction, Distance, Moving Speed	Forwarding	Not Discussed	Yes	Not Discussed	Not Discussed
DGR [6]	VANETs	2013	Moving Direction, Distance, Moving Speed	Replication	Not Discussed	Not Discussed	Not Discussed	Yes
TBHGR [9]	VANETs	2016	Moving Direction, Distance, Moving Speed, Visiting Preference	Replication	Yes	Yes	Not Discussed	Yes
——Mobile Destination via Realtime Geographic Information——								
DAER [12]	VANETs	2007	Distance	Replication	Yes	Yes	Not Discussed	Not Discussed
POR [27]	VANETs	2008	Distance, Message Size	Replication	Yes	Not Discussed	Not Discussed	Not Discussed
——Mobile Destination via Historical Geographic Information——								
LAROD-LoDiS [22]	ANs	2011	Distance, Replication Area	Replication	Yes	Not Discussed	Not Discussed	Not Discussed
geoDTN [29]	MSNs	2011	Distance, Movement Area	Forwarding	Not Discussed	Not Discussed	Yes	Yes
AaR [5]	VANETs	2014	Moving Speed, Moving Direction, Movement Area	Replication	Yes	Yes	Not Discussed	Yes
CaD [8]	VANETs	2013	Moving Speed, Moving Direction, Movement Area	Replication	Yes	Not Discussed	Not Discussed	Yes
★Hybrid Class★								
HVR [16]	VANETs	2009	Movement Area, Moving Speed, Encounter Angle	Replication	Not Discussed	Not Discussed	Not Discussed	Not Discussed
GeoSaR [7]	VANETs	2015	Moving Speed, Moving Direction, Movement Area, Encounter Angle	Replication	Yes	Yes	Yes	Yes

Consequently, there is a tradeoff between achieving the highest delivery ratio and the least redundancy.

The motivation of "Destination Unawareness" based protocols is to geometrically replicate messages, in case the destination cannot be tracked. In this branch, handling the "Local Maximum Problem" is "Not Applicable," as the qualification of relay node is unrelated to message destination. Here, the routing decision should focus more on estimating the delivery potential [37] for each message, rather than addressing the number of messages [15]. For example, messages may have different sizes where the one with a smaller size requires a shorter time for transmission and vice versa. Depending on this condition, a different moving direction or larger distance variation between pairwise encountered nodes enables a better chance for message replication.

In spite of this, tracking where the destination is (or approximately would be) plays a crucial role in driving high delivery ratio, as addressed in the "Destination Awareness" branch. Regarding those considering stationary destination and mobile destination via real-time geographic information, even if the destination can be tracked to guide routing decision, it is still essential to explore the mobility (e.g. jointly considering distance, moving direction and speed) of nodes and predict a future encounter opportunity. Relatively, those protocols applying historical geographic information tackle a practical concern, by taking the freshness of the location of destination into account. For example, relaying the message towards an obsolete location where mobile destination was located would degrade routing performance. Therefore, identifying the dynamic changed status of mobile destination should be further investigated.

In addition to the above two branches, integrating the efforts in the "Destination Unawareness" branch as a backup and in the "Destination Awareness" branch as the core intelligence can substantially enhance the reliability of the routing decision as investigated in the "Hybrid" class.

3.6 FUTURE DIRECTIONS

Thanks to the success of the Global Positioning System (GPS) technique, enabling geo-centric techniques for communication in ONs has a greater potential than those topological protocols [3], particularly for VANETs, UWSNs and ANs scenarios because of their highly dynamic characteristics. In this section, we detail a list of future directions that are worthwhile to study:

Handling the Local Maximum Problem: Although identified and initially tackled in literature, the local maximum problem is still an important issue for further investigation. In [34], a heuristic approach concerning the nodal mobility and message lifetime is proposed and has been proven through analysis and simulation. However, such an approach is locally estimated at each node, without an overview of other copies of a certain message in networks. On the one hand, the local maximum problem should be handled more greedily, if all message copies are close to an expiration deadline. On the other hand, more attention should be paid to nodal encounter prediction, if the message can still exist in a network for certain time. Therefore, apart from the knowledge of the message carrier itself, it is also essential to obtain an accurate (ideally) or approximate (practically) knowledge about how many copies of a message have been replicated. In order to further optimally handle the local maximum problem, an intelligent approach should be to jointly consider the number of copies of a message, its lifetime, as well as individual nodal mobility.

Concerning QoS Awareness: Since it is difficult to provide an end-to-end QoS support in ONs, appropriate message scheduling for transmission and buffer management is crucial given the limited bandwidth and buffer space. These two factors determine the number of messages that can be successfully transferred and received by relay nodes. Besides, since the nodes running out of energy cannot involve communication anymore, necessary energy saving approaches have been proposed, via an intelligent beacon control for nodal discovery [1]. Specifically, a frequent beacon broadcasting to discover neighbor nodes is energy costly, while that with infrequent broadcasting, however, may miss the communication opportunities with neighbor nodes. Note that the beacon control also has influence on the information updating (directly related to making routing decisions) that happens between pairwise nodes.

Combining with Coding Technique: As reviewed in [4], network coding [28] and erasure coding [40] techniques have been applied for routing protocols in ONs. In particular, the network coding enables efficient bandwidth usage, by encoding messages into a chunk block for transmission, while erasure coding compensates for the communication failure, by encoding the original message into a certain number of smaller size blocks for transmission. Since none of them has been explicitly applied for geographic routing protocols in ONs, these two well known coding techniques should be further investigated. Here, the nature of GeoSpray [35] could provide an initial guide on how to geometrically transmit those coded blocks using the erasure coding technique.

Assistance of Additional Infrastructure: Considering that the nodal mobility may be limited within certain areas, the assistance from an additional infrastructure, e.g. Message Ferry (MF)/gateway mentioned in [4], is able to help relay the message. Here, MF is a mobile entity that moves with a dedicated route, whereas the gateway is a deployed stationary entity. Both of them are able to bridge the communication among disconnected network islands, via trajectory controlling and location deployment. In this context, integrating them for a specific scenario is worthwhile to investigate.

Concerning Application Scenario: Combining routing protocols reviewed in this article with those conventionally designed for MANETs can adapt to the variation of network density. Initial observation in [10, 23] shows that it is intelligent to switch from MANETs to ONs based communication modes when networks become sparse. Although only the topological based routing protocol has been discussed in that work, the observation is also applicable to the geographic routing protocol. Such intelligence has been addressed by [10, 26] which combine the geographic routing intelligences in MANETs and ONs. Besides, given the application scenario highlighted in Table 3.1, the protocol design for MSNs is still inadequately investigated compared to others. Even if there has been an effort in linking geographic distance with users for fixed online social networks, the issue for mobile networks is still a challenge. Besides, the influence of heterogeneous mobility [9] should be addressed.

Concerning Security&Privacy: The information exchange for updating the historical geographic information requires security and privacy consideration. One major concern is the spoofing attack because the malicious node is able to create routing loops, generating false error information after information updates. Besides, overhearing the message passing through neighboring nodes might emulate selective forwarding by jamming the relayed message. Additionally it is also a privacy concern to release nodal geographic information to any encountered node. In particular, the location information should be released among friends who have common daily habits in MSNs.

3.7 CONCLUSION

Enabling geo-centric technologies is important for communication in ONs in line with dedicated use cases in Smart Cities. Different from MANETs, the sparse network density is the main challenge for communication in ONs. Motivated by the lack of attention to investigating geographic routing protocols in ONs, its challenges together with further reviews of the state of the art are presented in this chapter. Following the comparison and discussion of the literature with the, future directions are given, aimed at engaging continuing research interest in this field.

Bibliography

[1] D. Amendola, F. De Rango, K. Massri, and A. Vitaletti. Efficient Neighbor Discovery in RFID Based Devices Over Resource-Constrained DTN Networks. In *IEEE International Conference on Communications*, pages 3842–3847, Sydney, Australia, June 2014.

[2] F. Cadger, K. Curran, J. Santos, and S. Moffett. A Survey of Geographical Routing in Wireless Ad-Hoc Networks. *IEEE Communications Surveys Tutorials*, 15(2):1–33, 2013.

[3] Y. Cao, N. Wang, Z. Sun, and H. Cruickshank. A Reliable and Efficient Encounter-Based Routing Framework for Delay/Disruption Tolerant Networks. *IEEE Sensors Journal*, 15(7):1–15, 2015.

[4] Y. Cao and Z. Sun. Routing in Delay/Disruption Tolerant Networks: A Taxonomy, Survey and Challenges. *IEEE Communications Surveys Tutorials*, 15(2):654–677, 2013.

[5] Y. Cao, Z. Sun, H. Cruickshank, and F. Yao. Approach-and-Roam (AaR): A Geographic Routing Scheme for Delay/Disruption Tolerant Networks. *IEEE Transactions on Vehicular Technology*, 63(1):266–281, 2014.

[6] Y. Cao, Z. Sun, N. Wang, H. Cruickshank, and N. Ahmad. A Reliable and Efficient Geographic Routing Scheme for Delay/Disruption Tolerant Networks. *IEEE Wireless Communications Letters*, 2(6):603–606, 2013.

[7] Y. Cao, Z. Sun, N. Wang, M. Riaz, H. Cruickshank, and X. Liu. Geographic-Based Spray-and-Relay (GSaR): An Efficient Routing Scheme for DTNs. *IEEE Transactions on Vehicular Technology*, 64(4):1548–1564, 2015.

[8] Y. Cao, Z. Sun, N. Wang, F. Yao, and H. Cruickshank. Converge-and-Diverge: A Geographic Routing for Delay/Disruption-Tolerant Networks Using a Delegation Replication Approach. *IEEE Transactions on Vehicular Technology*, 62(5):2339–2343, 2013.

[9] Y. Cao, K. Wei, G. Min, J. Weng, X. Yang, and Z. Sun. A Geographic Multi-Copy Routing Scheme for DTNs with Heterogeneous Mobility. *IEEE Systems Journal*, 2016.

[10] P.-C. Cheng, J.-T. Weng, L.-C. Tung, K. C. Lee, M. Gerla, and J. Hårri. GeoDTN+Nav: A Hybrid Geographic and DTN Routing with Navigation Assistance in Urban Vehicular Networks. In *International Symposium on Visual Computing*, Las Vegas, Nevada, USA, November 2008.

[11] V. Erramilli, M. Crovella, A. Chaintreau, and C. Diot. Delegation Forwarding. In *ACM International Symposium on Mobile Ad Hoc Networking and Computing*, pages 251–259, Hong Kong, China, May 2008.

[12] H.-Y. Huang, P.-E. Luo, M. Li, D. Li, X. Li, W. Shu, and M.-Y. Wu. Performance Evaluation of SUVnet with Real-Time Traffic Data. *IEEE Transactions on Vehicular Technology*, 56(6):3381–3396, 2007.

[13] D. B. Johnson and D. A. Maltz. Dynamic Source Routing in Ad Hoc Wireless Networks. *Springer Mobile Computing*, 353(1):153–181, 1996.

[14] T. Jonson, J. Pezeshki, V. Chao, K. Smith, and J. Fazio. Application of Delay Tolerant Networking (DTN) in Airborne Networks. In *IEEE Military Communications Conference*, pages 1–7, San Diego, California, USA, November 2008.

[15] H. Kang and D. Kim. Vector Routing for Delay Tolerant Networks. In *IEEE Vehicular Technology Conference VTC-Fall*, pages 1–5, Calgary, Alberta, Canada, September 2008.

[16] H. Kang and D. Kim. HVR: History-Based Vector Routing for Delay Tolerant Networks. In *IEEE International Conference on Computer Communications and Networks*, pages 1–5, San Francisco, California, USA, August 2009.

[17] B. Karp. Challenges in Geographic Routing: Sparse Networks, Obstacles, and Traffic Provisioning. In *DIMACS Workshop on Pervasive Networking*, DIMACS Center, Rutgers University, Piscataway, New Jersey, USA, May 2001.

[18] B. Karp and H. T. Kung. GPSR: Greedy Perimeter Stateless Routing for Wireless Networks. In *ACM Mobile Computing and Networking*, pages 243–254, Boston, Massachusetts, USA, August 2000.

[19] J. Kartha and L. Jacob. Delay and Lifetime Performance of Underwater Wireless Sensor Networks with Mobile Element Based Data Collection. *International Journal of Distributed Sensor Networks*, 2015(6), 2015.

[20] M. Khabbaz, C. Assi, and W. Fawaz. Disruption-Tolerant Networking: A Comprehensive Survey on Recent Developments and Persisting Challenges. *IEEE Communications Surveys Tutorials*, 14(2):607–640, 2011.

[21] Y.-B. Ko and N. H. Vaidya. Location Aided Routing (LAR) in Mobile Ad Hoc Networks. *ACM Wireless Networks*, 6(4):307–321, 2000.

[22] E. Kuiper and S. Nadjm-Tehrani. Geographical Routing with Location Service in Intermittently Connected MANETs. *IEEE Transactions on Vehicular Technology*, 60(2):592–604, 2011.

[23] J. Lakkakorpi, M. Pitkänen, and J. Ott. Adaptive Routing in Mobile Opportunistic Networks. In *ACM International Conference on Modeling, Analysis, and Simulation of Wireless and Mobile Systems*, pages 101–109, Bodrum, Turkey, October 2010.

[24] J. LeBrun, C.-N. Chuah, D. Ghosal, and M. Zhang. Knowledge-Based Opportunistic Forwarding in Vehicular Wireless Ad Hoc Networks. In *IEEE Vehicular Technology Conference*, pages 2289–2293, Stockholm, Sweden, May 2005.

[25] I. Leontiadis and C. Mascolo. GeOpps: Geographical Opportunistic Routing for Vehicular Networks. In *IEEE International Symposium on a World of Wireless, Mobile and Multimedia Networks*, pages 1–6, Espoo, Finland, June 2007.

[26] F. Li, L. Zhao, X. Fan, and Y. Wang. Hybrid Position-Based and DTN Forwarding for Vehicular Sensor Networks. *International Journal of Distributed Sensor Networks*, 2012.

[27] X. Li, W. Shu, M. Li, H. Huang, and M.-Y. Wu. DTN Routing in Vehicular Sensor Networks. In *IEEE Global Telecommunications Conference*, pages 752–756, New Orleans, Louisiana, USA, December 2008.

[28] Y. Lin, B. Liang, and B. Li. Performance Modeling of Network Coding in Epidemic Routing. In *Proceedings of the First International MobiSys Workshop on Mobile Opportunistic Networking*, pages 67–74, San Juan, Puerto Rico, USA, June 2007.

[29] J. A. B. Link, D. Schmitz, and K. Wehrle. GeoDTN: Geographic Routing in Disruption Tolerant Networks. In *IEEE Global Telecommunications Conference*, pages 1–5, Houston, Texas, USA, December 2011.

[30] S. Ma, Y. Zheng, and O. Wolfson. Real-time City-scale Taxi Ridesharing. *IEEE Transactions on Knowledge and Data Engineering*, 27(7):1782–1795, July 2015.

[31] L. Moreira-Matias, J. Gama, M. Ferreira, J. Mendes-Moreira, and L. Damas. Predicting Taxi-passenger Demand Using Streaming Data. *IEEE Transactions on Intelligent Transportation Systems*, 14(3):1393–1402, September 2013.

[32] L. Pelusi, A. Passarella, and M. Conti. Opportunistic Networking: Data Forwarding in Disconnected Mobile Ad Hoc Networks. *IEEE Communications Magazine*, 44(11):134 –141, 2006.

[33] P. Pereira, A. Casaca, J. Rodrigues, V. Soares, J. Triay, and C. Cervello-Pastor. From Delay-Tolerant Networks to Vehicular Delay-Tolerant Networks. *IEEE Communications Surveys Tutorials*, 14(4):1166–1182, 2011.

[34] K. Peters, A. Jabbar, E.K. Cetinkaya, and J. P. G. Sterbenz. A Geographical Routing Protocol for Highly-Dynamic Aeronautical Networks. In *IEEE Wireless Communications and Networking Conference*, pages 492–497, Quintana Roo, Mescio, March 2011.

[35] V. Soares, J. Rodrigues, and F. Farahmand. GeoSpray: A Geographic Routing Protocol for Vehicular Delay-Tolerant Networks. *Information Fusion*, 15(1):102–113, 2014.

[36] W. Tong, W. Jiyi, X. He, Z. Jinghua, and C. Munyabugingo. A Cross Unequal Clustering Routing Algorithm for Sensor Network. *Measurement Science Review*, 13:200–205, August 2013.

[37] O. Türkes, H. Scholten, and P. Havinga. RoRo-LT: Social Routing with Next-Place Prediction from Self-Assessment of Spatiotemporal Routines. In *IEEE International Conference on Ubiquitous Intelligence and Computing and IEEE International Conference on Autonomic and Trusted Computing*, pages 201–208, Vietri sul Mare, Italy, December 2013.

[38] A. Vahdat and D. Becker. Epidemic Routing for Partially-Connected Ad Hoc Networks. Technical report, Duke University Technical Report, June 2000.

[39] T. Wang, Y. Cao, Y. Zhou, and P. Li. A Survey on Geographic Routing Protocols in Delay/Disruption Tolerant Networks. *Int. J. Distrib. Sen. Netw.*, 2016:8:8–8:8, January 2016.

[40] Y. Wang, S. Jain, M. Martonosi, and K. Fall. Erasure-coding Based Routing for Opportunistic Networks. In *ACM SIGCOMM 2005 Workshop on Delay-Tolerant Networking*, pages 220–236, Philadelphia, Pennsylvania, USA, August 2005.

[41] K. Wei, X. Liang, and K. Xu. A Survey of Social-Aware Routing Protocols in Delay Tolerant Networks: Applications, Taxonomy and Design-Related Issues. *IEEE Communications Surveys Tutorials*, 16(1):556–578, 2014.

[42] X. Xu, J. Zhou, Y. Liu, Z. Xu, and X. Zhao. Taxi-rs: Taxi-hunting recommendation System Based on Taxi GPS Data. *IEEE Transactions on Intelligent Transportation Systems*, 16(4):1716–1727, August 2015.

[43] J. Yuan, Y. Zheng, X. Xie, and G. Sun. T-drive: Enhancing Driving Directions with Taxi Drivers' Intelligence. *IEEE Transactions on Knowledge and Data Engineering*, 25(1):220–232, January 2013.

[44] A. Zanella, N. Bui, A. Castellani, L. Vangelista, and M. Zorzi. Internet of Things for Smart Cities. *IEEE Internet of Things Journal*, 1(1):22–32, February 2014.

[45] Z. Zhang. Routing in Intermittently Connected Mobile Ad Hoc Networks and Delay Tolerant Networks: Overview and Challenges. *IEEE Communications Surveys Tutorials*, 8(1):24–37, 2006.

[46] Y. Zheng. Trajectory Data Mining: An Overview. *ACM Trans. Intell. Syst. Technol.*, 6(3):29:1–29:41, May 2015.

[47] J. Zhu, J. Cao, M. Liu, Y. Zheng, H. Gong, and G. Chen. A Mobility Prediction-Based Adaptive Data Gathering Protocol for Delay Tolerant Mobile Sensor Network. In *IEEE Global Telecommunications Conference*, pages 730–734, New Orleans, LA, United States, December 2008.

CHAPTER 4

Routing Protocol for Low Power and Lossy IoT Networks

Xiyuan Liu

Beijing Laboratory of Advanced Information Network, Beijing University of Posts and Telecommunications

Zhengguo Sheng

Department of Engineering and Design, University of Sussex, UK

Changchuan Yin

Beijing Laboratory of Advanced Information Network, Beijing University of Posts and Telecommunications

CONTENTS

W ITH a growing need to better understand our environments, the Internet of Things (IoT) is gaining importance among information and communication technologies. IoT will enable billions of intelligent devices and networks, such as wireless sensor networks (WSNs), to be connected and integrated with computer networks. In order to support large-scale networks, IETF has defined the Routing Protocol for Low Power and Lossy Networks (RPL) to facilitate the multi-hop connectivity. In this chapter, we provide an overview of the working principle and latest research activities in developing effective RPL, which can serve as a reference for effective routing solutions in IoT use scenarios.

4.1 INTRODUCTION

The Internet of Things (IoT) has become a new focus for both industry and academia involving information and communication technologies (ICTs), and it is predicted that there will be almost 50 billion devices connected with each other through IoT by 2020 [13]. The concept of IoT that is initially linked to the new idea of using radio frequency identification in supply chains is now well known as a new ICT where the Internet is connected to the physical world via ubiquitous wireless sensor networks (WSNs) [3].

With the development of WSN technologies, a wide range of intelligent and tiny wireless sensing devices will be deployed in a variety of application environments. Generally, these sensing devices are constrained by limited energy resources (battery power), processing and storage capability, radio communication range and reliability, etc., and yet their deployment must cover a wide range of areas. In order to cope with those challenges, a number of breakthrough solutions have been developed, for example, efficient channel hopping in IEEE 802.15.4e TSCH [40], the emerging IPv6 protocol stack for connected devices [37] and improved bandwidth of mobile transmission.

Routing is always challenging for resource constrained sensor devices, especially in large scale networks. The IETF Routing Over Low-power and Lossy networks (ROLL) working group has been focusing on routing protocol design and is committed to standardize the IPv6 routing protocol for Low-power and Lossy Networks (LLN). RFC6550 [55], first proposed by ROLL group of IETF in the form of a draft to define Routing Protocol over Low Power and Lossy Networks (RPL), serves as a milestone in solving routing problems in LLNs. With an explosion of network scale and application deployments, RPL is designated to provide a viable solution to maintain connectivity and efficiency in a cost effective way.

In this chapter, we are focusing on the analysis of RPL performance in large scale networks; particularly, we would like to answer the following questions:

- What are the objective functions and metrics defined in RPL and how do they perform?

This work was supported in part by the National Science and Technology Major Project under Grant 2015ZX03003012-005 and the NSFC under Grant 61271257.

- How does RPL perform in a large scale network with multiple hops?

- Are there security issues that should be considered regarding RPL?

- What is the future prospect and potential of RPL?

We make a deep analysis of objective functions and metrics in RPL under varied scenarios, with references to the latest literature and studies, which can fundamentally contribute to the understanding of RPL performance and provide inspiration to raise more viable methods to further improve the network performance. Moreover, we also give an overview of current application deployments with RPL and the security issues it faces.

4.2 RPL: AN OVERVIEW AND ITS KEY MECHANISMS

IETF ROLL Working Group mainly focuses on the routing in LLNs and has proposed RPL in RFC6550 [55]. RFC6550 was first released in March 2012 and then a number of supplementary and supportive RFCs and Internet drafts progressed. For instance, RFC6997 [17] is aimed at clarifying the specified route discovery mechanism, while RFC7416 [50] is about the purpose of strengthening the security issues in RPL.

With the development of IoT, RPL is given new opportunities for the development of wireless sensor networks. It is able to meet the specific routing requirements of application areas including urban networks (RFC 5548) [11], building automation (RFC 5867) [36], industrial automation (RFC 5673) [41] and home automation (RFC 5826) [7]. Among those mechanisms standardized in RPL, routing and message control are two important mechanisms in establishing and maintaining an effective and reliable network, which will be highlighted in detail as follows.

4.2.1 Routing Mechanism of RPL

RPL is a distance vector routing protocol. It does not have predefined topology but will be generated through the construction of Destination-Oriented Directed Acyclic Graphs (DODAGs). Directed Acyclic Graphs (DAGs) describe tree shaped structures. However, a DAG is not a traditional tree structure in which one node is allowed to have multiple parent nodes. The DODAG, with sink node or the node providing default routing to the Internet as the root node, is a direction-oriented graph.

The construction of network topology is controlled by three types of control messages—DODAG Information Object (DIO), Destination Advertisement Object (DAO) and DODAG Information Solicitation (DIS) messages. DIO message is used for upward routing construction, which is essential for establishing communication from non-sink nodes (or multiple points) to the sink node (one point). Such a Multipoint-to-point (MP2P) mode is dominating the RPL applications. The construction of the upward route of RPL is realized by DIOs. The sink node will first broadcast DIOs, and the nodes receiving the DIO directly from the sink node become its neighbours. By setting the sink node as their parent nodes, those neighbour nodes will re-broadcast DIOs to further nodes. A similar step will repeat in such a way that the DODAG topology is constructed through handling DIOs and building parent sets.

DAO message is used for downward routing construction (Point-to-Point and Point-to-multipoint). There are two modes of downward routing—storing and non-storing modes—which indicate that the routing table information is stored in intermediate nodes (non-root and non-leaf nodes) and root node, respectively. DIS message is used for soliciting the sending of DIO in order to make immediate response to network inconsistency.

Objective Function (OF) defines the rule of selecting neighbours and parent nodes by rank computation. Routing metrics related to link or node characteristics (RFC6551 [47]) can be used by OF to make routing determination. One of the widely used OF0 is defined in RFC6552 with hop count as the routing metrics. OF determines Neighbour Set, Parent Set and Preferred Parents according to specified routing metrics and constraints. The node set selection is involved in the route discovery process and indicates the best path computation. The rank of a node must be larger than that of its parent node, in order to avoid routing loops.

It is worth noting that in order to construct a valid RPL routing, firstly, candidate neighbour node set must be the subset of nodes that can be reached through link local multicast. Secondly, parent set is the subset of candidate neighbour set which satisfies specific limitation conditions. Thirdly, preferred parents are those with optimal path characteristics. If there exist a group of nodes with equivalent rank and preferred extent regarding the metrics calculation, there can be more than one preferred parent node. Figure 4.1 illustrates logical relationships of candidate neighbour node set parent node set, and preferred parent node of the node.

4.2.2 Message Control Mechanism of RPL

It is obvious that any routing mechanism involves significant control overhead in a large-scale network. Particularly in a multi-hop network, an effective message control

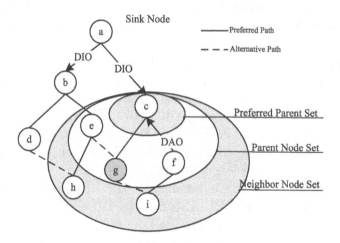

Figure 4.1 An example of DODAG and node set relationships.

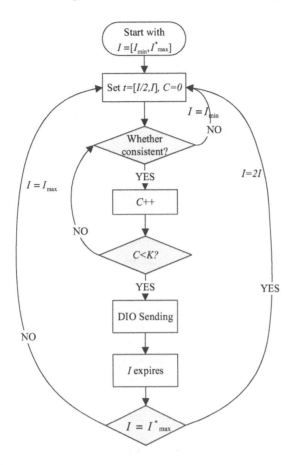

Start with
$I = [I_{min}, I^*_{max}]$

Set $t=[I/2,I]$, $C=0$

$I = I_{min}$

NO

Whether consistent?

YES

$I=2I$

$C++$

NO

$C<K$?

YES

YES

DIO Sending

NO

I expires

$I = I^*_{max}$

$I = I_{max}$

Figure 4.2 Trickle timer process.

mechanism is significantly important in reducing network overhead and balancing limited network resources.

Trickle timer mechanism [31], which is mainly used by DIO, has been emphasized as an important part of message control mechanism. A trickle timer is implemented based on a trickle algorithm and is able to detect and respond to network inconsistency and instability. Particularly, the inconsistency of RPL occurs in the following circumstances: detection of routing loops, first time joining a DODAG and rank change of a node. The fundamental mechanism of trickle time is shown in Figure 1.2. It is worth noting that the frequency of sending messages which is decided by the trickle timer can be dynamically adjusted to stabilize the network and govern the network status as well as improve the energy efficiency.

In the trickle timer process, t denotes the time for sending message, and C the counter indicates whether the network is consistent or not; predefined parameters include redundancy constant k, minimum time unit I_{min} and maximum time unit I_{max}. I^*_{max} denotes the maximum time period specified by the time units. Time consumption to transmit k packets is represented by t_K. Typically, we have

$$I^*_{max} = I_{min} \cdot 2^{I_{max}} \tag{4.1}$$

$$I_{min} = (2 \sim 3) \cdot t_K \tag{4.2}$$

Configuring the trickle timer with appropriate parameters is vital since it will influence the network reliability and stability [53], especially in large-scale networks. The redundancy constant k for each node should be carefully chosen in order to avoid mismatching values among all nodes in the network or being infinity which can lead to uneven load of traffic flow, depletion of energy or congestion in dense networks [31]. I_{min} also needs to be set accordingly to avoid congestion and high packet loss. In our study, we found out that the appropriate value of I_{min} falls into a fixed time period, which may be different in other settings. Particularly, with an inappropriate value of I_{min}, the packet delivery rate will be decreased.

4.2.3 RPL and Its Counterparts

The development of wireless sensor networks has contributed to proposals of a variety of routing protocols. LLNs have their specific requirements on routing. The commonly known routing protocols, such as Open Shortest Path First (OSPF) and Intermediate System to Intermediate System (IS-IS), are not suitable for the LLNs because they will lead to excessive control traffic in a constrained environment. Moreover, the large volume of routing traffic can also pose a threat to lossy links and rapid-in-change networks.

The comparison between RPL, LOAD and Geographical routing [24], in the case of advanced metering infrastructure (AMI), shows that LOAD fails to satisfy the requirements of LLNs regarding control overload, end-to-end delay and reliability. The next generation alternative, LOAD-ng [10], which is also raised by the IETF working group, is the representative of reactive routing while RPL is active routing. Under two cases of MP2P and P2MP traffic flows, in which the downward routing considers both storing and non-storing modes, both of the protocols perform closely in link quality and delay. RPL also suffers from instability in control overload, which is similar for LOAD-ng in the multicast situation. However, the reactive routing requires a larger cache. A brief comparison between RPL and LOAD-ng is shown in Table 4.1.

Table 4.1 Comparison between RPL and LOAD-NG

	RPL	**LOAD-ng**
Routing Mode	Active	Reactive
Delay	Shorter	Longer
Storage Requirement	Less	More
Complexity	More	Less
Control Overhead	More	Less

Compared with LOAD-ng and other routings in IoT, RPL is much more complex. The complicated types and options in control messages not only increase complexity in practice, but also elevate the hardware requirements in storage when it comes to the practical deployment.

4.3 RPL TOPOLOGY GENERATION METHODS

4.3.1 Objective Functions and Metrics

The topology of RPL is constructed according to specific OFs, which are configured according to metrics and constraints. OFs are responsible for constructing routing and providing optimal routing choice by determining DODAG topology and rank of each node. In the following, we summarize several typical OFs used in RPL.

1. Hop Count: It is one of the two well defined Objective Functions and also used as a routing metric [47]. Hop count is the most commonly used routing metric and it is deployed in the network routing calculation with the Hop Count OF.

2. ETX: Expected Transmission Count defined in RFC6551 [47] can also be used as a routing metric for OF in LLNs. The ETX metric is the number of transmissions a node is expected to a destination in order to successfully deliver a packet. With a higher value of ETX, the link quality may be worse. It is an addictive metric since it will add the ETX of each link along the path to the destination.

3. Per-Hop ETX: The combination and optimization of classic metrics can also bring better performance. Xiao *et al.* in [56] integrated the two traditional metrics—hop count and ETX into per-hop ETX. The new proposed metric is based on the addictive nature of ETX. It is demonstrated that calculating link metrics by dividing the aggregated ETX through the path using hop counts can improve packet delivery rate, delay and energy cost.

4. Stability Based OF: Iova *et al.* [21] offered an overview and outlook against RPL from aspects of reliability, end-to-end packet delivery rate, end-to-end delay and energy cost. Through comparisons and observations of OFs with metrics including hop count, ETX and link quality index (LQI), it reveals the tradeoff between network stability, which is mainly reflected by switching frequency of parent nodes, and the routing reliability. A deeper understanding of the issue regarding stability can be found in [57]. By taking numbers and frequency of control messages into consideration, the authors proposed a solution that combines DIO, DAO and DIS with given relative weights into one measurement for a specific node. Through this method, the packet delivery rate can be significantly improved, the control plane overload is largely deducted and the network stability is enhanced with reduced parent nodes switching times.

5. Energy Based OF: It is also interesting to consider energy based OF given that energy efficiency is highly required in large-scale sensor networks. Actually, the power supply of nodes is quite complex; therefore, in the structure

Table 4.2 Objective functions comparison

Routing Metrics Used	Observation Parameters	Key Features
Hop count [47]	Hop count between two nodes	Small end-to-end delay in sacrificing packet delivery rate
ETX [47]	Expected transmission count of data packet between two nodes	Packet Delivery Rate (PDR) is higher, increased delay
LQI [21]	Link quality data from the wireless chip after receiving data packets	End-to-end delay increases with an increase of PDR
Per-Hop ETX [56]	Expected transmission count of data packet per hop between two nodes	Delay and PDR are improved to some extent, the energy requirement is less in large-scale networks
Stability Index [57]	Numbers of DIO, DIS and DAO in the network	PDR is improved a lot while the number of control messages is largely reduced
Path loss metrics [26]	Remaining energy of nodes	Increased network longevity and evenly distributed energy consumption
ETX and energy composite metrics [8]	Energy parameter of nodes and ETX	Increased network longevity, given the same overall degree of network reliability
Expected Longevity [22]	Energy of nodes and ETX, forwarding according to specific probability	Increased network longevity

of routing metrics, there is a field indicating the power-supply type [47]. The power-supply sources include powered, battery and scavenger. Regarding different power-supply means, how to accommodate with various energy characters is a thought-provoking issue. Patrick *et al.* brought minimum path loss [26] into the definition of metrics, which is defined as the minimum node energy level that captures the energy-based path weight. It keeps the principle that parent nodes with maximum remaining energy are preferable and demonstrates satisfactory performance in network longevity and overload balance compared with ETX metrics. Existing literature also proposed to integrate node energy with other metrics. Capone *et al.* [8] combined node energy with ETX. By referring to the exponent of a ratio of transmission power and remaining energy and incorporating it with ETX, the method can gain improvement in network longevity and node energy.

Other solutions such as in RFC6551 [47], a series of metrics and constraints related to node and link attributes in RPL, are proposed. Table 4.1 summarizes classic and recently proposed OFs of RPL with metrics used, key observation parameters and performance of the metrics.

4.3.2 Multi-parents Consideration

One unique feature of the tree-based topology in RPL is that a node can possibly have multiple parent nodes. As described in RFC6550 [55], each node only chooses one preferred parent node to forward data packets to sink node, even though the node may store multiple parent nodes information in the parent set. RPL does not implement the parent switching mechanism; thus a node with a large number of child nodes will run out of energy easily. To increase the stability of a network and make full use of the candidate parent nodes to better balance the network overload, multi-parents selection is considered by some existing works.

The non-uniform flow distribution is likely to deplete some extensively selected nodes, which will be the bottlenecks and have significant impact on the longevity of the whole network. Additionally, nodes close to the sink or with lower rank tend to be more congested and with high energy cost. Capone et al. [8] proposed the expected longevity based metrics that considers both energy and ETX. The essence of the idea is that the network flow should be balanced when data packets are forwarded to different parent nodes according to a certain probability, which will help improve the network longevity.

ROLL group proposed an alternative approach by dividing RPL into clusters [48] within which nodes form to determine their parent nodes and construct the sub-topology. The RPL-based clustering scheme takes into account the remaining energy of cluster head based on the hierarchy of RPL topology as a sub-optimal parent for cluster member nodes. Therefore, the multi-parents issue can be transformed into a clusters problem in RPL [59]. The proposed solution with opportunistic packet forwarding and priority mechanism has been shown to obtain reduced delay and retransmission times compared with traditional RPL.

As shown in Figure 4.3, the data are obtained from the OMNeT++ simulation platform with basic settings in Section 4.3. We run the simulation with a duration of 100s during which the topology construction and packet forwarding have been finished and assume that nodes with zero energy will quit the topology immediately. According to the average energy consumption of 100 nodes using the two methods, the opportunistic forwarding method has a lower standard deviation (12.235) than the one-preferred-parent method (13.485) while the total consumption of both are closed, which indicates that more balanced energy consumption can be achieved in the opportunistic method.

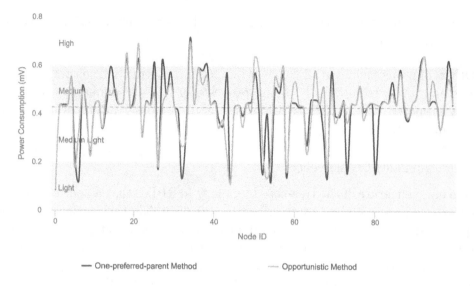

Figure 4.3 Energy consumption of 100 nodes with one-preferred-parent method and opportunistic method [59].

4.4 RPL APPLICATIONS

4.4.1 RPL Application Overview

RPL provides routing solutions for a wide scope of application areas including urban networks, building automation, industrial automation and home automation. In different use cases, adaptation of RPL needs to be considered to ensure optimized network performance.

Application-specific routing requirements vary according to the characteristics of different application use cases. We will take an overview here about the diversified use cases.

An urban low-power and lossy network will measure a wide gamut of physical data, including but not limited to: consumption data of water and gas, meteorological data such as temperature and pressure and pollution data such as sulfur dioxide. A prominent example is a "smart grid" application that consists of a citywide network of smart meters and distribution monitoring sensors. Besides advanced sensing functionalities such as measuring and automating, these meters may be capable of advanced interactive functionalities, which may invoke an actuator component, such as remote service disconnect or remote demand reset.

A building automation system is a hierarchical system of sensors, actuators, controllers and user interface devices that interoperate to provide a safe and comfortable environment while constraining energy costs. It is divided functionally across different but interrelated building subsystems such as heating, ventilation and air conditioning (HVAC); security; lighting; and elevator/lift control systems. The controllers are the most special part for the building management system. They are fed by sensor inputs to monitor the conditions within the building.

Historically, home automation used wired networks or power-line communication (PLC). Currently, the rapid development of wireless solutions enables homes to be more easily upgraded. The use cases of home automation include but are not limited to a "One event, Multiple actuators" case such as a "leaving home" key that will trigger all lights in the home to be turned off, remote control of home automation network, remote video surveillance and health-care devices for patients and elderly.

As for the industrial automation, it is not expected that wireless solutions will replace wired solutions in the foreseeable future, but it is believed that wireless cases will tremendously benefit its augmentation. The easy installation and maintenance is a significant consideration in this use case. Industrial automation can be divided into two categories—"process" and "discrete manufacturing." The product is typically a fluid (oil, gas, chemicals, etc.) in process control while the latter is an individual element (cars, dolls).

4.4.2 Application Scenarios

1. Smart Grid (SG): It has attracted much attention in both academia and industry. By monitoring energy usage and feedback responses automatically, SG is able to balance the energy distribution based on the power necessity. Countries including China, Japan, South Korea and Australia have invested extensive funding in the next-generation grid technology. The European Union set a target to deploy smart meters for more than 80% of customers by 2020. For Africa/Latin America, countries are directly investing in smart grid or indirectly utilizing renewable energy, which will ultimately require more advanced SG techonology [6].

   There is no denying that RPL plays a prominent role in smart grid deployment and is expected to be the standard routing protocol in AMI applications. The methods of rank computation as well as failures handling have been considered in AMI [1]. Ancillotti *et al.* [4] made a comprehensive elaboration and evaluation of RPL in AMI. They investigated the packet loss distribution of nodes in the network and pointed out that the scale of network and density of flows have significant impact on the network performance under AMI infrastructure applications. Ancillotti *et al.* in [5] made another research on RPL in AMI and proposed optimal methods for protocol deployment considering the presence of duty cycling with different RPL prototypes based on the Contiki simulation platform.

   SG is composed of a power system and smart grid communication network (SGCN). The latter can be partitioned into home area network, industrial area network and neighbour area network. Regarding smart grid under neighbour area network that involves devices at premises and utility monitors, Ho *et al.* [20] added positive parents switching functions in the RPL design which requires

---

AMI (Advanced Metering Infrastructure) is an important part of smart grid. AMI can be considered as an advanced version of Automated Meter Reading (AMR), which is capable of setting up two way communications with meter devices.

nodes to change their parent nodes proactively when packets are not received until a certain number of trials. The packets will be disposed if the switching times equal the number of candidate parent nodes. The proposed solution would definitely result in topology changes by providing dynamic updates.

2. Machine-to-Machine (M2M): It is able to realize autonomous communication and require no outer assistance to closed systems in a variety of fields. Aijaz *et al.* [2] summarized routing protocols design for M2M and proposed to modify RPL to adapt to cognitive radio. They also acknowledged the role of RPL as a standard routing protocol in future M2M development.

3. Agriculture Greenhouse: Quynh *et al.* [42] proposed a multi-path RPL protocol for the greenhouse environment monitoring system. According to the real-life greenhouse deployment and scale of the network, the proposed method can improve RPL with better energy balance and faster local repair, compared with the traditional hop count based RPL. The authors verified that RPL satisfies the requirements in greenhouse circumstances and can achieve better performance in packet delivery rate, time delay and packet error at the base station with multi-path improvements. The greenhouse scenarios provide decent results of RPL performance with consideration of hop count and residual energy.

4. Medical Applications: Gara *et al.* [15] considered such a use case with dynamic and hybrid topology and implemented a modified RPL in which the mobile nodes are implemented as leaf nodes and only send DIS to request parent without broadcasting DIO. The modified RPL shows better performance compared with native RPL in supporting low mobility nodes, which is indispensable in health care and medical applications.

In essence, the deployment of RPL should be adapted to real application scenarios, and further investigations of the diversified RPL deployments, especially those related with smart devices, are necessary in promoting the future development and applications of RPL. Table 4.3 summarizes the major RPL applications with topology features and metrics characteristics.

4.5 SECURITY ISSUES IN RPL

Security poses a serious challenge to RPL implementation. There are issues related to energy and link quality specified by LLNs [42]. LLNs require stable links maintenance and lower energy consumption beyond the common network circumstances and their limitation tends to have high impact on the effective design of security solutions. Especially in large-scale networks, security should be well considered in order to avoid large-scale contamination or information leakage.

Threats and attacks over RPL can lead to failures in authentication, maintenance of routing information and attacks on integrity or availability of the network operations [50]. Once an attacker captures a node, it is able to obtain the encrypted information and inject evil code to disturb the routing, which is quite difficult to be

Table 4.3 RPL applications and metrics

Application Scenarios	Topology Features	Metric/Objective Function Characteristics
Smart Grid [5, 20]	Large-scale and dense distribution	• Parents switching autonomously • Appropriate duty cycling
M2M (General Scope) [2]	Heterogeneous sensor system and involves a large number of devices	• Multiple next hops • Best forward selection
Agriculture Greenhouse [42]	Heterogeneous information system	• Multi-path • Hop count and residual Energy
Medical Applications [15]	Mobile nodes; Dynamic and hybrid topology	• Mobile nodes work as leaf node without broadcasting DIO

detected particularly when innocent nodes fail to know the attacks. Table 4.4 depicts the attack types in RPL.

As depicted in Figure 4.4, the attacks can lead to a non-optimal routing or even result in a worse situation such as routing loops or unreachable neighbours. For example, when node 3 chooses node 6 as its preferred parent, which has a larger rank, a rank attack happens with a formed loop of 3-6-5. Routing choice attack happens when node 7 detaches node 5 and chooses node 2 as its parent node. As for neighbour attack, node 4 can replicate messages from node 2 and deceives node 8 to choose node 2 as its parent, which is totally out of range for node 8.

To solve the above issues, an Intrusion Detection System (IDS) that is capable of analysing activities or processes in a network or in a node is proposed. The IDS normally deploys monitor nodes in finite state machine mode; every node in a network should be monitored under at least one of them. Such a method works well to efficiently detect rank attacks and local attacks [29]. Other IDS based methods,

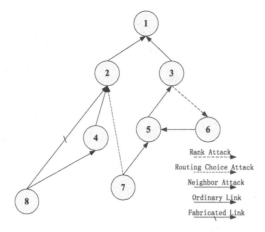

Figure 4.4 An illustration of attacks in RPL.

such as the one mainly focusing on the inner intrusion [58], can successfully solve routing choice attack by avoiding the optimal routing path failure caused by tampering options of DIOs. Le et al. [30] made a comprehensive analysis of rank attack, local repair stack, neighbour attack and DIS attack, and suggested that the handling models of the attacks can be developed through training of data.

Besides the intrusion detection based methods, the encryption of information in RPL is another option. Clark *et al.* [49] proposed a node-to-node encrypted authentication method by exchanging an encryption key. Seeber *et al.* [44] deployed a Trust Platform Model (TPM), which is able to provide cryptographic operations and node

Table 4.4 Category of attacks in RPL

Attack Type	Feature	Impact
Rank Attack [29, 30]	Choose non-preferred parent as parent node	Destroy routing or format loops
Local Repair Attack [29, 30]	Send local repair information untimely	Destroy routing, waste routing resources
Neighbour Attack [30]	Manipulate control information to deceive neighbour nodes	Forge and destroy routing, waste network resources
Routing Choice Attack [58]	Choose non-optimal routing path	Destroy routing, waste routing resources
Sinkhole Attack [49]	Route traffic to the node pretending to be a valid sink	Destroy routing and topology
Distance Spoofing Attack [49]	Route traffic to a node near the sink	Destroy routing and waste computation resources

authentication, to avoid evil routing information through related trust construction and key exchange mechanism.

As shown above, anti-attacks can be a challenging task for LLNs. ROLL WG analysed the security threats and attacks including authentication, access control, confidentiality, integrity and availability in [50]. Considering the different categories of threats and attacks, possible solutions have been offered, which mainly focus on establishing session keys, encapsulation during encryption and access control. It also points out that the sensor network limitations including energy, physical locations, directional traffic, etc., combined with use case requirements including urban networks [11], building automation [36], industrial automation [41] and home automation [7], can be the new motivation to design more effective RPL in real scenarios.

4.6 RPL PERFORMANCE EVALUATION IN LARGE-SCALE NETWORKS

4.6.1 Simulation Platforms

So far there are a number of software tools [28, 39, 46] that can be used for evaluating RPL performance. However, this is not always the case for a large-scale simulation. Table 4.5 summarises the key features and large-scale simulation capacity among major simulation platforms.

In our study, we consider to use OMNeT++, which is an event triggered, time discrete open source network simulator and based on module construction and realization. It is capable of implementing RPL simulation at a larger scale as well as with advantages in other aspects, such as easy access of OMNeT++ frameworks for different network scenarios and functional output API to obtain a series of targeted data.

4.6.2 Framework Integration for OMNeT++

The RPL simulation is developed based on the integration of INET 2.2.0 with MiXiM 2.3. The latest version INET 2.3 has already incorporated several functions from MiXiM. Both of them are the most prevalent frameworks in OMNeT++.

INET is a simulation framework with comparatively mature network layer realization. Its IPv6 network layer has been realized with diversified sub-modules, taking into account neighbour discovery functions and its message mechanism, including Neighbor Advertisement (NA), Neighbor Solicitation (NS), Router Advertisement (RA) and Router Solicitation (RS) handlings. Here we incorporate the IPv6 mechanism to construct RPL related functions. The routing table can store parents information, which is necessary in RPL DODAG construction.

---

https://omnetpp.org
https://inet.omnetpp.org
http://mixim.sourceforge.net

Table 4.5 Simulator comparisons for supporting RPL

Simulator	Support for RPL	Support for Large-scale Simulations	Supported Platforms and Programming Languages
JSim [51]	• Supports multiple protocols while the only MAC protocol that can be used is IEEE 802.11, which is a limitation in supporting RPL in JSim • Inactive since 2006	• Able to support simulation scale around 500 nodes while the execution takes a longer time • Complicated to use and less efficient	• Linux, Mac, and Windows • Java and tcl script language
Cooja [54]	• Fully supportive of RPL • Part of contiki OS • No specific energy consumption model	• Relatively low efficiency • Limited simulation scale with 200-500 nodes • Long processing time	• Linux, Mac, and Windows • Standard C
TOSSIM [32, 33]	• TinyRPL supports MP2P, P2P, P2MP traffic in RPL • However, TinyRPL is not supported on the TOSSIM simulator which requires a micaz binary. Therefore, it does not fully support RPL simulation	• Able to support thousands of nodes	• TOSSIM is designed specifically for TinyOS applications to be run on MICA Motes • C++ and python
Ns-2 [23]	• Object-oriented design which allows for straightforward creation and use of new protocols • Extensible for general WSN simulation • Fail to simulate problems of the bandwidth, power consumption or energy saving in WSN	• Only support less than 100 nodes • Rather complex and time-consuming • Only slightly maintained now	• General simulator and compatible with Linux, Mac, and Windows • C++ and OTcl
Ns-3 [9]	• Not backward compatible with Ns-2 • Modelling of Internet protocols and networks work • Weak in MAC and PHY layer development support	• Support of large scale but more nodes beyond 400+ may lead to unrealistic results	• General simulator and compatible with Linux, Mac, and Windows • C++ and python scripts
OMNeT++ [52]	• Offers various frameworks to deploy the network with RPL while the integration of available modules may introduce compatible problems • Extensible for general WSN simulation • Support energy consumption and mobility models	• Scale-free simulator	• General simulator and compatible with Linux, Mac, and Windows • C++ and NED language

MiXiM framework is well known for its realization of MAC and physical layers, especially IEEE 802.15.4. In our study, the CC2420 radio model is used for IEEE 802.15.4 MAC and PHY. Moreover, its battery module has been developed with a linear model which is more reliable in battery consumption observation. Here we deploy SimpleBattery module in MiXiM to model the energy consumption of networks and will mainly focus on the realistic results of Tx power consumption [18].

Moreover, INET framework provides several mobility models that can be easily utilized in the simulations, such as the mobility model in which the node randomness is controlled by the linear model, the Gauss–Markov model, etc. We only consider the stationary scenarios; therefore the StationaryMobility model is used as shown in Table 4.6.

With the integration of the above frameworks, we are able to run experiments with flexible parameters to observe the performance of large-scale RPL under various circumstances. Specifically, the integration offers an experimental basis to construct networks with specified functions, such as implementing new OFs, Metrics, Constraints, etc.

4.6.3 Configuration Details

We build our network layer based on IPv6 module in INET, the IEEE 802.15.4 MAC and PHY layer in MiXiM through 6LoWPAN adaptation. The upward routing has been realised with DIO and DIS messages mechanism. The DIO messages are implemented with a trickle timer. The parent and routing selections are decided by an extra class corresponding to the OFs we defined. The basic parameters of layers are defined in a .ini configuration file and the topology can also be preconfigured, which can be either randomly set or according to certain patterns. The source code is made available for further reference. With the node structure implemented above, the RPL mechanism can be implemented in the following three aspects.

1. The RPL message mechanism is defined and achieved in the **IPv6Neighbor Discovery** module. It replaces the default RA and RS message functions. The module is deployed with DIO and DIS messages handling and responsible for undertaking the update of preferred parents and path selection.

2. **RoutingTable** module plays a valid role of recording related routing information and making routing choice when forwarding packets. It mainly serves as a storing module that records the routing information and completes parent node selection.

3. The DODAG construction and rank computation obeys a certain Objective Function, which exists as an independent class completely performing the min-

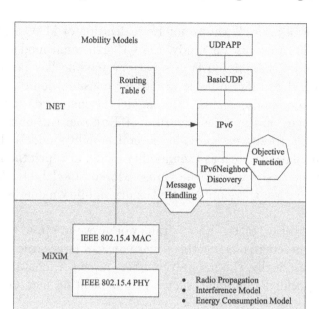

Figure 4.5 Simulation construction structure.

imum cost routing path selection. This paper mainly focuses on metrics analysis; therefore, multiple OFs with different metrics need to be deployed separately. We deploy **OF** as a single class file in the simulation, such that hard codes can be avoided and it is easy to be replaced and updated accordingly, which provides enough flexibility and extensibility. Figure 4.5 summarizes the simulation architecture that combines the frameworks, node structure and fundamental mechanisms.

We consider the messages that have been defined in RFCs—DIO and DIS for upward routing, which are the triggers for DODAG. The essential key options for routing selection are contained in messages. For example, Figure 4.6 shows the handling procedures across layers when a message uses received signal strength index (RSSI) as the key option. The RSSI information needs to be transmitted across layers and finally be utilized for path selection in the **IPv6NeighborDiscovey** module. It is worth noting that Figure 4.6 deploys RSSI based handling. Other key options may only involve the top two layers if there is nothing to do with the PHY or MAC layers.

Table 4.6 shows the parameter configuration in our simulation. The parameters in PHY and MAC are set according to the CC2420 datasheet. The trickle timer parameters have been explained in Section 4.2. Figure 4.7 depicts an example of random topology generated by OMNeT++ with 100 and 500 nodes, respectively.

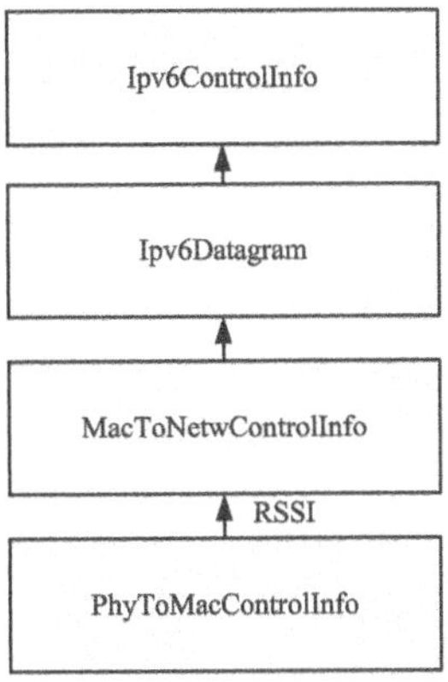

Figure 4.6 Message handling procedure cross layers for RSSI based handling.

4.6.4 Simulation of Cross-layer RPL Routing

RPL is compatible with a variety of MAC and physical protocols, especially IEEE 802.15.4. Since MAC and physical layer parameters have direct impacts on the link reliability as well as energy consumption, taking a cross-layer approach to incorporate low layer elements into metrics design may offer extra benefits for routing.

Sheng *et al.* in [45] proposed a novel method combining multi-path topology with duty cycle ratio in the MAC layer. It proves the sustainable network performance with the dynamic duty cycling adjustment. Di Marco *et al.* [34] proposed a reliability metric based on the Markov analysis model [35] and designed an algorithm with backwards and retransmission times of forward flow in IEEE 802.15.4. The forward flow contains flow generated by the node itself and relayed flow from child nodes. Compared with ETX, it takes the packet loss into account. Besides, in order to better balance the flow in the whole network, an optimized metric is also proposed to integrate itself flow and relayed flow with sending and receiving power, respectively. Sajan *et al.* [43] proposed a cognitive radio network (CRN) based RPL protocol by utilizing six frequency channels between nodes to represent the channel availability obtained through the efficient spectrum sensing algorithm. The main contribution lies in the routing repair functions regarding different channels with trickle timer of RPL.

To illustrate the cross-layer impact on the network performance, we develop a simulation experiment using OMNeT++. Particularly, under the same simulation settings, the OFs are compared among hop count, ETX and a tailored ETX with a correction factor-RSSI (RSSI-ETX), which is incorporated into the classic ETX to further rectify the deviation of link quality. RSSI is calculated as the maximum received signal strength in a time period from its last packet reception to the current reception. It will be logarithmically recorded and then combined with ETX. The role it plays is as a deviation controller to the ETX.

Table 4.6 Parameters of RPL simulation in OMNeT++

	MAC Layer	
	macTransPower	1 mW
	macMinBE	1
	macMaxBE	6
	macMaxCSMABackoffs	20
	rxSetupTime	0.1 s
MiXiM	macAckWaitDuration	0.000864 s
	PHY Layer	
	phySensitivity	-100 dBm
	phyMaxTXPower	1.1 mW
	AnalogueModel	LogNormalShadowing
	Connection Manager	
	carrierFrequency	2.4e9 Hz
	pMax	60 mW
	AttenuationThres	-84 dBm
	Trickle Timer	
	DIOIntMin (I_{min})	0.75 s
	DIORedun (t_K)	10
	DIOInetDoubl (I_{max})	8
	Topology Formation	
	Start Time	0 s
INET	Simulation End Time	300 s
	UDPApp (Packet Generation)	
	Size of Packet Payload	60~1000 Bytes
	StartTime (from)	60 s
	EndTime (stop at)	60.19 s
	Interval	0.02 s
	Destination Node Id	0
	Source Node Id	All nodes except 0
	Mobility	
	Mobility Type	StationaryMobility
	Playground Size	480 m×480 m

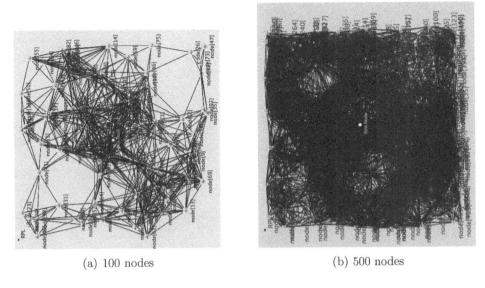

(a) 100 nodes (b) 500 nodes

Figure 4.7 A random topology with 100 and 500 nodes developed by OMNeT++.

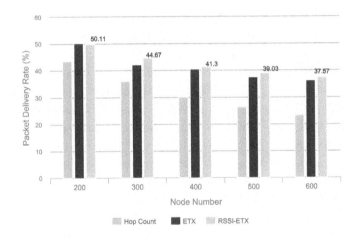

Figure 4.8 Packet delivery rate versus network scale under different Objective Functions.

We consider the simulation with the network size from 200 nodes to 600 nodes connected by a log-normal shadowing channel for a period of 300 s with a fixed UDPApp payload size of 60 bytes. The time schedule of data packet transmitting is shown in UDPApp parameters in Table 4.6. The data packets' generation is initiated from 60 s when a comparatively stable topology can be formed from the beginning of the simulation (0 s). The simulation result is averaged over 5 dependent trials with different random seeds. As depicted in Figures 4.8 and 4.9, packet delivery rate and

---

Random seeds are generated with a Mersenne Twister as a random sequence. The random seed can be set differently for each module. For example, the random seed will determine the random time unit generated in the trickle timer at the network layer.

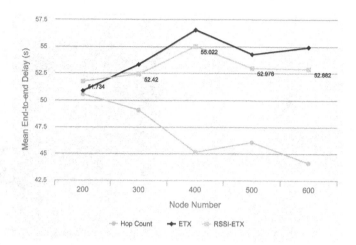

Figure 4.9 Mean end-to-end delay versus network scale under different Objective Functions.

mean end-to-end delay are shown, respectively. The maximum number of hops in the simulation is 12. It is worth noting that the general packet delivery rate in Figure 4.8 is lower compared with the simulation results in [12] because of the high packet loss in multi-hop networks and burst transmissions simultaneously in the same time frame, which causes significant congestion and interference. In Figure 4.9, we only consider the time delay of successful packet delivery. The retransmission and buffering have not been taken into account in our study. The increase of network size will lead to more hops when a node undertakes the parent selection process and therefore causes a longer delay. However, for the hop-based approach, the increasing density of nodes can lead to a better selection of a path with minimum hops; hence its mean end-to-end delay presents a decreasing trend. When an RSSI element is considered, the rectified OF performs better in a comprehensive view in packet delivery rate and mean end-to-end delay. Figure 4.10 shows the percentage of nodes with parent change which reflects the extent of dynamic adjustment in the network during the simulation. A higher change rate indicates a more dynamic network topology and prompt response to link quality. However, we should admit that the overhead imposed by dynamic change will be a bottleneck for large-scale deployment. In essence, the benefits by reflecting physical communication channels and signal behaviours on upper layers do play a vital role in routing communications; however, the performance trade-off between packet delivery rate, delay and maintenance overhead should be well considered in large-scale network design.

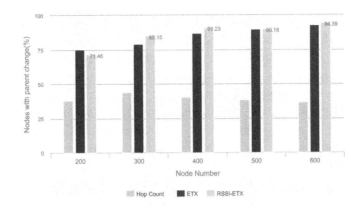

Figure 4.10 Percentage of nodes with parent change under different Objective Functions.

4.7 CHALLENGES AND PROSPECT

Although RPL is emerging as a comprehensive routing solution to general wireless sensor networks, there are still challenges as follows.

1. Transmission Mode

   Currently, the prominent transmission traffic type in RPL is MP2P, that is, the upward routing implemented by DIO, which is well defined in the standard. However, for the downward routing, the P2P and P2MP traffic modes that are mainly implemented by DAO are not precisely defined in literature. A complementary IETF standard protocol [17] has been proposed to solve the congestion and latency issues exposed by P2P traffic mode while the multicast protocol [27] has been taken into account for the MP2P mode. More efforts need to be done in DAO scheduling to relieve the congestion and buffer requirements.

   Furthermore, storage limitation is still a big challenge for large-scale routing. Considering the non-storing and storing modes in downward routing, with the network size increases, storing mode will lead to large memory consumption while the non-storing mode will introduce large communication overhead [14]. The challenge is to find a balanced solution by effectively integrating both models to reduce the memory overhead risk and improve the utilization of node capacity.

2. Diversification of OFs

   Existing literature investigated diversified influential elements in routing construction, including the control overhead, link quality, remaining energy, etc.

   Due to the nature of WSN and IoT applications, the network performance is not only limited to the packet delivery and time delay, but also energy efficiency and long term stability required by LLN. It has been verified that the combination of the influential elements can result in trade-off in routing performance. For example, a node's remaining energy and link quality can be considered jointly to create an optimal metrics in delivering long lifetime and reliable WSN. Kamgueu *et al.* [25] put forward a new perspective regarding the OF design, which introduced the fuzzy inference system (FIS) that is mainly defined for an uncertain system. The FIS is able to merge several metrics into one in a reasonable way. The qualitative approach is promising for RPL OF design. Additionally, the possibility of multi-parents in high dense networks can be further explored. Balancing the traffic load with multicast traffic or introducing parent switching in traffic routing can relieve the network load and prolong the network longevity.

In an emerging new IoT application domain, mobile nodes are allowed to connect to the static routing topology and thus the routing protocol to cope with node mobility is extremely challenging. Mobile RPL tends to lead to dynamic changes of topology and link failures. The technical question is how to react to a rapid change of preferred parent which has a significant impact on the reliability and stability of the network. The mobility influence should be considered in the OF for mobile based routing protocols. Hayes *et al.* [19] proposed a solution for mobile wireless sensor networks taking into account multiple paths' utilisation and blind forwarding technique, which is evaluated to be highly adaptable and robust. Mechanisms in the proposed routing protocol can be brought in by RPL and better support of mobile RPL.

In essence, an effective routing protocol design should consider the application environment. Thus, OFs need to be adjusted specifically to satisfy the characteristics of application scenarios.

3. Energy Issue

The energy consumption is always a concern in LLNs. Because of the differences of relative distances from the current node to the sink node in the network, energy consumption among nodes may be distinct and can lead to scenarios with emerging bottlenecks, which will affect the network reliability. Current studies make an effort to take nodes' energy depletion rate into the metrics and make predictions about the path that will consume energy at the lowest rate. An alternative method is to introduce backup nodes to take the place of the dead nodes with minimum network cost. However, the bottleneck nodes can be unavoidable to some extent; thus the critical question on how to balance the energy of nodes effectively is what needs to be looked into in the future.

In LLNs, especially in a large scale, equalizing the energy consumption is much more important than saving the energy in the network. Nurmio *et al.* [38] considered the energy of all parent nodes along the path towards the sink node, which resulted in an equalized energy consumption rate among nodes in the network.

Besides the scalability, the diversity of networks and distinction among nodes also have impacts on the energy consumption. Thus different Quality of Services requirements and power-supply types of nodes should also be considered in future work.

4. Cross-layer Issue

The cross-layer issue existing here is mainly related to the discrepancy between the payload in network layer and MAC layer. MTU of IPv6 network is 1280 bytes while that of IEEE 802.15.4 MAC is 127 bytes, thus an adaptive layer—6LoWPAN is indispensable to handle fragmentation and reassembly of data packets as well as head compression. Gardasevic *et al.* [16] has proved that with the increase of UDP payload, the routing performance including delay and PDR will be worse in both unicast and multicast scenarios.

The increasing size of the packet payload will impact the routing performance on consuming much more energy, decreasing packets' delivery and increasing network latency. The challenge of how the strategy of routing should be adjusted according to the packet payload needs to be further explored.

4.8 CONCLUSION

RPL has the potential to provide a viable solution for routing in IoT. We have discussed RPL in multiple aspects, including its principal mechanisms, key features, application scenarios and security issues. Moreover, a practical large-scale simulation investigation has been conducted. In essence, routing implementation needs to keep pace with the rapid development of IoT and we believe RPL will play an important role in IoT development.

Bibliography

[1] A. Aijaz, H. Su, and A. H. Aghvami. Corpl: A routing protocol for cognitive radio enabled AMI networks. *IEEE Transactions on Smart Grid*, 6(1):477–485, 2014.

[2] A. Aijaz and A. H. Aghvami. Cognitive machine-to-machine communications for internet-of-things: A protocol stack perspective. *IEEE Internet of Things Journal*, 2(2):103–112, 2015.

[3] C. Alcaraz, P. Najera, J. Lopez, and R. Roman. Wireless sensor networks and the internet of things: Do we need a complete integration? In *Proc. 1st International Workshop on the Security of the Internet of Things (SecIoT'10)*, Tokyo, Nov 2010.

[4] E. Ancillotti, R. Bruno, and M. Conti. The role of the RPL routing protocol for smart grid communications. *IEEE Communications Magazine*, 51(1):75–83, 2013.

[5] E. Ancillotti, R. Bruno, and M. Conti. Reliable data delivery with the ietf routing protocol for low-power and lossy networks. *IEEE Transactions on Industrial Informatics*, 10(3):1864–1877, 2014.

[6] S. Borlase. *Smart Grids: Infrastructure, Technology, and Solutions*. CRC Press, 2012.

[7] A. Brandt and G. Porcu. Home Automation Routing Requirements in Low-Power and Lossy Networks. RFC 5826, April 2010.

[8] S. Capone, R. Brama, N. Accettura, D. Striccoli, and G. Boggia. An energy efficient and reliable composite metric for rpl organized networks. In *Proc. 2014 12th IEEE International Conference on Embedded and Ubiquitous Computing (EUC)*, pages 178–184, Milano, Aug 2014.

[9] G. Carneiro. Ns-3: Network simulator 3. In *UTM Lab Meeting April*, volume 20, 2010.

[10] T. Clausen, A. Colin de Verdiere, J. Yi, A. Niktash, Y. Igarashi, H. Satoh, U. Herberg, C. Lavenu, T. Lys, C. Perkins, et al. The lightweight on-demand ad hoc distance-vector routing protocol-next generation (loadng). *draft-clausen-lln-loadng-09 (work in progress)*, 2013.

[11] M. Dohler, T. Watteyne, T. Winter, and D. Barthel. Routing Requirements for Urban Low-Power and Lossy Networks. RFC 5548, May 2009.

[12] S. Duquennoy, O. Landsiedel, and T. Voigt. Let the tree bloom: Scalable opportunistic routing with ORPL. In *Proceedings of the 11th ACM Conference on Embedded Networked Sensor Systems*, page 2, Roma, Nov 2013. ACM.

[13] D. Evans. The internet of things: How the next evolution of the internet is changing everything. *CISCO white paper*, 1:1–11, 2011.

[14] W. Gan, Z. Shi, C. Zhang, L. Sun, and D. Ionescu. Merpl: A more memory-efficient storing mode in RPL. In *Proc. 2013 19th IEEE International Conference on Networks (ICON)*, pages 1–5, Singapore, Dec 2013.

[15] F. Gara, L. Ben Saad, R. Ben Ayed, and B. Tourancheau. Rpl protocol adapted for healthcare and medical applications. In *Proc. 2015 International Wireless Communications and Mobile Computing Conference (IWCMC)*, pages 690–695, Dubrovnik, Aug 2015.

[16] G. Gardasevic, S. Mijovic, A. Stajkic, and C. Buratti. On the performance of 6lowpan through experimentation. In *Proc. 2015 International Wireless Communications and Mobile Computing Conference (IWCMC)*, pages 696–701, Dubrovnik, Aug 2015.

[17] M. Goyal, M. Philipp, A. Brandt, and E. Baccelli. Reactive Discovery of Point-to-Point Routes in Low-Power and Lossy Networks. RFC 6997, August 2013.

[18] S. K. Hammerseth. Implementing RPL in a mobile and fixed wireless sensor network with OMneT++. 2011.

[19] T. Hayes and F. H. Ali. Location aware sensor routing protocol for mobile wireless sensor networks. *IET Wireless Sensor Systems*, 6(2):49–57, 2016.

[20] Q. D. Ho, Y. Gao, G. Rajalingham, and T. Le-Ngoc. Robustness of the routing protocol for low-power and lossy networks (RPL) in smart grid's neighbor-area networks. In *Proc. 2015 IEEE International Conference on Communications (ICC)*, pages 826–831, London, June 2015.

[21] O. Iova, F. Theoleyre, and T. Noel. Stability and efficiency of RPL under realistic conditions in wireless sensor networks. In *Proc. 2013 IEEE 24th Annual International Symposium on Personal, Indoor, and Mobile Radio Communications (PIMRC)*, pages 2098–2102, London, Sept 2013.

[22] O. Iova, F. Theoleyre, and T. Noel. Using multiparent routing in RPL to increase the stability and the lifetime of the network. *Ad Hoc Networks*, 29:45–62, 2015.

[23] T. Issariyakul and E. Hossain. *Introduction to Network Simulator NS2*. Springer Science & Business Media, 2011.

[24] G. Iyer, P. Agrawal, and R. S. Cardozo. Performance comparison of routing protocols over smart utility networks: A simulation study. In *Proc. IEEE GLOBECOM Workshops*, pages 969–973, Atlanta, GA, Dec 2013.

[25] P. O. Kamgueu, E. Nataf, and T. Ndie Djotio. On design and deployment of fuzzy-based metric for routing in low-power and lossy networks. In *Proc. Local Computer Networks Conference Workshops (LCN Workshops), 2015 IEEE 40th*, pages 789–795, Clearwater Beach, FL, Oct 2015.

[26] P. O. Kamgueu, E. Nataf, T. D. Ndié, and O. Festor. *Energy-based routing metric for RPL*. PhD thesis, INRIA, 2013.

[27] R. Kelsey and J. Hui. Multicast Protocol for Low-Power and Lossy Networks (MPL). RFC 7731, February 2016.

[28] A. R. Khan, S. M. Bilal, and M. Othman. A performance comparison of open source network simulators for wireless networks. In *Proc. 2012 IEEE International Conference on Control System, Computing and Engineering (ICCSCE)*, pages 34–38, Penang, Nov 2012.

[29] A. Le, J. Loo, Y. Luo, and A. Lasebae. Specification-based IDS for securing RPL from topology attacks. In *Proc. Wireless Days (WD), 2011 IFIP*, pages 1–3, Niagara Falls, ON, Oct 2011.

[30] A. Le, J. Loo, Y. Luo, and A. Lasebae. The impacts of internal threats towards routing protocol for low power and lossy network performance. In *Proc. 2013 IEEE Symposium on Computers and Communications (ISCC)*, pages 000789–000794, Split, July 2013.

[31] P. Levis, T. Clausen, J. Hui, O. Gnawali, and J. Ko. The Trickle Algorithm. RFC 6206, March 2011.

[32] P. Levis and N. Lee. Tossim: A simulator for tinyos networks. *UC Berkeley, September*, 24, 2003.

[33] P. Levis, S. Madden, J. Polastre, R. Szewczyk, K. Whitehouse, A. Woo, D. Gay, J. Hill, M. Welsh, E. Brewer, et al. Tinyos: An operating system for sensor networks. In *Ambient Intelligence*, pages 115–148. Springer, 2005.

[34] P. Di Marco, C. Fischione, G. Athanasiou, and P. V. Mekikis. MAC-aware routing metrics for low power and lossy networks. In *INFOCOM, 2013 Proceedings IEEE*, pages 13–14, Turin, April 2013. IEEE.

[35] P. Di Marco, Pangun Park, C. Fischione, and K. H. Johansson. Analytical modeling of multi-hop IEEE 802.15.4 networks. *IEEE Transactions on Vehicular Technology*, 61(7):3191–3208, 2012.

[36] J. Martocci, Pieter De Mil, N. Riou, and W. Vermeylen. Building Automation Routing Requirements in Low-Power and Lossy Networks. RFC 5867, June 2010.

[37] G. Montenegro, C. Schumacher, and N. Kushalnagar. IPv6 over Low-Power Wireless Personal Area Networks (6LoWPANs): Overview, Assumptions, Problem Statement, and Goals. RFC 4919, August 2007.

[38] J. Nurmio, E. Nigussie, and C. Poellabauer. Equalizing energy distribution in sensor nodes through optimization of RPL. In *Proc. 2015 IEEE International Conference on Computer and Information Technology; Ubiquitous Computing and Communications; Dependable, Autonomic and Secure Computing; Pervasive Intelligence and Computing (CIT/IUCC/DASC/PICOM)*, pages 83–91, Liverpool, Oct 2015.

[39] A. K. Patil and Dr. P. M. Hadalgi. Evaluation of discrete event wireless sensor network simulators. *International Journal of Computer Science & Network*, 1(5), 2012.

[40] A. Paventhan, D. Darshini. B, H. Krishna, N. Pahuja, M. F. Khan, and A. Jain. Experimental evaluation of IETF 6TisCH in the context of smart grid. In *Proc. 2015 IEEE 2nd World Forum on Internet of Things (WF-IoT)*, pages 530–535, Milan, Dec 2015.

[41] K. Pister, P. Thubert, S. Dwars, and T. Phinney. Industrial Routing Requirements in Low-Power and Lossy Networks. RFC 5673, October 2009.

[42] T. N. Quynh, N. L. Manh, and K. N. Nguyen. Multipath RPL protocols for greenhouse environment monitoring system based on internet of things. In *Proc. 2015 12th International Conference on Electrical Engineering/Electronics, Computer, Telecommunications and Information Technology (ECTI-CON)*, pages 1–6, Hua Hin, June 2015.

[43] I. Sajan and E. M. Manuel. Cross layer routing design based on RPL for multi-hop cognitive radio networks. In *Proc. 2015 IEEE International Conference on*

Signal Processing, Informatics, Communication and Energy Systems (SPICES), pages 1–6, Kozhikode, Feb 2015.

[44] S. Seeber, A. Sehgal, B. Stelte, G. D. Rodosek, and J. SchÃűnwÃďlder. Towards a trust computing architecture for RPL in cyber physical systems. In *Proceedings of the 9th International Conference on Network and Service Management (CNSM 2013)*, pages 134–137, Zurich, Oct 2013.

[45] Z. Sheng, S. Yang, Y. Yu, and A. Vasilakos. A survey on the IETF protocol suite for the internet of things: Standards, challenges, and opportunities. *IEEE Wireless Communications*, 20(6):91–98, 2013.

[46] H. Sundani, H. Li, V. Devabhaktuni, M. Alam, and P. Bhattacharya. Wireless sensor network simulators a survey and comparisons. *International Journal of Computer Networks*, 2(5):249–265, 2011.

[47] Cisco Systems, M. Kim, K. Pister, N. Dejean, and D. Barthel. Routing metrics used for path calculation in low-power and lossy networks. *Heise Zeitschriften Verlag*, 2012.

[48] Y.-R. Tan. RPL-based Clustering Routing Protocol. Internet-Draft draft-tan-roll-clustering-00, Internet Engineering Task Force, December 2015. Work in Progress.

[49] C. Taylor and T. Johnson. Strong authentication countermeasures using dynamic keying for sinkhole and distance spoofing attacks in smart grid networks. In *Proc. 2015 IEEE Wireless Communications and Networking Conference (WCNC)*, pages 1835–1840, New Orleans, LA, March 2015.

[50] T. Tsao, R. Alexander, M. Dohler, V. Daza, A. Lozano, and M. Richardson. A Security Threat Analysis for the Routing Protocol for Low-Power and Lossy Networks (RPLs). RFC 7416, January 2015.

[51] H.-ying Tyan, Y. Gao, J. Hou, et al. Tutorial: Working with J-Sim, 2003.

[52] A. Varga et al. The OMNeT++ discrete event simulation system. In *Proceedings of the European Simulation Multiconference (ESMâĂŹ2001)*, volume 9, page 65, 2001.

[53] J.-P. Vasseur and A. Dunkels. *Interconnecting smart objects with IP: The next internet*. Morgan Kaufmann, 2010.

[54] T. Voigt. Contiki cooja crash course. *The International School on Cooperative Robots and Sensor Networks (RoboSense School 2012), Hammamet, Tunisia*, 2012.

[55] T. Winter. RPL: IPv6 Routing Protocol for Low-Power and Lossy Networks. RFC 6550, March 2012.

[56] W. Xiao, J. Liu, N. Jiang, and H. Shi. An optimization of the object function for routing protocol of low-power and lossy networks. In *Proc. Systems and Informatics (ICSAI), 2014 2nd International Conference on*, pages 515–519, Shanghai, Nov 2014.

[57] X. Yang, J. Guo, P. Orlik, K. Parsons, and K. Ishibashi. Stability metric based routing protocol for low-power and lossy networks. In *Proc. 2014 IEEE International Conference on Communications (ICC)*, pages 3688–3693, Sydney, NSW, June 2014.

[58] L. Zhang, G. Feng, and S. Qin. Intrusion detection system for RPL from routing choice intrusion. In *Proc. 2015 IEEE International Conference on Communication Workshop (ICCW)*, pages 2652–2658, London, June 2015.

[59] M. Zhao, H. Y. Shwe, and P. H. J. Chong. Cluster-parent based RPL for low-power and lossy networks in building environment. In *Proc. 2015 12th Annual IEEE Consumer Communications and Networking Conference (CCNC)*, pages 779–784, Las Vegas, NV, Jan 2015.

CHAPTER 5

Resource Allocation for Wireless Communication Networks with RF Energy Harvesting

Elena Boshkovska

Friedrich-Alexander-University Erlangen-Nürnberg (FAU), Germany

Derrick Wing Kwan Ng

The University of New South Wales, Australia

Robert Schober

Friedrich-Alexander-University Erlangen-Nürnberg (FAU), Germany

CONTENTS

5.1 INTRODUCTION

The successful development of wireless communication networks and technologies has triggered an exponential growth in the number of wireless communication devices worldwide. In the near future, devices embedded with multifunctional sensors and communication chip sets will be able to collect and exchange information via the Internet. Specifically, these smart devices will be connected to computationally powerful central computing systems to provide intelligent services for the daily life such as environmental monitoring, e-health, automated control, energy management, logistics, and safety management. This new concept of interconnecting a massive number of communication and sensing devices is known as the Internet of Things (IoT) [1].

It is predicted that in 2020, the number of devices interconnected via the Internet on the planet may reach up to 50 billion. Besides, the density of such networks will be around 1 million devices per km^2. Therefore, the wireless communication infrastructure is a key enabler of IoT. In fact, IoT requires energy-efficient and cost-effective wireless communications. Similar to conventional communication networks, the lifetime of IoT networks depends on the available energy at the transceivers. However, smart devices in IoT networks are ubiquitous with various levels of mobility. In other words, connecting these devices to fixed power grids to replenish their energy may not be a viable option. Therefore, most of the transceivers in IoT networks will be powered by batteries with limited energy storage which will reduce the lifetime of the networks significantly. Although the energy shortage can be alleviated by temporary battery replacements, such an intermediate solution may require frequent replacement of batteries which can be costly, time consuming, and cause interruption of service. This creates a serious performance bottleneck for providing stable communication, especially for delay sensitive services. On the other hand, a viable solution to extend the lifetime of wireless communication networks is to integrate wireless communication devices with energy harvesting (EH) technology to scavenge energy from the environment. In practice, wind, solar, and geothermal are the major renewable energy sources for generating electricity [2–4], thereby reducing substantially the reliance on the energy supply from the power grid. Yet, these conventional natural energy sources are usually climate and location dependent which restricts the mobility of smart devices. Besides, most of these energy sources are not available in indoor environments. More importantly, the uncontrollable and intermittent nature of these natural energy sources makes their use in IoT communication networks challenging.

Recently, wireless energy transfer (WET) has emerged as one of the technologies driving IoT networks and has attracted much attention from both academia and industry [5–27]. The existing WET technologies can be categorized into three classes: inductive coupling, magnetic resonant coupling, and radio frequency (RF)-based WET. The first two technologies rely on near-field electromagnetic (EM) waves. In particular, these two technologies can provide wireless charging over short distances only due to the required alignment of the magnetic field with the EH circuit. Therefore, in general, near-field techniques do not support the mobility of EH devices. In contrast, RF-based WET [5–24] exploits the far-field properties of EM waves facilitating long distance wireless charging. More importantly, EM waves not only serve as a vehicle for carrying energy, but also for carrying information which enables the possibility of simultaneous wireless information and power transfer (SWIPT) and wireless powered communication (WPC). Specifically, in SWIPT networks, a transmitter broadcasts both information and energy signals to provide information and energy delivery service simultaneously. In wireless powered communication networks (WPCNs), wireless communication devices first harvest energy, either from a dedicated power station or from ambient RF signals, and then use the harvested energy to transmit information signals. Compared to conventional EH, RF-based EH technology provides an on-demand energy replenishment which is suitable for smart wireless communication devices having strict quality of service (QoS) and energy requirements. On the other hand, various "last meter" wireless communication systems, such as Wi-Fi and small cell systems, can be potentially exploited for energy replenishment of battery constrained wireless devices. Nowadays, simple EH circuits are able to harvest microwatts to milliwatts of power over the range of several meters for a transmit power of 1 Watt and a carrier frequency of less than 1 GHz [28]. Although the development of WET technology is still in its infancy, there are already some preliminary practical applications of WET such as passive radio-frequency identification (RFID) systems. It is expected that the introduction of RF-based EH to smart communication devices will revolutionize the system architecture and resource allocation algorithm design.

Conventional wireless communication systems are required to provide different types of QoS requirements such as throughput, reliability, energy efficiency, fairness, and timeliness [29–32]. On top of this, efficient WET is expected to play an important role as an emerging QoS requirement for RF-based wireless EH communication networks. In practice, for a carrier frequency of 915 MHz, the signal attenuation is 50 dB for every 10-meter of free space propagation. Hence, the efficiency of WET will be unsatisfactory for long distance transmission unless advanced resource allocation and antenna technology are combined. As a result, various resource allocation algorithms exploiting multiple-antenna technology have been proposed [17–24]. Specifically, by utilizing the extra degrees of freedom offered by multiple transmit antennas, a narrow signal beam can be created and can be more accurately steered towards the desired receivers to improve the efficiency of WET. In this chapter, we study the resource allocation algorithm design for two specific RF-based multiple antenna EH communication networks.

The remainder of this chapter is organized as follows. In Section 5.2, we introduce various types of receiver structures for RF-based EH wireless communications. Sections 5.3 and 5.4 study the resource allocation algorithm design for SWIPT systems and WPCNs, respectively. In Section 5.5, we conclude with a brief summary of this chapter.

Notation

In this chapter, we adopt the following notations. $\mathbf{A}^H$, $\mathrm{Tr}(\mathbf{A})$, and $\mathrm{Rank}(\mathbf{A})$ represent the Hermitian transpose, trace, and rank of matrix $\mathbf{A}$; $\mathbf{A} \succeq \mathbf{0}$ indicates that $\mathbf{A}$ is a positive semidefinite matrix; matrix $\mathbf{I}_N$ denotes an $N \times N$ identity matrix.

vec($\mathbf{A}$) denotes the vectorization of matrix $\mathbf{A}$. $\mathbf{A} \otimes \mathbf{B}$ denotes the Kronecker product of matrices $\mathbf{A}$ and $\mathbf{B}$. $[\mathbf{B}]_{a:b,c:d}$ returns a submatrix of $\mathbf{B}$ including the a-th to the b-th rows and the c-th to the d-th columns of $\mathbf{B}$. $[\mathbf{q}]_{m:n}$ returns a vector with the m-th to the n-th elements of vector $\mathbf{q}$. A complex Gaussian random vector with mean vector $\boldsymbol{\mu}$ and covariance matrix $\boldsymbol{\Sigma}$ is denoted by $\mathcal{CN}(\boldsymbol{\mu}, \boldsymbol{\Sigma})$, and $\sim$ means "distributed as". $\mathbb{C}^{N \times M}$ denotes the space of all $N \times M$ matrices with complex entries. $\mathbb{H}^N$ represents the set of all N-by-N complex Hermitian matrices. $\mathcal{E}\{\cdot\}$ denotes statistical expectation. $|\cdot|$, $\|\cdot\|$, and $\|\cdot\|_{\mathrm{F}}$ denote the absolute value of a complex scalar, the Euclidean norm, and the Frobenius norm of a vector/matrix, respectively; $\mathrm{Re}\{\cdot\}$ denotes the real part of an input complex number.

5.2 RECEIVER STRUCTURE

Wireless communications via propagating EM waves in RF enables the possibility of SWIPT and WPC which is foreseen to be a key technology for facilitating the development of IoT communication networks with energy-limited wireless transceivers. Yet, the utilization of EM waves as a carrier for SWIPT and WPC poses many new research challenges for receiver design. Early studies on SWIPT and WPCNs were based on a pure information theoretical approach [5,33]. In particular, it was assumed in these works that information decoding and EH can be performed based on the same received signal and an ideal receiver. However, this is not possible in practice, yet. Specifically, existing EH circuits extract the energy of the received signal in the RF domain. The EH process destroys the information content embedded in the signal. Besides, conventional information decoding is performed in the digital baseband and frequency down converted signals cannot be used for EH. As a result, various types of practical EH receivers have been proposed to enable SWIPT. In particular, for SWIPT, the information decoding process and EH process have to be separated. A viable solution is to split the received RF power into two distinct parts, one for EH and one for information decoding. In the following, we discuss two commonly adopted techniques to achieve this signal splitting.

Time Switching (TS) Receiver:
With TS receivers, each transmission block is divided into two orthogonal time slots, one for transferring wireless power and the other one for transmitting information, cf. Figure 5.1(a). The co-located energy harvester and information receiver switch between harvesting energy and decoding in two time slots [17]. In practice, by taking into account the channel statistics and QoSs for power transfer, the time durations for wireless information transfer and energy transfer can be optimized to achieve different system design objectives. Although the TS receiver structure allows for a simple hardware implementation, it requires accurate time synchronization and information/energy scheduling, especially in multi-user systems.

Power Splitting (PS) Receiver:
A power splitting (PS) receiver splits the signal received at the antenna into two streams at different power levels using a PS unit, cf. Figure 5.1(b). In particular, one stream is sent to the RF energy harvester for EH, and the other one is converted to baseband for information decoding [17,19]. The PS process incurs a higher receiver

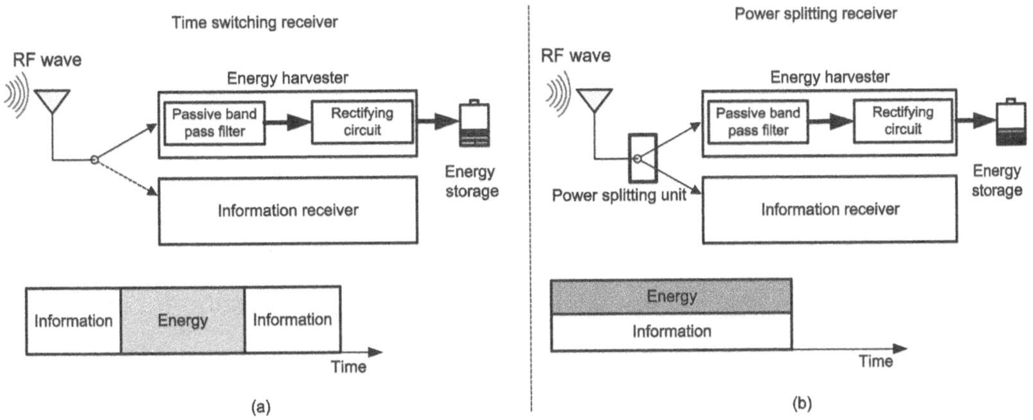

Figure 5.1 Simple receiver structures for wireless information and power transfer; (a) Time switching receiver; (b) Power splitting receiver.

complexity compared to the TS process. Besides, optimization of the ratio of the two power streams is needed in order to achieve a balance between the performances of information decoding and EH. Furthermore, additional noise may be introduced due to the adopted PS process [14]. Nevertheless, this receiver structure achieves SWIPT, as the signal received in one time slot is exploited for both information decoding and power transfer. Therefore, it is more suitable than the TS receiver for applications with critical information/energy or delay constraints [6].

In the sequel, we study the resource allocation algorithm design for two practical wireless information and power transfer networks based on the TS receiver structure, due to its simpler hardware implementation. Since the unit of "Joule-per-second" is used for energy consumption in this chapter, the terms "power" and "energy" are interchangeable.

5.3 SWIPT COMMUNICATION NETWORKS

In this section, we outline the adopted system model for the considered SWIPT systems.

5.3.1 Channel Model

A frequency flat fading communication channel is considered. The SWIPT system comprises a transmitter, an information receiver (IR), and J EH receivers (ER), cf. Figure 5.2. The transmitter is equipped with $N_T \geq 1$ antennas and serves both the IR and the ERs simultaneously in the same frequency band. We assume that the IR is a single-antenna device for assuring low hardware complexity. Each ER is equipped with $N_R \geq 1$ receive antennas to facilitate wireless EH. The received signals at the

---

The considered system can be treated as having $J + 1$ TS receivers where one of the receivers is in the IR mode and the remaining J receivers are in the ER mode.

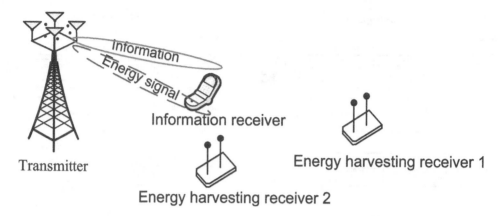

Figure 5.2 A simple SWIPT system model with one information receiver and $J = 2$ EH receivers (ERs), e.g., wireless sensors. The ERs harvest energy from the received RF signals to extend their lifetimes.

IR and ER $j \in \{1, \ldots, J\}$ are given by

$$y = \mathbf{h}^H \mathbf{w} s + \mathbf{w}_E + n, \text{ and} \tag{5.1}$$

$$\mathbf{y}_{ER_j} = \mathbf{G}_j^H \mathbf{w} s + \mathbf{w}_E + \mathbf{n}_{ER_j}, \ \forall j \in \{1, \ldots, J\}, \tag{5.2}$$

respectively, where $s \in \mathbb{C}$ and $\mathbf{w} \in \mathbb{C}^{N_T \times 1}$ are the data symbol and the information beamforming vector, respectively. Without loss of generality, we assume that $\mathcal{E}\{|s|^2\} = 1$. The channel vector between the transmitter and the IR is denoted by $\mathbf{h} \in \mathbb{C}^{N_T \times 1}$ and the channel matrix between the transmitter and ER j is denoted by $\mathbf{G}_j \in \mathbb{C}^{N_T \times N_R}$. $n \sim \mathcal{CN}(0, \sigma_s^2)$ and $\mathbf{n}_{ER_j} \sim \mathcal{CN}(\mathbf{0}, \sigma_s^2 \mathbf{I}_{N_R})$ are the additive white Gaussian noises (AWGN) at the IR and ER j, respectively, where σ_s^2 denotes the noise power at the receiver. $\mathbf{w}_E \in \mathbb{C}^{N_T \times 1}$ is a Gaussian pseudo-random sequence generated by the transmitter to facilitate efficient wireless power transfer. In particular, $\mathbf{w}_E$ is modelled as a complex Gaussian random vector with

$$\mathbf{w}_E \sim \mathcal{CN}(\mathbf{0}, \mathbf{W}_E), \tag{5.3}$$

where $\mathbf{W}_E \in \mathbb{H}^{N_T}, \mathbf{W}_E \succeq \mathbf{0}$, denotes the covariance matrix of the pseudo-random energy signal.

5.3.2 Non-linear Energy Harvesting Model

In this section, we discuss two mathematical models used in the literature to capture the characteristic of practical RF EH circuits. To this end, we first study a basic approach for extracting electrical energy from the received RF signals. In practice, after the transmitted RF signal is received at the antenna(s) of an ER, a passive bandpass filter is employed before the received RF signal is passed on to a rectifying circuit, cf. Figure 5.1. In fact, the rectifying circuit is the core element of RF EH circuits. In particular, it is a passive electronic circuit comprising diodes, resistors, and capacitors that converts the incoming RF power to direct current (DC) power. Then, the converted power can be stored in the energy storage unit of the receiver.

The RF-to-DC energy conversion efficiency depends greatly on the characteristics of the rectifying circuit. In general, rectifiers can be implemented using different non-linear circuits, starting from the simplest half-wave rectifiers, cf. Figure 5.3, to complicated circuits that offer N-fold increase of the circuit output power so as to improve the efficiency of the circuit, cf. Figure 5.4. A half-wave rectifier, as depicted in Figure 5.3, passes either the positive or negative half of the alternating current (AC) wave, while the other half is blocked [34]. Although half-wave rectifiers result in a lower output voltage compared to other types of rectifiers, a half-wave rectifier requires only a single diode and is a very simple design. Thus, half-wave rectifiers are suitable for cheap and small mobile devices such as wireless sensors for IoT applications. On the other hand, Figure 5.4 depicts an array of voltage doubler circuits, where each part of the circuit consists of two diodes and other corresponding elements. Depending on the number of stages required for a particular rectifier, the circuit parts can be repeated until the N-th element is reached. This configuration offers an increase of the conversion efficiency of the circuit.

In general, one can derive mathematical equations to describe the input-output characteristic of an EH circuit based on its schematic, e.g., Figures 5.3 and 5.4. However, they usually lead to complicated expressions which are intractable for resource allocation algorithm design. More importantly, such an approach relies on specific implementation details of EH circuits and the corresponding mathematical expressions may differ significantly across different types of EH circuits. In the following, we discuss two general tractable models proposed in the literature for characterizing the aforementioned RF EH process. Mathematically, the total received RF power at ER j is given by

$$P_{\mathrm{ER}_j} = \mathrm{Tr}\left((\mathbf{w}\mathbf{w}^H + \mathbf{W}_{\mathrm{E}})\mathbf{G}_j\mathbf{G}_j^H\right). \tag{5.4}$$

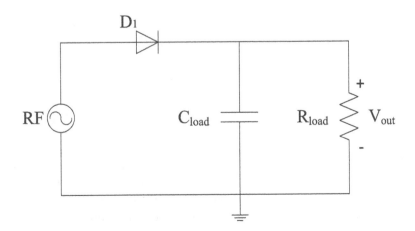

Figure 5.3 A schematic of a half-wave rectifier [35] where C_{load}, R_{load}, D_1, and V_{out} denote a load capacitance, load resistance, diode, and the output voltage, respectively.

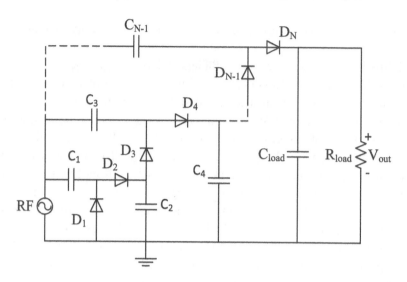

Figure 5.4 A schematic of a Dickson charge pump [35] with N stages, where D_i, and C_i, $i \in \{1, \dots, N\}$, denote the diode and the capacitor in the i-th stage.

In the SWIPT literature [36–44], the total harvested power at ER j, $\Phi_{\text{ER}_j}^{\text{Linear}}$, is typically modelled by the following linear equation:

$$\Phi_{\text{ER}_j}^{\text{Linear}} = \eta_j P_{\text{ER}_j}, \tag{5.5}$$

where $0 \leq \eta_j \leq 1$ is the constant power conversion efficiency of ER j. In other words, the total harvested power at the ER is linearly and directly proportional to the received RF power. Besides, the total harvested power increases with the amount of received power without bound.

Yet, practical RF-based EH circuits introduce non-linearities into the end-to-end WET and the conventional linear model fails to capture this important characteristic, as shown by experimental results [35, 45, 46]. Recently, a parametric non-linear EH model was proposed in [24, 47] to facilitate the design of resource allocation algorithms for practical SWIPT systems. Here, the total harvested power at ER j, Φ_{ER_j}, is modelled as:

$$\Phi_{\text{ER}_j} = \frac{[\Psi_{\text{ER}_j} - M_j \Omega_j]}{1 - \Omega_j}, \quad \Omega_j = \frac{1}{1 + \exp(a_j b_j)}, \tag{5.6}$$

$$\text{where } \Psi_{\text{ER}_j} = \frac{M_j}{1 + \exp\left(-a_j(P_{\text{ER}_j} - b_j)\right)} \tag{5.7}$$

is a logistic function which has the received RF power, P_{ER_j}, as the input. In particular, three parameters, i.e., M_j, a_j, and b_j, are introduced to describe the shape of the logistic function which depends on various physical properties of the RF EH circuit. Specifically, M_j is a positive constant denoting the maximum harvestable power at ER j, when the EH circuit is saturated due to an exceedingly large input power. Parameters a_j and b_j are constants which capture the joint effects of resistance, capacitance, and circuit sensitivity. Specifically, a_j denotes the non-linear charging rate

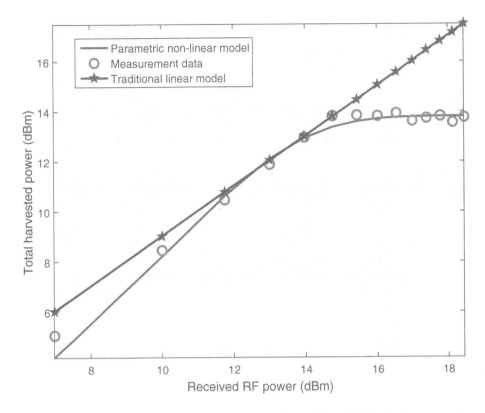

Figure 5.5 A comparison between experimental data from [45], the harvested power for the non-linear model in (5.6), and the linear EH model with $\eta_j = 0.8$ in (5.5).

with respect to the input power and b_j is related to the minimum turn-on voltage of the EH circuit.

In practice, for a given EH hardware circuit, the values of parameters a_j, b_j, and M_j of the proposed model in (5.6) can be estimated by using a standard curve fitting algorithm. In Figure 5.5, we show an example for the curve fitting for the non-linear EH model in (5.6) with parameters $M = 0.024$, $b = 0.014$, and $a = 150$. As can be observed, the parametric non-linear model matches the experimental result provided in [45] closely for the RF power harvested by a practical EH circuit. For comparison, Figure 5.5 also illustrates the total harvested power predicted by the linear model in (5.5). It can be seen that the conventional linear RF energy harvesting model fails to capture the non-linear characteristics of practical EH circuits, especially in high and low received RF power regimes.

5.3.3 Channel State Information

We assume that only imperfect channel state information (CSI) is available at the transmitter for resource allocation due to the slow time varying nature of the communication channels. To capture the impact of the CSI imperfection on resource allocation design, we adopt a commonly used deterministic model [19, 20]. In particular, the CSI of the links between the transmitter and the information receiver as well as EH receiver j can be modelled as:

$$\mathbf{h} = \widehat{\mathbf{h}} + \Delta\mathbf{h}, \tag{5.8}$$

$$\boldsymbol{\Upsilon} \triangleq \left\{\Delta\mathbf{h} \in \mathbb{C}^{N_{\mathrm{T}} \times 1} : \|\Delta\mathbf{h}\|_2^2 \leq \rho^2\right\}, \tag{5.9}$$

$$\mathbf{G}_j = \widehat{\mathbf{G}}_j + \Delta\mathbf{G}_j, \forall j \in \{1, \ldots, J\}, \text{ and} \tag{5.10}$$

$$\boldsymbol{\Xi}_j \triangleq \left\{\Delta\mathbf{G}_j \in \mathbb{C}^{N_{\mathrm{T}} \times N_{\mathrm{R}}} : \|\Delta\mathbf{G}_j\|_{\mathrm{F}}^2 \leq v_j^2\right\}, \forall j, \tag{5.11}$$

respectively, where $\widehat{\mathbf{h}}$ and $\widehat{\mathbf{G}}_j$ are the estimates of channel vector $\mathbf{h}$ and channel matrix $\mathbf{G}_j$, respectively. $\Delta\mathbf{h}$ and $\Delta\mathbf{G}_j$ represent the channel uncertainty due to channel estimation errors. In (5.9) and (5.11), sets $\boldsymbol{\Upsilon}$ and $\boldsymbol{\Xi}_j$ define the continuous spaces spanned by all possible channel uncertainties, respectively. Constants ρ and v_j denote the maximum value of the norm of the CSI estimation error vector $\Delta\mathbf{h}$ and the CSI estimation error matrix $\Delta\mathbf{G}_j$, respectively.

Remark 1 *In practical systems, the values of ρ^2 and v_j^2 depend not only on the adopted channel estimation method, but also on the packet duration and the coherence time of the associated communication channel.*

5.3.4 Achievable System Data Rate

The energy signal $\mathbf{w}_{\mathrm{E}}$ is a Gaussian pseudo-random sequence which is known to all the transceivers. Hence, interference cancellation can be performed at the IR to facilitate information decoding. As a result, given perfect CSI at the receiver for coherent information decoding, the achievable rate (bit/s/Hz) between the transmitter and the IR is given by

$$R = \log_2\left(1 + \frac{|\mathbf{h}^H\mathbf{w}|^2}{\sigma_{\mathrm{s}}^2}\right), \tag{5.12}$$

where the interference caused by the energy signal, i.e., $\mathrm{Tr}(\mathbf{h}^H\mathbf{W}_{\mathrm{E}}\mathbf{h})$, has been removed.

5.3.5 Problem Formulation and Solution

In the considered SWIPT system, we aim to maximize the total achievable data rate of the system while guaranteeing a minimum total harvested power at multiple ERs. The resource allocation algorithm design is formulated as the following optimization problem:

Problem 1 Robust Resource Allocation for SWIPT:

$$\underset{\mathbf{w}, \mathbf{W}_{\mathrm{E}} \in \mathbb{H}^{N_{\mathrm{T}}}}{\text{maximize}} \; \underset{\Delta\mathbf{h} \in \boldsymbol{\Upsilon}}{\min} \quad \log_2\left(1 + \frac{|\mathbf{h}^H\mathbf{w}|^2}{\sigma_{\mathrm{s}}^2}\right) \tag{5.13}$$

subject to $\quad$ C1 : $\|\mathbf{w}\|_2^2 + \mathrm{Tr}(\mathbf{W}_{\mathrm{E}}) \leq P_{\max}$,

$\qquad\qquad$ C2 : $\underset{\Delta\mathbf{G}_j \in \boldsymbol{\Xi}_j}{\min} \; \Phi_{\mathrm{ER}_j} \geq P_{\mathrm{req}_j}, \forall j \in \{1, \ldots, J\}$.

The objective function in (5.13) takes into account the CSI uncertainty set $\boldsymbol{\Upsilon}$ to provide robustness against CSI imperfection. Constants $P_{\max}$ and P_{req_j} in constraints C1 and C2 are the maximum transmit power from the power station and the required minimum harvested power at ER j, respectively. It can be observed that there are infinitely many possibilities in both the objective function and constraint C2, due to the CSI uncertainties. In order to design a computationally efficient resource allocation algorithm, we first define $\mathbf{W} = \mathbf{w}\mathbf{w}^H$ and transform the considered problem into the following equivalent rank-constrained semi-definite program (SDP):

Problem 2 Rank-constrained Robust Resource Allocation for SWIPT:

$$\underset{\mathbf{W},\mathbf{W}_{\mathrm{E}} \in \mathbb{H}^{N_{\mathrm{T}}}, \tau, \boldsymbol{\beta}}{\text{maximize}} \quad \tau \tag{5.14}$$

subject to

$$\text{C1}: \ \mathrm{Tr}(\mathbf{W} + \mathbf{W}_{\mathrm{E}}) \leq P_{\max},$$

$$\text{C2}: \ M_j \geq \Theta_j\Big(1 + \exp\big(-a_j(\beta_j - b_j)\big)\Big), \forall j \in \{1, \ldots, J\},$$

$$\text{C3}: \ \underset{\Delta\mathbf{h} \in \boldsymbol{\Upsilon}}{\min}\ \mathrm{Tr}(\mathbf{W}\mathbf{H}) \geq \tau,$$

$$\text{C4}: \ \underset{\Delta\mathbf{G}_j \in \boldsymbol{\Xi}_j}{\min}\ \mathrm{Tr}((\mathbf{W} + \mathbf{W}_{\mathrm{E}})\mathbf{G}_j\mathbf{G}_j^H) \geq \beta_j, \forall j \in \{1, \ldots, J\},$$

$$\text{C5}: \ \mathrm{Rank}(\mathbf{W}) \leq 1,$$

$$\text{C6}: \ \mathbf{W} \succeq \mathbf{0},$$

$$\text{C7}: \ \mathbf{W}_{\mathrm{E}} \succeq \mathbf{0},$$

---

where

$$\Theta_j \ = \ P_{\mathrm{req}_j}(1 - \Omega_j) + M_j\Omega_j \quad \text{and} \tag{5.15}$$

$$\mathbf{H} \ = \ \mathbf{h}\mathbf{h}^H. \tag{5.16}$$

$\boldsymbol{\beta} = \{\beta_1, \ldots, \beta_j, \ldots, \beta_J\}$ and τ are auxiliary optimization variables. We note that $\mathbf{W} \succeq \mathbf{0}$, $\mathbf{W} \in \mathbb{H}^{N_{\mathrm{T}}}$, and $\mathrm{Rank}(\mathbf{W}) = 1$ in (5.14) are imposed to guarantee that $\mathbf{W} = \mathbf{w}\mathbf{w}^H$ after optimization. Now, the transformed problem in (5.14) involves infinitely many constraints only in C3 and C4. Besides, the rank constraint in C5 is non-convex. To further facilitate the solution, we first transform constraints C3 and C4 into linear matrix inequalities (LMIs) using the following lemma:

Lemma 1 (S-Procedure [48]) *Let a function* $f_m(\mathbf{x}), m \in \{1, 2\}, \mathbf{x} \in \mathbb{C}^{N \times 1}$*, be defined as*

$$f_m(\mathbf{x}) = \mathbf{x}^H\mathbf{A}_m\mathbf{x} + 2\mathrm{Re}\{\mathbf{b}_m^H\mathbf{x}\} + c_m, \tag{5.17}$$

where $\mathbf{A}_m \in \mathbb{H}^N$*,* $\mathbf{b}_m \in \mathbb{C}^{N \times 1}$*, and* $c_m \in \mathbb{R}$*. Then, the implication* $f_1(\mathbf{x}) \leq 0 \Rightarrow f_2(\mathbf{x}) \leq 0$ *holds if and only if there exists a* $\delta \geq 0$ *such that*

$$\delta\begin{bmatrix} \mathbf{A}_1 & \mathbf{b}_1 \\ \mathbf{b}_1^H & c_1 \end{bmatrix} - \begin{bmatrix} \mathbf{A}_2 & \mathbf{b}_2 \\ \mathbf{b}_2^H & c_2 \end{bmatrix} \succeq \mathbf{0}, \tag{5.18}$$

provided that there exists a point $\hat{\mathbf{x}}$ *such that* $f_m(\hat{\mathbf{x}}) < 0$*.*

Exploiting Lemma 1, the original constraint C3 holds if and only if there exists a $\delta \geq 0$, such that the following LMI constraint holds:

$$\text{C3: } \mathbf{S}_{\text{C}_3}\left(\mathbf{W}, \delta, \tau\right) = \begin{bmatrix} \delta\mathbf{I}_{N_T} & \mathbf{0} \\ \mathbf{0} & -\delta\rho^2 - \tau \end{bmatrix} + \mathbf{U}_{\hat{\mathbf{h}}}^H \mathbf{W} \mathbf{U}_{\hat{\mathbf{h}}} \succeq \mathbf{0}, \qquad (5.19)$$

where $\mathbf{U}_{\hat{\mathbf{h}}} = \begin{bmatrix} \mathbf{I}_{N_T} & \hat{\mathbf{h}} \end{bmatrix}$. Similarly, constraint C4 can be equivalently written as

$$\text{C4: } \mathbf{S}_{\text{C}_{4_j}}\left(\mathbf{W}, \mathbf{W}_{\text{E}}, \boldsymbol{\nu}, \boldsymbol{\beta}\right) \qquad (5.20)$$

$$= \begin{bmatrix} \nu_j\mathbf{I}_{N_T N_R} & \mathbf{0} \\ \mathbf{0} & -\beta_j - \nu_j v_j^2 \end{bmatrix} + \mathbf{U}_{\widetilde{\mathbf{g}}_j}^H (\boldsymbol{\mathcal{W}} + \boldsymbol{\mathcal{W}}_{\text{E}}) \mathbf{U}_{\widetilde{\mathbf{g}}_j} \succeq \mathbf{0}, \forall j,$$

for $\boldsymbol{\nu} = \{\nu_1, \ldots, \nu_j, \ldots, \nu_J\}$, $\nu_j \geq 0$, $\boldsymbol{\mathcal{W}} = \mathbf{I}_{N_R} \otimes \mathbf{W}$, $\boldsymbol{\mathcal{W}}_{\text{E}} = \mathbf{I}_{N_R} \otimes \mathbf{W}_{\text{E}}$, $\mathbf{U}_{\widetilde{\mathbf{g}}_j} = [\mathbf{I}_{N_T N_R} \quad \widetilde{\mathbf{g}}_j]$, and $\widetilde{\mathbf{g}}_j = \text{vec}(\hat{\mathbf{G}}_j)$. Then, the considered optimization problem can be rewritten as

---

Problem 3 Rank-constrained SDP for SWIPT:

$$\underset{\mathbf{W}, \mathbf{W}_{\text{E}} \in \mathbb{H}^{N_T}, \tau, \boldsymbol{\nu}, \delta, \boldsymbol{\beta}}{\text{maximize}} \quad \tau \qquad (5.21)$$

subject to $\quad$ C1 : $\text{Tr}(\mathbf{W} + \mathbf{W}_{\text{E}}) \leq P_{\text{max}}$,

$\qquad$ C2 : $M_j \geq \Theta_j\left(1 + \exp\left(-a_j(\beta_j - b_j)\right)\right), \forall j \in \{1, \ldots, J\}$,

$\qquad$ C3 : $\mathbf{S}_{\text{C}_3}\left(\mathbf{W}, \delta, \tau\right) \succeq \mathbf{0}$,

$\qquad$ C4 : $\mathbf{S}_{\text{C}_{4_j}}\left(\mathbf{W}, \mathbf{W}_{\text{E}}, \boldsymbol{\nu}, \boldsymbol{\beta}\right) \succeq \mathbf{0}, , \forall j \in \{1, \ldots, J\}$,

$\qquad$ C5 : $\text{Rank}(\mathbf{W}) \leq 1$,

$\qquad$ C6 : $\mathbf{W} \succeq \mathbf{0}$,

$\qquad$ C7 : $\mathbf{W}_{\text{E}} \succeq \mathbf{0}$,

---

where δ and $\boldsymbol{\nu}$ are the non-negative auxiliary optimization variables introduced in Lemma 1 for handling constraints C3 and C4, respectively. We note that constraints C3 and C4 involve only a finite number of LMI constraints which facilitates the resource allocation algorithm design. However, the rank constraint in C5 is still an obstacle in solving the considered optimization problem due to its combinatorial nature. As a result, we adopt SDP relaxation by removing constraint C5 from the problem formulation which yields:

Problem 4 SDP relaxation of (5.21)

$$\underset{\mathbf{W},\mathbf{W}_\mathrm{E} \in \mathbb{H}^{N_\mathrm{T}}, \tau, \boldsymbol{\nu}, \delta, \boldsymbol{\beta}}{\text{maximize}} \quad \tau \tag{5.22}$$

subject to

C1 : $\mathrm{Tr}(\mathbf{W} + \mathbf{W}_\mathrm{E}) \leq P_{\max}$,

C2 : $M_j \geq \Theta_j\left(1 + \exp\left(-a_j(\beta_j - b_j)\right)\right), \forall j \in \{1, \ldots, J\}$,

C3 : $\mathbf{S}_{\mathrm{C}_3}\left(\mathbf{W}, \delta, \tau\right) \succeq \mathbf{0}$,

C4 : $\mathbf{S}_{\mathrm{C}_{4_j}}\left(\mathbf{W}, \mathbf{W}_\mathrm{E}, \boldsymbol{\nu}, \boldsymbol{\beta}\right) \succeq \mathbf{0}, ,\, \forall j \in \{1, \ldots, J\}$,

C5 : ~~Rank(W) ≤ 1~~.

C6 : $\mathbf{W} \succeq \mathbf{0}$,

C7 : $\mathbf{W}_\mathrm{E} \succeq \mathbf{0}$.

The rank relaxed problem is a convex optimization problem and can be solved efficiently by standard numerical solvers such as CVX [49]. Yet, the constraint relaxation may not be tight when $\mathrm{Rank}(\mathbf{W}) > 1$ and in that case the result of the relaxed problem serves as a performance upper bound for the original problem. Therefore, we study the tightness of the adopted SDP relaxation in the following theorem.

Theorem 1 *Assuming the considered problem is feasible for $P_{\max} > 0$, a rank-one solution of (5.22) can always be constructed.*

Proof: Please refer to Appendix 5.6.1.

In other words, (5.21) can be solved optimally. In particular, information beamforming is optimal for the maximization of achievable rate, despite the imperfection of the CSI and non-linearity of the RF EH circuits.

Table 5.1 Simulation parameters

Carrier center frequency	915 MHz
Bandwidth	200 kHz
Transceiver antenna gain	10 dBi
Number of receive antennas N_R	2
Noise power σ^2	−95 dBm
Maximum transmit power P_{max}	36 dBm
Transmitter-to-ER fading distribution	Rician with Rician factor 3 dB
Transmitter-to-IR fading distribution	Rayleigh

5.3.6 Numerical Example

In this section, we evaluate the IoT system performance of the proposed optimal resource allocation algorithm via simulations. We summarize the important simulation parameters in Table 5.1. We assume that the IR and the J ERs are located at 100 meters and 5 meters from the transmitter, respectively. In particular, the IR is an IoT device connecting to the transmitter for information transfer while the J ERs are idle IoT receivers requesting wireless energy to extend their lifetimes. Unless further specified, we adopt the normalized maximum channel estimation errors of ER j and the IR as $\sigma^2_{\text{est}_G} = 1\% \geq \frac{v_j^2}{\|\mathbf{G}_j\|_F^2}, \forall j$, and $\sigma^2_{\text{est}_h} = 1\% \geq \frac{\rho^2}{\|\mathbf{h}\|_2^2}$. For the non-linear EH circuits, we set $M_j = 24$ mW which corresponds to the maximum harvested power per wireless powered device. Besides, we adopt $a_j = 150$ and $b_j = 0.014$. We solve the optimization problem in (5.22) and obtain the average system performance by averaging over different channel realizations.

In Figure 5.6, we show the average achievable rate of the system versus the average total harvested energy in a downlink system for the optimal beamforming scheme. In particular, a transmitter equipped with N_T antennas serves a single-antenna IR and $J = 1$ ER. As can be observed, there is a non-trivial trade-off between the achievable system data rate and the total harvested energy. In other words, system data rate maximization and total harvested energy maximization are two conflicting system design objectives. Besides, for the optimal resource allocation, the trade-off region of the system achievable rate and the harvested energy is enlarged significantly with N_T and N_R. This is due to the fact that the extra degrees of freedom offered by multiple transmit antennas help the transmitter to focus the energy of the information signal and thus improve the beamforming efficiency. On the other hand, increasing the number of receive antennas N_R can significantly improve the total harvested energy at the ER. In fact, the extra receiver antennas act as additional energy collectors which enables a more efficient energy transfer. Furthermore, it is verified by simulation that Rank($\mathbf{W}$) = 1 can be obtained/construsted for all the considered channel realizations which confirms the correctness of Theorem 1.

In Figure 5.7, we study the average achievable data rate versus the number of ERs for different maximum normalized channel estimation error variances. The maximum transmit power is $P_{max} = 36$ dBm and $N_R = 2$. Besides, the maximum normalized

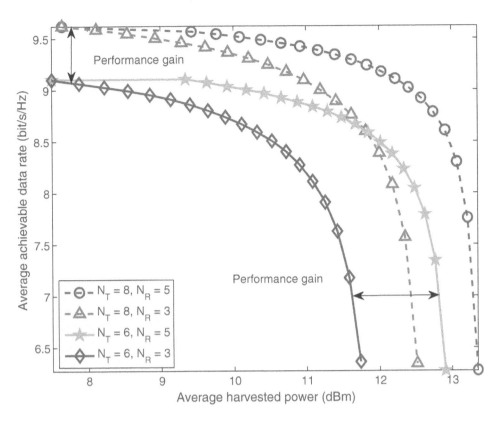

Figure 5.6 Average achievable data rate $(bit/s/Hz)$ versus the average harvested power (dBm) for different numbers of antennas.

channel estimation error variance of the transmitter-to-IR link and the transmitter-to-ERs links are set to be identical, i.e., $\sigma^2_{est_G} = \sigma^2_{est_h} = \sigma^2_{est}$. As can be observed, the average achievable data rate decreases with an increasing number of ERs. In fact, constraints C4 become more stringent when there are more ERs in the system which reduces the flexibility of the transmitter in resource allocation. In particular, for a large number of ERs in the system, the transmitter is forced to steer the transmit direction towards the ERs to improve the efficiency of wireless power transfer which reduces the received signal strength at the IR. On the other hand, the achievable data rate decreases with increasing σ^2_{est}, since the CSI quality degrades with increasing σ^2_{est}. In particular, for a larger value of σ^2_{est}, it becomes more difficult for the transmitter to focus the transmitter energy for improving the efficiency of SWIPT.

5.4 WIRELESS POWERED COMMUNICATION NETWORKS

In the last section, we studied the robust resource allocation algorithm design for systems where a transmitter provides information and wireless energy simultaneously to IR and ERs, respectively. In this section, we focus on a second line of research in WET: WPCN, where the wireless communication devices are first powered by WET and then use the harvested energy to transmit data. For instance, dedicated

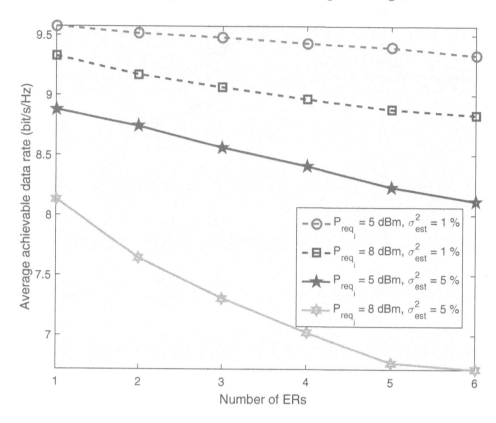

Figure 5.7 Average achievable data rate (bit/s/Hz) versus the number of ERs for $N_T = 8$.

power beacons or power stations can be deployed in the system for WET. Compared to conventional base stations, power stations/beacons do not require data backhaul connections and can be installed in an ad hoc or on-demand manner. This kind of system setup has various IoT applications for energy-limited wireless communication sensors which need to first harvest enough energy from the environment before sending information to an information receiver. In the following, we discuss a resource allocation design to improve the system performance of such a WPCN.

5.4.1 Channel Model

A simple WPCN is considered in this section. We assume that there is a power station transferring wireless energy to J wireless powered mobile users in the downlink to facilitate their information transfer in the uplink, cf. Figure 5.8. We assume that both the power station and each of the wireless powered mobile users are equipped with $N_T > 1$ and $N_R > 1$ antennas, respectively, to facilitate efficient energy and information transfer. On the other hand, there is a single-antenna IR receiving the uplink information from the J wireless powered mobile users. In the considered network, we adopt the "harvest-then-transmit" protocol [12,50,51] for WET and information transmission. Specifically, the transmission is divided into two orthogonal time periods, namely the WET period and wireless information transfer (WIT) period, cf.

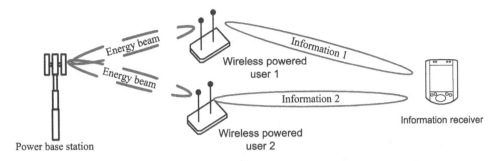

Figure 5.8 A WPCN with $J = 2$ multiple-antenna wireless powered users harvesting energy from a dedicated power base station. The harvested energy will be exploited for future information transmission.

Figure 5.9. In the WET period, the power station sends an energy signal to the J wireless powered users for EH. The instantaneous received signal at mobile user $j \in \{1, \ldots, J\}$ is given by

$$\mathbf{y}_{\mathrm{EH}_j} = \mathbf{G}_j^H \mathbf{v} + \mathbf{n}_{\mathrm{EH}_j}, \tag{5.23}$$

where $\mathbf{v} \in \mathbb{C}^{N_\mathrm{T} \times 1}$ is the beamforming vector in the downlink for WET. The channel matrix between the power station and mobile user j is denoted by $\mathbf{G}_j \in \mathbb{C}^{N_\mathrm{T} \times N_\mathrm{R}}$. Vector $\mathbf{n}_{\mathrm{EH}_j} \sim \mathcal{CN}(\mathbf{0}, \sigma_{\mathrm{s}_j}^2 \mathbf{I}_{N_\mathrm{R}})$ is the AWGN at mobile user j. Then, in the WIT period, the J wireless powered mobile users exploit the energy harvested in the RF to transmit independent information signals in the uplink to the information receiver in a time division manner. In particular, mobile user J is allocated τ_j amount of time for uplink transmission. The instantaneous received signal at the information receiver from mobile user j is given by

$$y_j^{\mathrm{IR}} = \mathbf{h}_j^H \mathbf{w}_j s_j + n, \ \forall j \in \{1, \ldots, J\}, \tag{5.24}$$

where $\mathbf{h}_j \in \mathbb{C}^{N_\mathrm{R} \times 1}$ is the channel vector between wireless powered user j and the information receiver. Scalar $s_j \in \mathbb{C}$ is the information signal of mobile user j, $\mathbf{w}_j \in \mathbb{C}^{N_\mathrm{R} \times 1}$

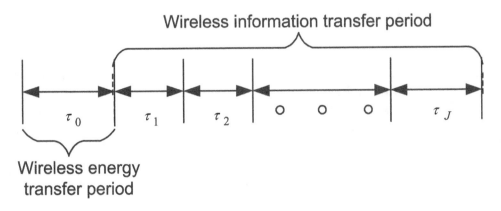

Figure 5.9 Wireless energy and information transfer protocol.

is the precoding vector adopted by user j intended for WIT, and $n \sim \mathcal{CN}(0, \sigma_n^2)$ is the AWGN at the information receiver. Without loss of generality, we assume that $\mathcal{E}\{|s_j|^2\} = 1, \forall j \in \{1, \ldots, J\}$.

Channel State Information

In practice, a power station is expected to be a simple device with limited signal processing capability. As a result, the estimates of the CSI of the communication links between the power station and the J wireless powered users may not be perfect. To capture the imperfectness of the CSI for resource allocation, we adopt equations (5.10) and (5.11). In contrast, a sophisticated information receiver can be implemented in WPCNs for signal processing. Therefore, we assume that the CSI of the communication links between the J wireless powered users and the information receiver is perfectly known for resource allocation design.

5.4.2 Problem Formulation and Solution

The resource allocation policy, $\{\boldsymbol{\tau}, \mathbf{V}, \mathbf{w}_j\}$, for maximizing the total system throughput can be obtained by solving the following problem:

Problem 5 Robust Resource Allocation for WPCN:

$$\underset{\mathbf{V} \in \mathbb{H}^{N_T}, \mathbf{w}_j, \tau_j}{\text{maximize}} \quad \sum_{j=1}^{J} \tau_j \log_2 \left(1 + \frac{|\mathbf{h}_j^H \mathbf{w}_j|^2}{\sigma_s^2}\right) \tag{5.25}$$

subject to $\quad$ C1 : $\mathrm{Tr}(\mathbf{V}) \leq P_{\max}$,

$$\text{C2}: \tau_0 + \sum_{j=1}^{J} \tau_j \leq T_{\max},$$

$$\text{C3}: \tau_j \|\mathbf{w}_j\|^2 \leq \min_{\Delta \mathbf{G}_j \in \Xi_j} \tau_0 \frac{\dfrac{M_j}{1 + \exp\left(-a_j (\mathrm{Tr}(\mathbf{V}\mathbf{G}_j\mathbf{G}_j^H) - b_j)\right)} - M_j \Omega_j}{1 - \Omega_j}, \forall j,$$

$$\text{C4}: \tau_r \geq 0, \forall r \in \{0, 1, \ldots, J\},$$

$$\text{C5}: \mathbf{V} \succeq \mathbf{0}.$$

Constants $P_{\max}$ and $T_{\max}$ in constraints C1 and C2 are the maximum transmit power for the power station and the maximum duration of a time slot, respectively. Constraint C3 is imposed such that for a given CSI uncertainty set Ξ_j, the maximum energy available for information transmission at wireless powered user j is limited by the total harvested RF energy during the wireless EH period τ_j. In particular, the right-hand side of constraint C3 denotes the total harvested power at ER j if a practical non-linear RF EH circuit is assumed. C4 is the non-negativity constraint

---

Here, we assume that the circuit power consumption of each wireless powered user is negligibly small compared to the transmit power consumption and thus is not taken into account.

for information scheduling variable τ_j. Constraint C5 and $\mathbf{V} \in \mathbb{H}^{N_T}$ constrain matrix $\mathbf{V}$ to be a positive semi-definite Hermitian matrix.

The optimization problem in (5.25) is a non-convex optimization problem which involves infinitely many constraints in C3. Besides, inequality constraint C3 involves the coupling of optimization variables τ_j and $\mathbf{w}_j$. Furthermore, the right-hand side of constraint C3 is a quasi-concave function. In general, there is no systematic approach for solving non-convex optimization problems. In order to obtain a computationally efficient resource allocation algorithm design, we introduce several transformations of the optimization problem. First, to handle the quasi-concavity of constraint C3, we solve the optimization problem for a fixed constant τ_0 and obtain an optimal solution for one instance of the optimization problem. Then, we repeat the procedure for all possible values of τ_0 and record the corresponding achieved system objective values. At the end, we select that τ_0 as the optimal time allocation for WET from all the trials which provides the maximum system objective value. Therefore, in the sequel, we assume that τ_0 is given by its optimal value for the design of the resource allocation algorithm.

Next, we introduce a change of variable to decouple the optimization variables in constraint C3. Specifically, we define a new optimization variable $\tilde{\mathbf{w}}_j = \sqrt{\tau_j} \mathbf{w}_j$ and rewrite the optimization problem as

---

Problem 6 Transformed Problem for WPCN:

$$\underset{\mathbf{V} \in \mathbb{H}^{N_T}, \tilde{\mathbf{w}}_j \in \mathbb{H}^{N_U}, \tau_j, \beta_j}{\text{maximize}} \quad \sum_{j=1}^{J} \tau_j \log_2 \left(1 + \frac{|\mathbf{h}_j^H \tilde{\mathbf{w}}_j|^2}{\tau_j \sigma_s^2} \right) \tag{5.26}$$

subject to

$$\text{C1}: \text{Tr}(\mathbf{V}) \leq P_{\max},$$

$$\text{C2}: \tau_0 + \sum_{j=1}^{J} \tau_j \leq T_{\max},$$

$$\text{C3}: \|\tilde{\mathbf{w}}_j\|^2 \leq \tau_0 \frac{\frac{M_j}{1+\exp\left(-a_j(\beta_j - b_j)\right)} - M_j \Omega_j}{1 - \Omega_j}, \forall j,$$

$$\text{C4}: \tau_r \geq 0, \forall r \in \{0, 1, \ldots, J\},$$

$$\text{C5}: \mathbf{V} \succeq \mathbf{0},$$

$$\text{C6}: \min_{\Delta \mathbf{G}_j \in \Xi_j} \text{Tr}(\mathbf{V} \mathbf{G}_j \mathbf{G}_j^H) \geq \beta_j, \forall j \in \{1, \ldots, J\}.$$

To handle the infinitely many constraints in C6, we can apply Lemma 1 for (5.26). In particular, constraint C6 can be equivalently written as

$$\text{C6}: \mathbf{S}_{\text{C}_{6_j}}\left(\mathbf{V}, \boldsymbol{\mu}, \boldsymbol{\beta}\right) \tag{5.27}$$

$$= \begin{bmatrix} \nu_j \mathbf{I}_{N_T N_R} & \mathbf{0} \\ \mathbf{0} & -\beta_j - \nu_j v_j^2 \end{bmatrix} + \mathbf{U}_{\tilde{\mathbf{g}}_j}^H \boldsymbol{\mathcal{V}} \mathbf{U}_{\tilde{\mathbf{g}}_j} \succeq \mathbf{0}, \forall j,$$

Table 5.2 Simulation parameters

Carrier center frequency	915 MHz
Bandwidth	200 kHz
Transceiver antenna gain	10 dBi
Noise power (including quantization noise) σ^2	−47 dBm
Power station-to-wireless powered user distance	5 meters
Power station-to-wireless powered user fading distribution	Rician with Rician factor 3 dB
Wireless powered user-to-IR fading distribution	Rayleigh
Maximum duration of a communication slot, $T_{\max}$	1 unit

for $\boldsymbol{\nu} = \{\nu_1, \ldots, \nu_j, \ldots, \nu_J\}$, $\nu_j \geq 0$, $\boldsymbol{\mathcal{V}} = \mathbf{I}_{N_{\mathrm{R}}} \otimes \mathbf{V}$, $\mathbf{U}_{\tilde{\mathbf{g}}_j} = [\mathbf{I}_{N_{\mathrm{T}}N_{\mathrm{R}}} \quad \tilde{\mathbf{g}}_j]$, and $\tilde{\mathbf{g}}_j = \mathrm{vec}(\hat{\mathbf{G}}_j)$.

Problem 7 Transformed Problem for WPCN:

$$\underset{\mathbf{V} \in \mathbb{H}^{N_{\mathrm{T}}}, \tilde{\mathbf{w}}_j \in \mathbb{H}^{N_{\mathrm{U}}}, \tau_j, \beta_j, \mu_j}{\text{maximize}} \quad \sum_{j=1}^{J} \tau_j \log_2 \left(1 + \frac{|\mathbf{h}_j^H \tilde{\mathbf{w}}_j|^2}{\tau_j \sigma_{\mathrm{s}}^2} \right) \tag{5.28}$$

$$\text{subject to} \quad \mathrm{C1} - \mathrm{C5},$$

$$\mathrm{C6:} \ \ \mathbf{S}_{\mathrm{C}_{6_j}}\left(\mathbf{V}, \boldsymbol{\mu}, \boldsymbol{\beta} \right) \succeq \mathbf{0}, \forall j \in \{1, \ldots, J\}.$$

The above transformed problem is jointly concave with respect to the optimization variables and can be solved efficiently via standard numerical solvers for convex programs.

5.4.3 Numerical Example

In this section, we evaluate the IoT system performance of the proposed resource allocation algorithm via simulations. We summarize the relevant simulation parameters in Table 5.2. We assume that a dedicated power station is deployed for wireless charging of IoT devices. There are $J = 4$ ERs in the IoT network requiring energy for WIT. For the non-linear EH circuits, we set $M_j = 24$ mW which corresponds to the maximum harvested power per ER. Besides, we adopt $a_j = 150$ and $b_j = 0.014$. To obtain the average system performance, we solve the optimization problem in (5.28) for each channel realization and average the result over different channel realizations.

In Figure 5.10, we study the average total system throughput versus the maximum transmit power from the power station, $P_{\max}$, for different numbers of antennas equipped at the power station, N_{T}, and at the wireless powered users, N_{R}. We set the normalized maximum channel estimation errors of wireless powered user j as $\sigma_{\mathrm{est}_G}^2 = 1\% \geq \frac{v_j^2}{\|\mathbf{G}_j\|_F^2}, \forall j$. As can be observed, the average total system throughput

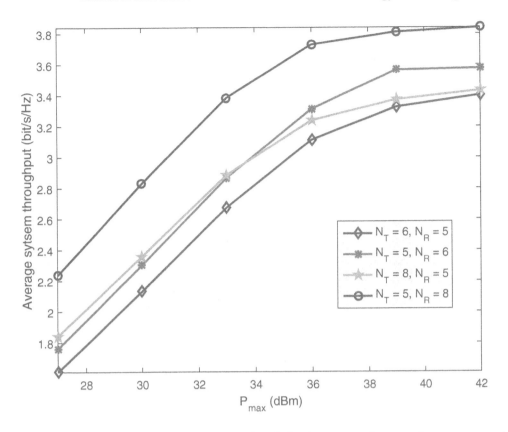

Figure 5.10 Average system throughput (bit/s/Hz) versus the maximum transmit power at the power base station (dBm).

increases with increasing P_{max}. Indeed, with a higher value of P_{max}, the wireless powered users are able to harvest more energy for information transmission. However, there is a diminishing return in performance as P_{max} increases in the high transmit power regime. This is due to the fact that the high transmit power from the power station causes saturation in practical non-linear EH circuits which limits the available harvested power for WIT. On the other hand, when the number of antennas equipped at the power base station increases, a higher system throughput can be achieved by the proposed optimal scheme. In fact, the extra antennas provide extra spatial degrees of freedom which facilitates a more flexible resource allocation, since the power station can steer the energy signal towards the wireless powered users more accurately to improve the efficiency of WET. Besides, the system throughput increases rapidly with the number of antennas equipped at the wireless powered users. In fact, the extra antennas equipped at the wireless powered users act as additional wireless energy collectors which increase the amount of total harvested energy. Furthermore, the extra antennas at the wireless powered users would also provide extra spatial degrees of freedom which improves the transmit beamforming gain in the WIT phase.

Figure 5.11 shows the time allocation ratio for the proposed algorithm with respect to the WET and WIT periods for the case of $N_T = 6$ and $N_R = 5$ in Figure 5.10. As can be observed, the WET period for the proposed scheme becomes shorter

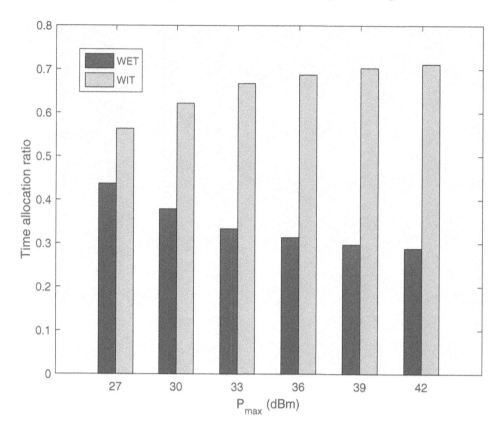

Figure 5.11 Time allocation ratio for WET and WIT versus the maximum transmit power at the power base station (dBm).

as the value of P_{max} increases. In fact, for a higher maximum transmit power from the power station, the wireless powered users can harvest the amount of energy required for information transmission in a shorter period of time. In contrast, the WIT period becomes longer for an increasing value of P_{max}. This is due to the fact that the achievable throughput of each wireless powered user is an increasing function with respect to the time allocation for information transmission, i.e., τ_j, for a fixed amount of total transmit energy. In the extreme case, for a sufficiently large P_{max}, one can expect that $\tau_0 \to 0$ since an infinitesimal amount of time is enough to provide sufficient energy to fully charge the wireless powered users.

5.5 CONCLUSION

In this chapter, we studied resource allocation algorithms for two RF-based EH wireless communication network architectures, which are of interest for IoT applications. We first discussed a parametric non-linear EH model which facilitates the resource allocation algorithm design to enable efficient wireless powered IoT communication networks. The algorithm designs were formulated as two non-convex optimization problems for maximizing the sum-throughput in SWIPT and WPCN systems, respectively. The problem formulations took into account the imperfectness of the CSI and the non-linearity of the EH circuits in order to ensure robust resource allocation. The proposed resource allocation design optimization problems were optimally solved by advanced signal processing techniques. Numerical results showed the potential gains in harvested power enabled by the proposed optimization and the benefits in adopting multiple-antenna technology for IoT communication networks.

Acknowledgements

This work was supported in part by the AvH Professorship Program of the Alexander von Humboldt Foundation and by the Australian Research Council (ARC) Linkage Project LP 160100708.

5.6 APPENDIX

5.6.1 Proof of Theorem 1

We provide a method for constructing an optimal rank-one solution for (5.22) when $\text{Rank}(\mathbf{W}) > 1$ is obtained from (5.22). For a given optimal τ^* from the solution of (5.22), we solve the following auxiliary convex optimization problem [52, 53]:

Auxiliary Convex Optimization Problem

$$\underset{\mathbf{W}, \mathbf{W}_E \in \mathbb{H}^{N_T}, \, \boldsymbol{\nu}, \delta, \boldsymbol{\beta}}{\text{minimize}} \quad \text{Tr}(\mathbf{W}) \qquad (5.29)$$

$$\text{subject to} \quad \text{C1}, \text{C2}, \text{C4}, \text{C6}, \text{C7},$$

$$\text{C3}: \ \mathbf{S}_{\text{C}_3}\left(\mathbf{W}, \delta, \tau^*\right) \succeq \mathbf{0}.$$

We note that the optimal resource allocation policy obtained from the above auxiliary convex optimization problem is also an optimal resource allocation policy for (5.22), since both problems have the same feasible solution set and τ^* is fixed for (5.29).

Now, we aim to show that (5.29) admits a rank-one beamforming matrix. In this context, we first need the Lagrangian of problem (5.29):

$$
\begin{aligned}
L \ = \ & \text{Tr}(\mathbf{W}) + \lambda(\text{Tr}(\mathbf{W} + \mathbf{W}_E) - P_{\max}) - \text{Tr}(\mathbf{W}\mathbf{Y}) \\
& - \sum_{j=1}^{J} \text{Tr}(\mathbf{S}_{\text{C}_{4_j}}\left(\mathbf{W}, \mathbf{W}_E, \boldsymbol{\nu}, \boldsymbol{\beta}\right)\mathbf{D}_{\text{C}_{4_j}}) \\
& - \text{Tr}(\mathbf{S}_{\text{C}_3}\left(\mathbf{W}, \delta, \tau\right)\mathbf{D}_{\text{C}_3}) - \text{Tr}(\mathbf{W}_E\mathbf{Z}) + \boldsymbol{\Delta},
\end{aligned}
\qquad (5.30)
$$

where $\lambda \geq 0$, $\mathbf{D}_{\text{C}_3} \succeq \mathbf{0}$, $\mathbf{D}_{\text{C}_{4_j}} \succeq \mathbf{0}, \forall j \in \{1, \ldots, J\}$, $\mathbf{Y} \succeq \mathbf{0}$, and $\mathbf{Z} \succeq \mathbf{0}$ are the dual variables for constraints C1, C3, C4, C6, and C7, respectively. $\boldsymbol{\Delta}$ is a collection of primal and dual variables and constants that are not relevant to the proof.

Now, we focus on those Karush–Kuhn–Tucker (KKT) conditions which are needed for the proof.

KKT conditions:

$$\mathbf{Y}^*, \mathbf{Z}^*, \mathbf{D}_{C_3}^*, \mathbf{D}_{C_{4_j}}^* \succeq \mathbf{0}, \quad \lambda^* \geq 0, \tag{5.31a}$$

$$\mathbf{Y}^*\mathbf{W}^* = \mathbf{0}, \quad \mathbf{Q}^*\mathbf{V}^* = \mathbf{0}, \tag{5.31b}$$

$$\mathbf{Y}^* = (1 + \lambda^*)\mathbf{I}_{N_T} - \mathbf{U}_{\hat{\mathbf{h}}}\mathbf{D}_{C_2}\mathbf{U}_{\hat{\mathbf{h}}}^H - \mathbf{\Xi}, \tag{5.31c}$$

$$\mathbf{Z}^* = \lambda^*\mathbf{I}_{N_T} - \mathbf{\Xi}, \tag{5.31d}$$

$$\mathbf{S}_{C_3}\left(\mathbf{W}, \delta, \tau\right)\mathbf{D}_{C_3} = \mathbf{0}, \tag{5.31e}$$

---

where $\mathbf{\Xi} = \sum_{j=1}^{J} \sum_{l=1}^{N_R} \left[\mathbf{U}_{\tilde{\mathbf{g}}_j}\mathbf{D}_{C_{4_j}}\mathbf{U}_{\tilde{\mathbf{g}}_j}^H\right]_{a:b,c:d}, a = (l-1)N_T + 1, b = lN_T, c = (l-1)N_T + 1$, and $d = lN_T$. The optimal primal and dual variables of the SDP relaxed version are denoted by the corresponding variables with an asterisk superscript.

Subtracting (5.31d) from (5.31c) yields:

$$\mathbf{Y}^* + \mathbf{U}_{\hat{\mathbf{h}}}\mathbf{D}_{C_3}\mathbf{U}_{\hat{\mathbf{h}}}^H = \mathbf{Z}^* + \mathbf{I}_{N_T}. \tag{5.32}$$

Next, we multiply both sides of (5.31c) by $\mathbf{W}^*$ leading to

$$\mathbf{W}^*\mathbf{U}_{\hat{\mathbf{h}}}\mathbf{D}_{C_3}\mathbf{U}_{\hat{\mathbf{h}}}^H = \mathbf{W}^*(\mathbf{Z}^* + \mathbf{I}_{N_T}). \tag{5.33}$$

From (5.33), we can deduce that

$$\begin{aligned}
\text{Rank}(\mathbf{W}^*) &= \text{Rank}(\mathbf{W}^*\mathbf{U}_{\hat{\mathbf{h}}}\mathbf{D}_{C_3}\mathbf{U}_{\hat{\mathbf{h}}}^H) \\
&\leq \min\{\text{Rank}(\mathbf{W}^*), \text{Rank}(\mathbf{U}_{\hat{\mathbf{h}}}\mathbf{D}_{C_3}\mathbf{U}_{\hat{\mathbf{h}}}^H)\}.
\end{aligned} \tag{5.34}$$

Therefore, if $\text{Rank}(\mathbf{U}_{\hat{\mathbf{h}}}\mathbf{D}_{C_3}\mathbf{U}_{\hat{\mathbf{h}}}^H) \leq 1$, then $\text{Rank}(\mathbf{W}^*) \leq 1$. To show $\text{Rank}(\mathbf{U}_{\hat{\mathbf{h}}}\mathbf{D}_{C_3}\mathbf{U}_{\hat{\mathbf{h}}}^H) \leq 1$, we pre-multiply and post-multiply (5.31e) by $[\mathbf{I}_{N_T}\ \mathbf{0}]$ and $\mathbf{U}_{\hat{\mathbf{h}}}^H$, respectively. After some mathematical manipulations, we have the following equality:

$$(\delta\mathbf{I}_{N_T} + \mathbf{W}^*)\mathbf{U}_{\hat{\mathbf{h}}}\mathbf{D}_{C_3}\mathbf{U}_{\hat{\mathbf{h}}}^H = \delta[\mathbf{0}\ \hat{\mathbf{h}}]\mathbf{D}_{C_3}\mathbf{U}_{\hat{\mathbf{h}}}^H. \tag{5.35}$$

Besides, it can be shown that $\delta\mathbf{I}_{N_T} + \mathbf{W}^* \succ \mathbf{0}$ and $\delta > 0$ hold for the optimal solution such that the dual optimal solution is bounded from above. Therefore, we have

$$\text{Rank}(\mathbf{U}_{\hat{\mathbf{h}}}\mathbf{D}_{C_3}\mathbf{U}_{\hat{\mathbf{h}}}^H) = \text{Rank}(\delta[\mathbf{0}\ \hat{\mathbf{h}}]\mathbf{D}_{C_3}\mathbf{U}_{\hat{\mathbf{h}}}^H) \leq \text{Rank}([\mathbf{0}\ \hat{\mathbf{h}}]) \leq 1. \tag{5.36}$$

By combining (5.34) and (5.36), we can conclude that $\text{Rank}(\mathbf{W}^*) \leq 1$. On the other hand, $\mathbf{W}^* \neq \mathbf{0}$ is not optimal for $P_{\max} > 0$ and thus $\text{Rank}(\mathbf{W}^*) = 1$. ■

Bibliography

[1] M. Zorzi, A. Gluhak, S. Lange, and A. Bassi, "From today's INTRAnet of things to a Future INTERnet of Things: a Wireless- and Mobility-Related View," *IEEE Wireless Commun.*, vol. 17, pp. 44–51, Dec. 2010.

[2] D. W. K. Ng, E. S. Lo, and R. Schober, "Energy-Efficient Resource Allocation in OFDMA Systems with Hybrid Energy Harvesting Base Station," *IEEE Trans. Wireless Commun.*, vol. 12, pp. 3412–3427, Jul. 2013.

[3] M. Zhang and Y. Liu, "Energy Harvesting for Physical-Layer Security in OFDMA Networks," *IEEE Trans. Inf. Forensics Security*, vol. 11, pp. 154–162, Jan. 2016.

[4] I. Ahmed, A. Ikhlef, D. W. K. Ng, and R. Schober, "Power Allocation for an Energy Harvesting Transmitter with Hybrid Energy Sources," *IEEE Trans. Wireless Commun.*, vol. 12, pp. 6255–6267, Dec. 2013.

[5] P. Grover and A. Sahai, "Shannon Meets Tesla: Wireless Information and Power Transfer," in *Proc. IEEE Intern. Sympos. on Inf. Theory*, Jun. 2010, pp. 2363–2367.

[6] I. Krikidis, S. Timotheou, S. Nikolaou, G. Zheng, D. W. K. Ng, and R. Schober, "Simultaneous Wireless Information and Power Transfer in Modern Communication Systems," *IEEE Commun. Mag.*, vol. 52, no. 11, pp. 104–110, Nov. 2014.

[7] Z. Ding, C. Zhong, D. W. K. Ng, M. Peng, H. A. Suraweera, R. Schober, and H. V. Poor, "Application of Smart Antenna Technologies in Simultaneous Wireless Information and Power Transfer," *IEEE Commun. Mag.*, vol. 53, no. 4, pp. 86–93, Apr. 2015.

[8] X. Chen, Z. Zhang, H.-H. Chen, and H. Zhang, "Enhancing Wireless Information and Power Transfer by Exploiting Multi-Antenna Techniques," *IEEE Commun. Mag.*, no. 4, pp. 133–141, Apr. 2015.

[9] X. Chen, D. W. K. Ng, and H.-H. Chen, "Secrecy Wireless Information and Power Transfer: Challenges and Opportunities," *IEEE Commun. Mag.*, vol. 23, no. 2, pp. 54–61, Apr. 2016.

[10] X. Chen, J. Chen, and T. Liu, "Secure Transmission in Wireless Powered Massive MIMO Relaying Systems: Performance Analysis and Optimization," *IEEE Trans. Veh. Technol.*, vol. 65, pp. 8025–8035, Oct. 2016.

[11] C. Zhong, X. Chen, Z. Zhang, and G. K. Karagiannidis, "Wireless-Powered Communications: Performance Analysis and Optimization," *IEEE Trans. Commun.*, vol. 63, pp. 5178–5190, Dec. 2015.

[12] Q. Wu, M. Tao, D. Ng, W. Chen, and R. Schober, "Energy-Efficient Resource Allocation for Wireless Powered Communication Networks," *IEEE Trans. Wireless Commun.*, vol. 15, pp. 2312–2327, Mar. 2016.

[13] X. Chen, X. Wang, and X. Chen, "Energy-Efficient Optimization for Wireless Information and Power Transfer in Large-Scale MIMO Systems Employing Energy Beamforming," *IEEE Wireless Commun. Lett.*, vol. 2, pp. 1–4, Dec. 2013.

[14] D. W. K. Ng, E. S. Lo, and R. Schober, "Wireless Information and Power Transfer: Energy Efficiency Optimization in OFDMA Systems," *IEEE Trans. Wireless Commun.*, vol. 12, pp. 6352–6370, Dec. 2013.

[15] M. Zhang, Y. Liu, and R. Zhang, "Artificial Noise Aided Secrecy Information and Power Transfer in OFDMA Systems," *IEEE Trans. Wireless Commun.*, vol. 15, pp. 3085–3096, Apr. 2016.

[16] Y. Liu, "Wireless Information and Power Transfer for Multirelay-Assisted Cooperative Communication," *IEEE Wireless Commun. Lett.*, vol. 20, pp. 784–787, Apr. 2016.

[17] R. Zhang and C. K. Ho, "MIMO Broadcasting for Simultaneous Wireless Information and Power Transfer," *IEEE Trans. Wireless Commun.*, vol. 12, pp. 1989–2001, May 2013.

[18] S. Leng, D. W. K. Ng, N. Zlatanov, and R. Schober, "Multi-Objective Resource Allocation in Full-Duplex SWIPT Systems," in *Proc. IEEE Intern. Commun. Conf.*, May 2016.

[19] D. W. K. Ng, E. S. Lo, and R. Schober, "Robust Beamforming for Secure Communication in Systems with Wireless Information and Power Transfer," *IEEE Trans. Wireless Commun.*, vol. 13, pp. 4599–4615, Aug. 2014.

[20] D. W. K. Ng and R. Schober, "Secure and Green SWIPT in Distributed Antenna Networks with Limited Backhaul Capacity," *IEEE Trans. Wireless Commun.*, vol. 14, no. 9, pp. 5082–5097, Sep. 2015.

[21] M. Khandaker and K.-K. Wong, "Robust Secrecy Beamforming with Energy-Harvesting Eavesdroppers," *IEEE Wireless Commun. Lett.*, vol. 4, pp. 10–13, Feb. 2015.

[22] D. W. K. Ng, E. S. Lo, and R. Schober, "MultiObjective Resource Allocation for Secure Communication in Cognitive Radio Networks with Wireless Information and Power Transfer," *IEEE Trans. Veh. Technol.*, vol. 65, pp. 3166–3184, May 2016.

[23] Q. Wu, W. Chen, and J. Li, "Wireless Powered Communications with Initial Energy: QoS Guaranteed Energy-Efficient Resource Allocation," *IEEE Wireless Commun. Lett.*, vol. 19, pp. 2278–2281, Dec. 2015.

[24] E. Boshkovska, D. Ng, N. Zlatanov, and R. Schober, "Practical Non-Linear Energy Harvesting Model and Resource Allocation for SWIPT Systems," *IEEE Commun. Lett.*, vol. 19, pp. 2082–2085, Dec. 2015.

[25] S. Kisseleff, I. F. Akyildiz, and W. Gerstacker, "Beamforming for Magnetic Induction Based Wireless Power Transfer Systems with Multiple Receivers," in *Proc. IEEE Global Telecommun. Conf.*, Dec. 2015, pp. 1–7.

[26] S. Kisseleff, X. Chen, I. F. Akyildiz, and W. Gerstacker, "Wireless Power Transfer for Access Limited Wireless Underground Sensor Networks," in *Proc. IEEE Intern. Commun. Conf.*, May 2016.

[27] S. Kisseleff, X. Chen, I. F. Akyildiz, and W. H. Gerstacker, "Efficient Charging of Access Limited Wireless Underground Sensor Networks," *IEEE Trans. Commun.*, vol. 64, no. 5, pp. 2130–2142, May 2016.

[28] Powercast Coporation, "RF Energy Harvesting and Wireless Power for Low-Power Applications," 2011. [Online]. Available: http://www.mouser.com/pdfdocs/Powercast-Overview-2011-01-25.pdf

[29] D. W. K. Ng, E. S. Lo, and R. Schober, "Energy-Efficient Resource Allocation in Multi-Cell OFDMA Systems with Limited Backhaul Capacity," *IEEE Trans. Wireless Commun.*, vol. 11, pp. 3618–3631, Oct. 2012.

[30] D. W. K. Ng, E. Lo, and R. Schober, "Energy-Efficient Resource Allocation in OFDMA Systems with Large Numbers of Base Station Antennas," *IEEE Trans. Wireless Commun.*, vol. 11, pp. 3292–3304, Sep. 2012.

[31] Q. Wu, W. Chen, M. Tao, J. Li, H. Tang, and J. Wu, "Resource Allocation for Joint Transmitter and Receiver Energy Efficiency Maximization in Downlink OFDMA Systems," *IEEE Trans. Commun.*, vol. 63, pp. 416–430, Feb. 2015.

[32] Q. Wu, M. Tao, and W. Chen, "Joint Tx/Rx Energy-Efficient Scheduling in Multi-Radio Wireless Networks: A Divide-and-Conquer Approach," *IEEE Trans. Wireless Commun.*, vol. 15, pp. 2727–2740, Apr. 2016.

[33] L. Varshney, "Transporting Information and Energy Simultaneously," in *Proc. IEEE Intern. Sympos. on Inf. Theory*, Jul. 2008, pp. 1612–1616.

[34] C. W. Lander, *Power Electronics: Principles and Practice*. McGraw-Hill, 1993, 3rd ed.

[35] C. Valenta and G. Durgin, "Harvesting Wireless Power: Survey of Energy-Harvester Conversion Efficiency in Far-Field, Wireless Power Transfer Systems," *IEEE Microw. Mag.*, vol. 15, pp. 108–120, Jun. 2014.

[36] X. Zhou, R. Zhang, and C. K. Ho, "Wireless Information and Power Transfer: Architecture Design and Rate-Energy Tradeoff," in *Proc. IEEE Global Telecommun. Conf.*, Dec. 2012.

[37] D. W. K. Ng, E. S. Lo, and R. Schober, "Energy-Efficient Resource Allocation in Multiuser OFDM Systems with Wireless Information and Power Transfer," in *Proc. IEEE Wireless Commun. and Netw. Conf.*, 2013.

[38] S. Leng, D. W. K. Ng, and R. Schober, "Power Efficient and Secure Multiuser Communication Systems with Wireless Information and Power Transfer," in *Proc. IEEE Intern. Commun. Conf.*, Jun. 2014.

[39] D. W. K. Ng, L. Xiang, and R. Schober, "Multi-Objective Beamforming for Secure Communication in Systems with Wireless Information and Power Transfer," in *Proc. IEEE Personal, Indoor and Mobile Radio Commun. Sympos.*, Sep. 2013.

[40] D. W. K. Ng, R. Schober, and H. Alnuweiri, "Secure Layered Transmission in Multicast Systems with Wireless Information and Power Transfer," in *Proc. IEEE Intern. Commun. Conf.*, Jun. 2014, pp. 5389–5395.

[41] D. W. K. Ng and R. Schober, "Resource Allocation for Coordinated Multipoint Networks with Wireless Information and Power Transfer," in *Proc. IEEE Global Telecommun. Conf.*, Dec. 2014, pp. 4281–4287.

[42] M. Chynonova, R. Morsi, D. W. K. Ng, and R. Schober, "Optimal Multiuser Scheduling Schemes for Simultaneous Wireless Information and Power Transfer," in *23rd European Signal Process. Conf. (EUSIPCO)*, Aug. 2015.

[43] Q. Wu, M. Tao, D. W. K. Ng, W. Chen, and R. Schober, "Energy-Efficient Transmission for Wireless Powered Multiuser Communication Networks," in *Proc. IEEE Intern. Commun. Conf.*, Jun. 2015.

[44] D. Ng and R. Schober, "Max-Min Fair Wireless Energy Transfer for Secure Multiuser Communication Systems," in *IEEE Inf. Theory Workshop (ITW)*, Nov. 2014, pp. 326–330.

[45] J. Guo and X. Zhu, "An Improved Analytical Model for RF-DC Conversion Efficiency in Microwave Rectifiers," in *IEEE MTT-S Int. Microw. Symp. Dig.*, Jun. 2012, pp. 1–3.

[46] T. Le, K. Mayaram, and T. Fiez, "Efficient Far-Field Radio Frequency Energy Harvesting for Passively Powered Sensor Networks," *IEEE J. Solid-State Circuits*, vol. 43, pp. 1287–1302, May 2008.

[47] E. Boshkovska, "Practical Non-Linear Energy Harvesting Model and Resource Allocation in SWIPT Systems," Master's thesis, University of Erlangen-Nuremberg, 2015. [Online]. Available: http://arxiv.org/abs/1602.00833.

[48] S. Boyd and L. Vandenberghe, *Convex Optimization*. Cambridge University Press, 2004.

[49] M. Grant and S. Boyd, "CVX: Matlab Software for Disciplined Convex Programming, version 2.0 Beta," [Online]. https://cvxr.com/cvx, Sep. 2013.

[50] H. Ju and R. Zhang, "Throughput Maximization in Wireless Powered Communication Networks," *IEEE Trans. Wireless Commun.*, vol. 13, pp. 418–428, Jan. 2014.

[51] E. Boshkovska, D. W. K. Ng, N. Zlatanov, and R. Schober, "Robust Resource Allocation for Wireless Powered Communication Networks with Non-linear Energy Harvesting Model," *submitted for possible publication*, 2016.

[52] E. Boshkovska, A. Koelpin, D. W. K. Ng, N. Zlatanov, and R. Schober, "Robust Beamforming for SWIPT Systems with Non-linear Energy Harvesting Model," in *Proc. IEEE Intern. Workshop on Signal Process. Advances in Wireless Commun.*, Jul. 2016.

[53] Y. Sun, D. W. K. Ng, J. Zhu, and R. Schober, "Multi-Objective Optimization for Robust Power Efficient and Secure Full-Duplex Wireless Communication Systems," *IEEE Trans. Wireless Commun.*, vol. 15, pp. 5511–5526, Aug. 2016.

II

Data Era: Data Analytics and Security

CHAPTER 6

Distributed Machine Learning in Big Data Era for Smart City

Yuan Zuo

College of Engineering, Mathematics and Physical Sciences, University of Exeter, UK

Yulei Wu

College of Engineering, Mathematics and Physical Sciences, University of Exeter, UK

Geyong Min

College of Engineering, Mathematics and Physical Sciences, University of Exeter, UK

Chengqiang Huang

College of Engineering, Mathematics and Physical Sciences, University of Exeter, UK

Xing Zhang

Beijing Advanced Innovation Center for Future Internet Technology, Beijing University of Technology (BJUT) and Beijing University of Posts and Telecommunications (BUPT), China

CONTENTS

BIG DATA and Internet of Things are playing indispensable roles in human's daily life. In the era of information, all things are linked and enhanced by the Internet, from original virtual data to daily essentials, known as Internet of Things, from national strategy to urban development, known as smart city. Abundant digital sources bred by a variety of applications are not only a feature but also a challenge of big data analysis. This chapter focuses on large-scale data processing methods suitable for big data analysis scenarios in smart city, mainly the large-scale machine learning. The chapter presents six prevalent directions for classification distributed optimization evolving from standalone mode working to cluster mode. The classical classification algorithms are intrinsically sequential violating the parallel framework; therefore, several types of parallelization approaches are proposed to adapt stale mechanisms to advanced distributed fashions. These approaches include a variety of improvements facing distinct issues. In addition, this chapter discusses the basic statistical learning problems, in which classification conforms to a generic paradigm, and details theoretical analysis for each method. Additionally, a brief case study is given to gain deep insight of practical applications.

6.1 INTRODUCTION

As a part of digital revolution, the Internet has been promoting information science and technology over the last decades, eventually being considered as the symbol of the information revolution. A universal view was that although the information technology (IT) reinforced human interaction and communication, there was still a lack of connection between the digital and the physical world. Recently, however, Internet of Things (IoTs), commonly equipped with sensors, embedded systems and wireless access, arose to link objects, machines and even human bodies [1] [2] [3]. Since the IoT, associated with ubiquitous networks, enhances the interaction with the surrounding environment, it provides new and cost-efficient services for smart city [4], where numerous frameworks or architectures have been proposed [5] [6].

On these frameworks, numerous techniques can be adopted to accomplish a sequence of tasks. Among them, classification is one of the pervasive tools in a majority of applications. Take "traffic" as an example; it is necessary for drivers to acquire the road condition in a particular road segment, especially in rush hour to avoid traffic congestion. Collecting data from sensors placed along the roadside to detect the vehicle flows and data from sensors in vehicles recording running status, intelligent systems can calculate and predict the traffic condition to generate a driving recommendation ranking. The condition can be simply depicted as "good" and "bad" corresponding to "recommended" and "not recommended," respectively. Such a paradigm is generally called binary-classification or classification problem [7], which is a significant branch of machine learning (ML). At the end of this chapter, there will be a brief case study to demonstrate the idea.

Ubiquitous wireless sensors, smart items and smooth access, in one way, enrich the data sources enhancing critical and insightful understanding of actions in smart city, while, in other ways, substantially raise the volume of raw data. In the era of big data, the underlying trend is rooted in the five Vs (Volume, Velocity, Variety, Veracity, Value) and this vision is provided by the IoT and smart city [2] [4] using high performance computing (HPC), especially Cloud computing. In this sense, Cloud computing is widely thought to be a highly potential approach to overcome the large-scale data size issues [8]. In HPC, although techniques are distinct from each other [9], they inherently share some attributes where Parallelism is one of them.

With the incremental requirement of large-scale processing, a considerable amount of methods including machine learning algorithms need to be modified or even redesigned to fit in a parallel mode, because many of them are inherently sequential methods. This chapter will focus on parallelization of several main-stream optimization algorithms applied in binary classification as well as other machine learning theories, for instance, stochastic gradient descent (SGD) and Newton methods.

As for parallel computing, there is a common belief that a complicated problem can be divided into small segments which will be executed simultaneously by multiple workers commonly in a distributed environment (including multiple machines and multiple processors/cores). The basic tutorial and concepts can be found in [10] [11]. One may confuse the concept of "parallel" and "distributed" systems, though there exist some differences [12], e.g., "parallel" emphasizes the concurrent processing, while "distributed" refers to separate computing nodes with either synchronous patterns or asynchronous patterns. To avoid the confusion, in this chapter, the distributed system is regarded as a platform to implement parallel algorithms.

At present, there exist some distributed systems (e.g., Hadoop, Spark and Storm) for big data processing, which are prevalent under some circumstances. These distributed systems are rooted in the very idea that a small, middle or large cluster can only be deployed by off-the-shelf x86 hardware, instead of ad-hoc costly servers or workstations. As open source communities grow, the above popular systems are basically developed in an open source manner. Inspired by Google's MapReduce paradigm [13], Hadoop implements its own mechanism with the Hadoop distributed file system [14]. The very name of MapReduce suggests that it consists of two phases: *Mapping*, which performs splitting and sorting, and *Reducing*, which performs aggregation. Although MapReduce provides a host of merits when coping with large scale datasets, the mechanism is not as suitable as Spark in terms of machine learning [15]. With its special data structure, namely resilient distributed dataset (RDD), Spark is developed for data stream and machine learning. There also exists a well-behaved machine learning library called MLlib (Machine Learning Library). This chapter will involve some distributed-like systems; as such, the above description may provide an overview of the present big data processing fashion.

As for the general machine learning, it derived from a sub-sphere of artificial intelligence in the 1960s, but evolved into a pivotal toolset for the modern information technology. From Deep Blue of IBM in 1997 beating the world chess champion to AlphaGo of Google recently beating Lee Sedol, machine learning has implied a promising dawn of artificial intelligence to convey insight of every aspect on this planet. High performance computers equipped with distributed machine learning and ubiquitous sensors will boost Big Data and IoT applications in smart cities.

According to IBM Research THINKLab, which is an advanced institution targeted at technologies in business challenges, it is expected that active sensors for city management and daily life assistance will reach more than 50 billion. Abundant sensors will produce TB magnitude structured and unstructured data every day in the world, which triggers an urgent need for information retrieval. Effective methods such as classification, clustering and regression in machine learning are introduced to facilitate smart city. For instance, moving trace data of human, traffic monitoring and air delivers vital messages for assembling crowds, traffic congestion, air quality and other crucial issues associated with citizens' life quality. In section 6.7, an application for traffic monitoring is given to build a model for navigation service, which classifies traffic conditions with data patterns collected by sensors. Another example could be urban infrastructure maintenance.

In New York City, the infrastructure of a power grid over 100 years old is considerably outdated, which has become the largest challenge to grid reliability. To gradually replace and maintain new smart grid components, researchers access historical data from failure records, repairing records and other grid source records incorporating supervised machine learning algorithms for ranking potential objects, failure proactive action and decision, and utility plan decision [16]. It optimistically illustrates that intelligent management in smart grid through knowledge discovery and machine learning analysing historical data can enhance its resilience, decision making and effectiveness. Readers who have interests in machine learning can acquire further information in [17].

In this chapter, the concern is about a part of basic theories. As we mainly consider that the root of a set of machine learning algorithms lies in the study of statistical learning theory adopted by, for instance, pattern recognition, regression estimation and density estimation [18], one of the principal problems is the optimization of learning algorithms. In [18], it illustrates an explicit explanation from theoretical derivation to pragmatic implementation.

The first assumption is that in a supervised scenario, there is an unknown sample vector with distribution $P(x)$, associated with label y, paired (x_i, y_i). The target is to train a model (in practice, training is to acquire the optimal parameter vector ω) to predict the label of future input points, and the operation of training is to measure the discrepancy between the real response y_i and the learning result y_l, denoted as $L(y_l, (x_i, y_i))$. We then have a risk function

$$R = \int L(y_l, (x_i, y_i))dP(x) \qquad (6.1)$$

Here, the pair (x_i, y_i) can be depicted as a function $f(\omega)$ using the model ω. The problem is to obtain an optimal $f(\omega)$ from a function set to minimize the risk function. However, there is no prior experience in the unknown distribution. Therefore, the problem will be transferred into minimizing an empirical risk function

$$R_{empr} = \frac{1}{l} \sum L(y_l, (x_i, y_i)) \tag{6.2}$$

where l represents the number of samples x. Obviously, the form of loss functions is case dependent; more details can be found in [18].

The subsequent part of the chapter intends to elucidate several prevalent optimization methods reducing a concrete risk function. The rest of the chapter is organised as follows. In Section 6.2 and Section 6.3, two prevalent convex optimization methods and their parallelization are introduced, namely, the SGD and the Newton method. Section 6.4 presents the Petuum system, a specific parallel computing platform for machine learning. In Section 6.5, an architecture for decomposing optimization processing is introduced. Section 6.6 summarises two other parallel improvements. In Section 6.7, we conduct a brief case study for an application in the smart city by using basic classification. Finally, Section 6.8 concludes this chapter.

6.2 THE STOCHASTIC GRADIENT DESCENT (SGD) IN PARALLELIZATION

Owing to the basic statistical learning scenario, the follow-up problem is narrowed down to picking out an optimal solution f_{opt} in a set of functions, which is also known as optimization (training model). A general learning algorithm named SGD is proven to be an efficient tool for a large training dataset based on a series of recent contributions [19] [20] [21] [22] [23] [24], which also introduce several advanced parallelization implementations.

Figure 6.1 demonstrates the gradient of sinc function with a 2-D contour. Sinc function is given in a 3-D graph. Generally, gradient implies the greatest increasing direction of the function, which is represented by blue arrows in (b). The length of an arrow denotes its magnitude and the colors of vertical lines are associated to the location in the function in (a). The gradient descent was first introduced in [25]. Simply put, minimizing the empirical risk function is achieving the optimal value of ω. One way to do this is to iteratively update ω along with the direction of the function descent with the entire set of training data and an initialization. As such, the descent direction is opposed to the gradient of loss function. At the time for the next iteration, we have

$$\omega_{t+1} = \omega_t - \alpha \sum \nabla L(y_l, (x_i, y_i)) \tag{6.3}$$

where α means the manual descent step. In the present work, generally, ∇ represents the gradient of a function. Furthermore, if substituting the above step for the Hessian matrix (the matrix of the second-order partial derivative of a function) of the loss function, the gradient descent is turned to be the 2nd order gradient descent. It is easy to see that the gradient descent is intrinsically a batch method and has to buffer

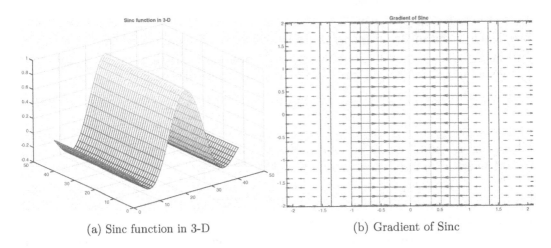

(a) Sinc function in 3-D　　　　　　　(b) Gradient of Sinc

Figure 6.1 Gradient of sinc function. The gradient usually indicates the direction of function increasing.

the whole data. In the meantime, the volume of data is seeing a sharp increase at an unprecedented rate, which triggers serious large-scale assignments problems.

To overcome this problem, SGD randomly draws one single point for each iteration. It can be denoted as follows

$$\omega_{t+1} = \omega_t - \eta \nabla L(y_t, (x_t, y_t)) \tag{6.4}$$

where, at time t, it only uses one sample with the step size η. Note that SGD is inherently sequential and can be easily employed in a fast online fashion. Some recommendations and attributes are provided in [19]. Admittedly, SGD fits in the large-scale environment. Nevertheless, the latest high-performance computing structures are composed of multi-nodes or multi-processors, making it difficult to be compatible with the pure sequential approaches. Consequently, researchers recently have paid more attention to the parallelized paradigm. In what follows, some recent endeavours are outlined.

In practice, the gradient descent is frequently employed in convex optimization, which prevails in classification and regression problems. An essential difference between classification and regression is the object type, discrete value for classification and continuous value for regression. In smart city, classification promotes prediction and recognition for behaviours, activities and status, such as transportation condition, crowding and security, based on data collected from weather, entry-exit inspection, the stock market, etc. On the other hand, urban administrators take advantage of regression to estimate the incoming economy trend, the future population, electricity consumption rate, etc. In the case of crowding and human flow, security departments are highly prone to concern about accidents, terrorism and criminal, which will usually be discerned by the gathering and movement of crowds. Handling urban sensor data, an administration can track the status of crowds in real-time to discover some hidden anomaly symptoms and predict an emergency, allowing the local police force in advance to reduce losses and even prevent: chaos by evacuating people ahead of time.

6.2.1 Parallelized SGD Based on MapReduce

In [20], it sticks to the classic SGD methods because of its consistent convergence and easy implementation. There are two major branches in the state of the art of parallel architectures in high-performance computing in light of recent work, parallel computing in a multi-core system and distributed computing in a multi-machine platform. As it should be clarified, here we concentrate on the tasks being completed in a distributed scenario. The multi-core structure in the optimization of existing binary-classification problems, also known as the two-classification problem, will be discussed in subsequent statements. As mentioned before, due to several merits, the SGD method is continuously broadly regarded as one of the most effective and efficient methods of numerical optimization in large-scale machine learning domain [26].

According to [20], the proposed algorithm is devised to fit in a MapReduce scheme, which has relatively higher latency of data access and lower interaction between machines than multi-core platforms. Apparently, it seems to be a natural process that each local machine trains part of data of the entire collection of samples. Nevertheless, this working mode leads to a heavy communication pressure as well as low parameter convergence. Alleviating communication overhead is one of the most pertinent issues in the sense of simultaneous processing. Since each iteration of MapReduce needs access data in the local disk, the training procedure brings about a waste of computation. Therefore, the authors adopt sub-problems segregation and local solution aggregation [27] to dramatically reduce communication cost while they replace the inner loop with a parallel-SGD step. In addition, another contribution is integrating an asymptotic convergence property of online learning into their parallel-SGD algorithm. This property ensures all local parameter vectors will end up with the same limit value with a fixed learning rate η.

The main discussion is a limit at general convex cost function with regularization following minimization of real risk function.

$$(Cost Function with L_2 Regularization) f(x, y, \omega, b) = \frac{1}{2} \|\omega\|^2 + L(x, y, \omega, b) \quad (6.5)$$

where b and ω are the bias and the parameter, respectively, and $\frac{1}{2} \|\omega\|^2$ is the L_2 regularization. It is noteworthy that in this paper, rigorous mathematical proofs have been presented to guarantee contractive mapping through distribution algorithms and low errors in the sense of a stationary point, based on general statistical learning theories and functional analysis. Detailed analysis can be found in its appendix.

6.2.2 Online SGD in Round-robin

The online SGD version is also a hot spot in the high-performance computing sphere. Parallel computing can be represented as a low processor-level or thread-level paradigm, for instance, multi-core CPU, GPU and multi-processors system with shared memory, opposed to a distributed model, such as MapReduce or Spark. Aside from the existing ideas, online learning is another way to tackle general large-scale and dataflow problems, which characterises samples input as streams.

This contribution [28] helps online learning to develop a parallel property along with its advantages in convergence and flexibility in spite of being intrinsically consecutive. Apparently, according to [28], traditional online measurements encounter some bottlenecks and low efficiency problems based on some high speed applications and large amounts of data cube in effect. The authors claim there exist a few factors, low efficient usage of multi-core processors, incompatible speed between CPUs and peripherals, namely memory access, disk storage and network, which are unable to fit in pervasive cloud architectures. Admittedly, they propose two delayed update algorithms targeted at the shared memory multi-core structure and ordinary machine clusters, but it is uncertain that the method is suitable for distributed models.

The two algorithms share one property of delay updates. The first one, called asynchronous optimization, makes each core calculate gradients independently using a shared parameter variable; the second one, called pipelined optimization, makes a family of linear functions be calculated independently on different cores. During the whole procedure, each core updates parameters based on its own timeline, which could generate the delay up to $\tau = n - 1$ (suppose the parameter $\omega \in \mathbb{R}^n$ if the multi-core is formed in a round-robin fashion. From the pseudo-code, the update step would be 1) $\frac{\delta}{t-\tau}$ and 2) $\frac{\delta}{\sqrt{t-\tau}}$ (τ is a constant specified by heuristics). Based on the two delay coefficients, the subsequent theoretical analysis presents corresponding regret bounds. More details can be found in the version [29]. Additionally, three other bounds are inferred under assumptions, namely linearity of functions and Lipschitz continuity of gradients, because the impact of the correlation between adjacent points in a data stream is not negligible.

To face sequential challenges, this literature provides theoretical proofs with evaluation results, which indicate that slight delayed updates affect the convergence little but accelerate the computation significantly.

6.2.3 HOGWILD! for "Lock-free"

In terms of the basic inner loop of SGD, the synchronization step, also called locking, is of necessity for robustness and convergence, which in turn heavily affects its parallelized formalization. Thereupon, Niu et al. [21] proposed an approach without locking to parallelizing the SGD, called "HOGWILD!", due to an ill-behaved MapReduce-like pattern for large-scale and web-scale data sets.

In this work, HOGWILD! introduces a concept of sparsity and keeps its target at three sparse separable cost examples: Sparse SVM based on bi-classification, Matrix Completion based on collaborative filtering, and Graph Cuts. It is assumed that the fractional functions would only act over a small part of data because the data volume is huge. From the perspective of sparsity, it guarantees the linear convergence with a $\frac{1}{k}$ rate for a constant step and the corresponding robustness. In a nutshell, HOGWILD! makes it available that processors can arbitrarily access and revise decision variables stored in a shared memory regardless of writing conflicts amongst them. Evidently, the risks because of overwriting are negligible. Specifically, an additional operation is seen as atomic as a single element step while numerous elements usually occupy synchronization running. To update the decision, at the outset, HOGWILD! randomly pick samples in the sets. Then, it follows the general formalization while with one normal basis in the vector space.

$$x_\nu \leftarrow x_\nu - \eta b_\nu^T G_e(x) \tag{6.6}$$

where η is the constant or variable step, b_ν denotes one basis, ν ranges from 1 to n and also $\nu \in e$, and G_e is the sub-gradient of the corresponding risk functions multiplied by the size of datasets $|E|$ (here, $e \in E$). It is evident that some risk functions are zeros with the corresponding samples outside the subset e.

As a consequence, each processor utilizes the above formula randomly and iteratively computes the parameter vector. Theoretically, the authors derive some conditions and present adequate proofs to ensure its convergence, even nearly within the same steps as the sequential one. In their experiments, compared with an online method [28] [29] and Vowpal Wabbit [30], HOGWILD! achieves better speedup and converges nearly linearly.

6.2.4 AsySVRG for asynchronous SGD Variant

In contrast to the gradient descent dealing with batch loads, the SGD resembles a stream process updating parameters once per sample until the empirical risk function decreases to a sufficiently small value. Abundant literature indicate that SGD fits in high-speed large-scale computing.

In [22], it proposes an asynchronous parallel SGD approach coupled with the parallelization of the stochastic variance reduced gradient (SVRG) [31], named AsySVRG. Although recent computer clusters have gained popularity in high-performance computing, each single machine incorporates a multi-core system to handle sub-problems. Also, a single node is facing a data flood trend. Admittedly, increasing physical workstations can alleviate the pressure, however costly. On the contrary, increasing cores for parallel threads in one machine is likely to be a plausible way.

The proposed algorithm firstly allocates a shared memory storing the parameter ω (only one thread can access it at a time) and inconsistent reading ω (reading is free while updating is restricted). Then, the step below is taken in parallel.

$$\nu_m = \nabla f(u_k) - \nabla f(u_0) + \nabla f(u_0) \tag{6.7}$$

$$u_{m+1} = u_m - \eta \nu_m \tag{6.8}$$

where $u_0 = \omega_t$, $u_m = \omega_{t+m}$, k is the updating number of ν_m and m is the updating number of ω. More details can be found in [31]. After the parallel phrase, there are two options, replacing the current solution in memory with the latest result or averaging the result within the loop. This depends on the subtraction τ, from $k - m < \tau$.

Convergence upon two conditions is analysed, which presents the linearity. The experiments evidently show AsySVRG outperforms the previous Hogwild! paradigm in terms of runtime and speedup.

6.2.5 ASGD with Single-sided Communication

Inspired by [20], ASGD (Asynchronous Parallel Stochastic Gradient Descent) [23] is proposed by two researchers still attempting to further improve synchronization efficiency, also called "locking" in article [21], induced during the parallelization procedure. As it asserts, since parallel machine learning is becoming increasingly significant, many practitioners endeavour to tailor traditional machine learning algorithms into the MapReduce framework. Obviously, among them, the SGD is a desirable one. It is also widely accepted that MapReduce yields high interaction overhead between nodes. As such, differing from previous work, the authors target replacing the ordinary communication mode by an asynchronous single-sided scheme so as to achieve the "lock-free" goal.

The authors notice that the general communication of two-sided protocols is the main reason that expensive links happen. Therefore, the proposal of Asynchronous Parallel Stochastic Gradient Descent (ASGD) is raised, which first adopts the mini-Batch update approach [32], and then enables the asynchronous single-sided strategy [33]. To be more specific, after taking the same initialization and parameter set-ups as article [20], this method treats the online gradient descent update step as their key leverage point. There exists an external update term $\Delta(\omega)$

$$\Delta'(\omega_{t+1}) = \omega_t - \frac{1}{2}(\omega_t + \omega_{t'}) + \Delta(\omega_{t+1}) \qquad (6.9)$$

where ω_t denotes a parameter vector at iteration t. Additionally, in order to overcome the drawback of data races mentioned in article [21], the ASGD introduces a decision function to distinguish "bad" or "good" state parameters. According to its practice results, the effect of data races is negligible.

Unfortunately, as claimed, ASGD just integrates existing algorithms and will degrade to parallel if the communication interval is infinite. However, it still provides fast convergence and is more scalable than previous works.

6.3 THE NEWTON METHOD IN PARALLELIZATION

In real-world, there are substantial situations where we cannot solve a problem ending up with an exact formulization, which is an analytical solution. A majority of cases are prone to acquiring an approximately numerical solution. The Newton method, also called the Newton–Raphson method, is a commonly used technique to efficiently find an approximate solution to a real-valued function in light of numerical analysis. A brief introduction of the Newton method will be demonstrated as follows:

For the sake of convenience, we consider one single variable in functions, which are assumed second-order differentiable. A basic and simple idea for approximation is a linear function. From the perspective of Taylor's expansion about the point x_0,

$$f(x) = f(x_0) + f'(x_0)(x - x_0) + O(f'')(1st\,order\,Taylor\,polynomial) \qquad (6.10)$$

Omitting the remainder term $O(f'')$, we have

$$f(x) \approx f(x_0) + f'(x_0)(x - x_0) \qquad (6.11)$$

Let $f(x) = 0$, then

$$x = x_0 - \frac{f(x_0)}{f'(x_0)} \tag{6.12}$$

Substituting this result into the objective function $f(x)$, it would check if it is sufficiently close to 0. If not, we simply iteratively continue this updating process to obtain

$$x_{n+1} = x_n - \frac{f(x_n)}{f'(x_n)} \tag{6.13}$$

until the value of $f(x_{n+1})$ is almost 0. Numerically, there is still a tiny residual while it can be negligible under the accuracy requirement.

The Newton method can be also interpreted in a geometrical sense. Turning back to the aforementioned formula $f(x) \approx f(x_0) + f'(x_0)(x - x_0)$, then

$$f(x_{n+1}) \approx f(x_n) + f'(x_n)(x - x_n) \tag{6.14}$$

It can be regarded as a tangent line of the curve $f(x)$ at the point $(x_n, f(x_n))$, and the root or approximation of it, x_{n+1}, is the x-intersect of the tangent line. Now, as x_{n+1} is the next initial of the true root, the same step is repeated for the next approximation x_{n+2}. This is an intriguing procedure because each result will asymptotically be close to the true solution at the rate of quadratic convergence. As such, the Newton method prevails in the sphere of convex optimization.

As we can see from above, classification largely narrows down to convex optimization, which can be accelerated by the Newton method. The typical machine learning algorithm plays a crucial part in smart city infrastructure applications. Urbanization, tremendous lifestyle changing and rapid work pace heavily rely on sustainable urban infrastructures, including hospital, police, electricity and sanitization, which urban service providers have been requesting for many years. Abundant data from energy efficiency, serviceability and functionality have attracted increasing attention, where clustering is adopted to merge similar fields to identify invisible troubles and classification is applied to guide utility planning and malfunction diagnosis. Furthermore, society and public security issues have been raised to an unprecedented high level owing to global terrorism and regional instability. Based on public data collected from social media, blog posts or video comments, global security departments are expecting a reliable classification method to assist in locating potential dangers.

The Newton method is a very effective and efficient tool with quadratic convergence, although it may consume considerable resources and may not have convergence guarantee. It heavily relies on the initial point. If the initial guess is close enough to the optimum, it may perform ideally, otherwise slowly, even divergently. The basic theory and details can be found in [34].

Recently, the parallelization of the Newton method attracted researchers' attention due to data acquisition in big data. Lin and his team make great contributions to it [35] [36] [37] [38]. Their work is presented below.

6.3.1 A truncated Newton Method: The Trust Region Newton Method (TRON)

In either logistic regression (LR) or support vector machines (SVM), the convex optimization always attracts the most attention. At the outset of the chapter, we have introduced that, technically, one of the overriding issues of machine learning is optimization. Take typical statistical learning for example. The primitive purpose is to minimize the "real expected risk," $E_r(f)$, while the truly concerned and obtained term is the function f^* which finally makes the value of the risk smallest.

In the article [35], Lin et al. modified the bound-constrained conditions to unconstrained occasions based on their previous work in truncated Newton approaches. The Trust Region Newton Method (TRON) is also aiming at the huge volume dataset with thousands of, even millions of, dimensions, namely features. Because the LR and SVM problem are alike, the proposed TRON algorithm is suitable to both with L-1 or L-2 regularization. In a nutshell, a basic assumption is that the objective function is convex. The optimal solution will be searched along the Newton direction and the parameter will be updated until it reaches stopping condition.

$$d = -\frac{\nabla f}{\nabla^2 f}(Newton Direction), \omega_{t+1} = \omega_t + d(Update) \tag{6.15}$$

where $\nabla^2 f$ is the Hessian matrix if it contains multi-variables and ∇f is the gradient. TRON intends to consider two concerns: 1) how to determine the appropriate direction and guarantee the convergence quality; 2) owing to the high dimensionality, the Hessian matrix is so large that storing it and calculating its inverse become nearly infeasible. In general, to tackle issue 1), it defines a variable region to bound the updating length. Within it, the calculated direction shall be trusted and be applied to the parameter, otherwise the parameter remains. Meanwhile, the region itself would be updated based on an approximation ratio between practical reduction and expected reduction. To address 2), it makes an assumption that the Hessian matrix is sparse while the product of the Hessian and data vector is denser and smaller. Therefore, the inverse is avoided.

Specifically, let q_{rs} be the quadratic Taylor's expansion,

$$q_{rs} = \nabla f d + \frac{1}{2}\nabla^2 f d^2, \tag{6.16}$$

$$ratio = \frac{f(\omega + d) - f(\omega)}{q_{rs}}. \tag{6.17}$$

The region is denoted by δ_{trust}, which is also updated at the end of each iteration. From now on, the problem is transferred to minimize q_{rs}, subject to $\|d\| \leq \Delta_{trust}$. At the following stage, it is to solve q_{rs} by the conjugate gradient method. Moreover, as for the product,

$$\nabla^2 f d = d + CX^T(D(Xd)) \tag{6.18}$$

where C is a constant matrix, D denotes a diagonal matrix and X represents the data matrix. To update the ω and δ_{trust} is to check whether the actual reduction ratio is large enough.

Last but not least, with six test-sets containing distinctive numbers of features, TRON presents well-behaved performance and a high convergence rate. Based on the algorithm, their future work focuses on distributed improvement while still maintaining high quality.

6.3.2 The Distributed TRON Based on Spark and MPI

As one of the most fashionable platforms for Cloud computing and big data analytics, Spark has attracted much interest from fellow researchers. Since MapReduce is broadly viewed as being slow with iterative computation, Spark is quite prevalent among researchers. The aforementioned TRON inspires them to come up with a distributed implementation using Spark. It is well-interpreted because the basic idea still adheres to the universal master-slave mode and only maps matrix-vector multiplication into slave nodes.

Technically, the research [36] follows the main scenario of the TRON while the heavy computation occurs in the approximation stage of minimizing the Taylor's expansion. Taking the product (6.18) out as an exclusive step, the data matrix is split into parts. Therefore, it becomes a component-wise operation.

$$f = \frac{1}{2}\omega^T\omega + C\sum f_k \rightarrow \triangledown f = \omega + C\sum \triangledown f_k \rightarrow \triangledown^2 fd = d + C\sum \triangledown^2 f_k d \quad (6.19)$$

From the above theory, it is evident that the computation takes place locally if the divided data are stored in slave nodes; then the algorithm aggregates the intermediate results into the master node. Undoubtedly, two phrases are introduced in this article: 1) Partitioning data, aggregating intermediate and updating the parameter vectors; 2) Shipping the current parameter vectors and updating step vectors to slave nodes.

The authors also dissect the implementation details to figure out to what extent they would affect the ultimate convergence and speedup. The details include the loop structure in the program languages, the data encapsulation, the mapping and reducing style and the temporal variables. In the end, experimental results show the distributed TRON is of fault-tolerance.

6.3.3 General Distributed Implementation of TRON

A successor study of the distributed Newton method [37]. Differing from the employment of the TRON in Spark, the authors intend to adjust it to a universal parallel framework without loss of generality. They adhere to the parallelization of matrix-vector multiplication to handle the challenges of parameter communication cost. Meanwhile, to facilitate the computation speed, the separation format of training data has been mainly discussed.

As can be seen, the matrix-vector product indeed decreases the delivery pressure between each iteration, while in this article, the separation format of samples also exerts a significant influence on the computing phrase if the dimensionality outnumbers the magnitude of data. To simplify the idea, the update step is denoted as

$$\triangledown f(\omega) = \omega + C(YX)^T(\delta(YX\omega)^{-1} - E) \quad (6.20)$$

where E denotes all the elements of the matrix are 1. For instance, it is similar to the ordinary operation, while feature-wise, each node will contain a part of the divided parameter vector and the whole set of label matrix Y. The formalization is as follows

$$X\omega = \sum X_j\omega_j \tag{6.21}$$

where ω_j represents the sub-vector of ω. The evaluation manifests the instance-wise split and feature-wise split are suitable for the two conditions, magnitude of data being larger and smaller than dimensionality, respectively. Also, compared to a decomposition framework, ADMM, and a machine learning project, Vowpal Wabbit [30], the distributed TRON illustrates that its iteration maintains the same efficiency ignoring the cluster scale and it achieves comparable, even faster and scalable, performance.

6.3.4 Matrix-vector Product Improvement for Inner Mechanism

Further performance improvement in the built-in matrix-vector product mechanism with respect to the previous Newton method is studied in the paper. Obviously, in [38], the authors assume that the Newton method performs well while the bottleneck is the sparse matrix-vector multiplication. They primarily consider a multi-processor environment. First, a matrix split type [39] is adopted for products' uncorrelation. Second, several ways are conducted to parallelize the matrix-vector multiplication, including OpenMP, MKL and an effective format, called Recursive Sparse Blocks (RSB) [40] [41], which is employed for multi-core thread parallel sparse matrix-vector multiply (SpMV). The outcome demonstrates the implementation of OpenMP outperforms other recent methods over the parallel Newton method.

6.4 THE PETUUM FRAMEWORK

It is natural to think that distributed machine learning plainly inclines people towards optimization of a variety of methods in a sense of isolation. In other words, one may purely emphasise its superiority upon some particular occasions or in a restricted field instead of a universal area. It turns out that almost each branch of machine learning tends to evolve gradually to parallelization along with its preliminary solvable direction. Indeed, the above practice is effective and makes significant breakthroughs, while little work has been done devising an ad hoc mechanism for general machine learning algorithms. Although the common big data platforms have their own ML extension libraries, for instance, Mahout [42], MLlib [43], Vowpal Wabbit [30], and GraphLab [44], the drawbacks still occur inevitably [45]. The problem discovered in [45] is the migration of classic algorithms from a single computer to a cluster or cloud environment. To put it simply, these systems may fail to offer proper built-in blocks or take advantage of flexible iterations.

Xing et al. [45] attempted to acquire in-depth insight into the fine-grained operations to extract several intrinsic properties of the ML methods without loss of generality. They believe that the two keys underlying the collection of ML are data and model, which inspire data-parallel and model-parallel approaches, repsectively. Consequently, three crucial and fundamental shared-properties are extracted: 1) error tolerance, 2) dynamic structural dependency and 3) non-uniform convergence, which are supposed to be treated equally, carefully and efficiently. Petuum, proposed in this work, sets its goal as a general-purpose platform to systematically address parallelism challenges with the observed inner properties. As can be seen, it still adopts the centralized formation.

Alongside the three properties, three corresponding system objectives are introduced: 1) converging parameters simultaneously at low communication cost, 2) scheduling the parameters' updating policy considering their dependencies and 3) prioritizing communication for non-convergent parameter components. From a perspective of statistical learning theory, Petuum first views the ML algorithms as a process of iterative-convergence. Given D is the dataset and δ is the updating function, M represents the model with a particular stopping condition

$$M^t = F(M^{t-1}, \Delta(M^{t-1}, D)) \tag{6.22}$$

where t denotes the order of iteration and F aggregates all intermediate values. Then, the data-parallel and model-parallel mechanisms dissect the following stages, respectively.

In data-parallelism, the training set D is separated and stored in the individual nodes. Thereby, let the function δ be applicable to a subset of D,

$$M^t = F(M^{t-1}, \sum \Delta(M^{t-1}, D)) \tag{6.23}$$

At this phrase, the master worker will aggregate the temporal partial variables taking on the additive operation. This data separated scenario is rooted in the feasibility of the additive operation over independent and identical distributed (IID) data.

In mode-parallelism, the mode S is the one partitioned,

$$M^t = F(M^{t-1}, [\Delta(M^{t-1}, M^{t-1}S_p^{t-1})]_p), \tag{6.24}$$

where p is the pth node in the system and S denotes a scheduling function which determines the updating components of the parameter vector in local workers. That is to say, with a well-designed scheduling function, the whole elements in the parameter are free to be updated individually over the same IID dataset. Significantly, the key foundation of this scenario is the independency or ill-correlation of the elements [46].

Based on the above principles, Petuum is ultimately designed with three principal modules: scheduler, workers and parameter server. As stated, the scheduler takes responsibility for choosing specified elements to update upon the customized standards. The workers execute the regular computational tasks simultaneously ignorant of data existence format, for example, the file system HDFS or Yarn. The third one, a parameter server, makes it convenient to access the shared vector; in the meantime, it restricts the consumption of network resources.

It is noteworthy that Petuum is an open source library including a bunch of prevalent ML algorithms despite not mentioning in detail. Additionally, Petuum successfully exploits three intrinsic properties to speed up ML algorithms and the rigorous theoretical analysis can be found in [45].

On top of the state-of-the-art distributed ML frameworks, it is inevitable in communication that the parameters will be shared, transmitted or asynchronously overwritten. Based on the learning theory, the core of optimization is to adjust the parameter vector iteratively to a stable and convergent value. Empirical experiments show numerical models are massively matrix-parameterized. Thereby the communication pressure arises. Taking a platform example of Petuum, one of its main modules is the parameter server (PS). Based on the public server, partial components are modified during one iteration, then aggregated. A Sufficient Factor Broadcasting (SFB) structure is proposed in [47], aiming at reducing the communication cost of matrix-parameterized models to solve SGD and stochastic dual coordinate ascent (SDCA) [48] [49]. Typically, the models can be abstracted as follows

$$\min_{A} L(Ax) + R(A) \tag{6.25}$$

where A is the parameter matrix, L denotes a set of loss functions and R represents the regularization function.

The SFB takes a basic idea from non-negative matrix factorization (NMF), which factorizes a tense large matrix into two sparse and low rank non-negative matrices. Simply put, the model consists of three primary phrases: 1) decomposing the raw matrix; 2) shipping the core "Sufficient Factors" to workers; and 3) reconstructing with two factors. During the whole runtime, the full matrix will never be shipped. Since the factorized low-rank matrices are much smaller, the cost will be reduced drastically. It is worth mentioning that to avoid full broadcasting, the authors introduced the Halton sequence [50] idea of having a subset of machines connected. Messages can only be sent within a particular circle. Evidently in experiments, the convergence quality is slightly worse than the full matrix scheme in the runtime, while it asymptotically converges to the optimum. Details and the potential extension can be found in [47].

Specified devised distributed machine learning overcomes a series of common shortcomings existing in general-purpose distributed architectures. A wise idea would be making two types of processing frames connect seamlessly to construct a Big Data analysis flow. The future countless sensors will inevitably be embedded in a variety of smart devices for numerous types of purposes, producing excessively high volume unstructured data with tedious noise, which has to be taken seriously ahead of relation and dependence analysis. Those general-purpose systems are extremely suitable for data cleaning, batch relational transformation, bulk outputs, precise index and storing for data warehouses, whereas not for distributed machine learning platforms. Seamless merging guarantees data integrity and ordering for machine learning iterative and consistent access, which underpins the outcomes for end system administrators and also consolidates the core role of machine learning.

6.5 THE CONVEX OPTIMIZATION DECOMPOSITION METHOD

For general distributed implementation, the synchronous SGD or the distributed Newton method is an effective measure to cope with noticeable challenges. However, there is a different point of view. The ADMM [51] recently has been introduced to build a general framework for distributed learning [52]. Essentially, it is a way to decompose a complex convex problem into plenty of solvable sub-problems, blending merits of dual decomposition and augmented Lagrangian methods. For better understanding the ADMM, we start from primal decomposition and dual decomposition.

Based on a simple idea, once there is a large project, there is a hierarchical form from higher to lower and division of labour for divided tasks. For convenience and efficiency sake, directly divided tasks make the responsibility clear. In large-scale computing, once there is a large problem, splitting it into mutually independent pieces appears to be plausible and rational. For the primal problems, assume that the primal convex unconstrained function could be depicted as:

$$\min f(x_1, x_2...x_n, y_1, y_2...y_m) = \min f_x(x_1, x_2...x_n) + \min f_y(y_1, y_2...y_m) \quad (6.26)$$

Here, one may have the formula summed by two totally decoupled sub-functions with mutually independent n inputs $\mathbf{x}$ and m inputs $\mathbf{y}$, or with several fixed variables. Particular techniques can be taken to solve the subsequent minimization issues with respect to separation. This is called primal decomposition.

Unfortunately, in reality, there are less idealistic occasions than having problems without constraints. Now, consider the following primal problem:

$$\min_x f(\mathbf{x}, \mathbf{y}) = \min(f_x(x) + f_y(y)) \quad (6.27)$$

$$Subjecttog_i(\mathbf{x}, \mathbf{y}) \le G_i, h_j(\mathbf{x}, \mathbf{y}) = H_j \quad (6.28)$$

where $g_i(x)$ and $h_j(x)$, respectively, represent inequality and equality constraints, G_i and H_j are the corresponding constants and $\mathbf{x}$ and $\mathbf{y}$ refer to the previous parameter vectors. A feasible way is to adopt Lagrangian multipliers.

First, for equality constraints, it will be written as $y_1 = y_2$. By introducing the Lagrangian multiplier τ, the Lagrangian can be expressed as

$$L(\mathbf{x}, \mathbf{y}, \tau) = f_1(\mathbf{x}, y_1) + f_2(\mathbf{x}, y_2) + \tau(y_1 - y_2) \quad (6.29)$$

$$L(\mathbf{x}, \mathbf{y}, \tau) = (f_1(\mathbf{x}, y_1) + \tau y_1) + (f_2(\mathbf{x}, y_2) - \tau y_2) \quad (6.30)$$

Thereupon, the dual function is

$$F(\tau) = F_1(\tau) + F_2(\tau) \quad (6.31)$$

$$F_1(\tau) = \inf(f_1(\mathbf{x}, y_1) + \tau y_1), F_2(\tau) = \inf(f_2(\mathbf{x}, y_2) - \tau y_2) \quad (6.32)$$

which is illustrated as separable, followed by its Lagrangian dual formalization.

Second, for the inequality, suppose it could be expressed as

$$g_1(x) + g_2(y) \le G \quad (6.33)$$

The Lagrangian would be

$$L(\mathbf{x}, \mathbf{y}, \lambda) = (g_1(\mathbf{x}) + \lambda x) + (g_2(\mathbf{t}) - \lambda y) \quad (6.34)$$

Therefore, it looks similar to the equality condition solving in parallel. The two conditions refer to the dual composition. Note that, in effect, the primal function is continuous but not necessarily strictly differential, in which the sub-gradient notation can be helpful. One can find more details and rigorous mathematical explanations in [51] [52] [53].

As for ADMM, it can be considered as two-step evolution, augmented Lagrangian and the method of multipliers, and alternating direction. In augmented Lagrangian, there is a penalty term,

$$L_{aug} = L(\mathbf{x}, \mathbf{y}, \lambda) + (\frac{\rho}{2}) \|h(\mathbf{x}, \mathbf{y})\|_2^2 \tag{6.35}$$

where ρ is called the penalty parameter. Meanwhile, the update policy yields

$$\mathbf{x}, \mathbf{y} \leftarrow \arg\min L_{aug}(\mathbf{x}, \mathbf{y}, \lambda^k) \tag{6.36}$$

$$\lambda^{k+1} \leftarrow \lambda^k + \rho h(\mathbf{x}, \mathbf{y}) \tag{6.37}$$

which is viewed as the method of multipliers.

The only difference between the ADMM and the augmented one is that the equality constraint can be decoupled with respect to the variables. To be precise,

$$\min f_1(\mathbf{x}) + f_2(\mathbf{y}) \tag{6.38}$$

$$s.t. h_1(\mathbf{x}) + h_2(\mathbf{y}) = 0 \tag{6.39}$$

$$L_{ADMM} = L(\mathbf{x}, \mathbf{y}, \lambda) + (\frac{\rho}{2}) \|h_1(\mathbf{x}) + h_2\mathbf{y}\|_2^2 \tag{6.40}$$

where $h(\cdot)$ and $L(\cdot)$ represent the equality and the loss function, respectively.

Now that the decoupled step is presented, the upcoming iterations are

$$\mathbf{x}^{k+1} \leftarrow \arg\min L_{ADMM}(\mathbf{x}, \mathbf{y}^k, \lambda^k) \tag{6.41}$$

$$\mathbf{y}^{k+1} \leftarrow \arg\min L_{ADMM}(\mathbf{y}, \mathbf{x}^k, \lambda^k) \tag{6.42}$$

$$\lambda^{k+1} \leftarrow \lambda^k + \rho(h_1(\mathbf{x}^{k+1}) + h_2(\mathbf{y}^{k+1})) \tag{6.43}$$

by which it alternates the direction of variables $\mathbf{x}$, $\mathbf{y}$ during optimization. This becomes the basic idea that ADMM has the dual problem solved in parallel. The literature [52] details the inner theory and analyses its properties.

6.5.1 An Implementation Example of ADMM

Since the ADMM method provides an efficient framework to decompose the dual problem, research over ADMM has gained popularity in recent years. It is natural that several existing algorithms are embedded in the sub-problems under distinct circumstances and assumption-oriented various applications. This paper [1] is a typical exemplification of the ADMM structure. It proposed to use a dual coordinate descent method and the aforementioned TRON to tackle disk loading and dense matrix challenges in dual and primal model, respectively.

$$\omega = \arg\min_{\omega} C \sum \max(1 - f(\omega), 0)^2 + prox(\omega) \tag{6.44}$$

where C is a constant and $prox(\cdot)$ is the proximal operator [1]. Solving the above sub-problem achieves linear runtime and spatial complexity, and the data need to be loaded only once.

In the dual coordinate descent method, the authors describe the normal sub-problem in a quadratic form, in which the optimization is achieved by updating one variable at a time. In a nutshell, the basic idea is to minimize the objective function over one element each iteration in the parameter vector ω, while others remain the same value as the last iteration. In light of the ADMM super-linear convergence, the framework might process it slowly. Nevertheless, the empirical work shows ω converges within only tens of iterations. Also, some contemporary solvers are compared to infer its competitive performance.

6.5.2 Other Work Relevant to Decomposition

As we have claimed in the previous part, due to the advent of huge volume data sets, these data sets can no longer be handled effectively and efficiently by typical global methods. Decentralised and decomposed methods have to be employed in theoretical analysis, one of which could be characterized as primal problems and dual problems decomposition [53]. This work [55] adopted general dual decomposition framework to formulate its own specific sphere in the aforementioned communication and network: distributed model predictive control (DMPC) [56], network utility maximization (NUM) [57] and direct current optimal power flow for a power system (DC-OPF) [58].

It focuses on minimization of a separable convex problem with numerous linear constraints. As usual, a general primal decomposition representation is transferred into a dual composition formulation via the method of Lagrange multipliers.

Suppose that x represents input samples $x \in \mathbb{R}^n, i \in \mathbb{N}$ with n dimensions,

$$
\begin{aligned}
(Primal)P^* &= \min_{x\in\mathbb{R}^n} f(x) = \min_{x\in\mathbb{R}^n, f_i\in F} \sum f_i(x) \Rightarrow \\
(Dual)D^* &= \max_{\nu_i,\omega_i\in\mathbb{R}, d_i\in D} d_i(\nu_i,\omega_i) = \max_{x\in\mathbb{R}^n} L(x,\nu,\omega)
\end{aligned}
\tag{6.45}
$$

where $f_i(x)$, $d_i(\nu_i,\omega_i)$ and $L(x,\nu,\omega)$ are the original function, dual function and loss function, respectively, and ν_i, ω_i are the parameters. The primal problem is subjected to linear constraints,

$$
Ax = b, Ex \le e, A \in \mathbb{R}^{m\times n}, b \in \mathbb{R}^m, E \in \mathbb{R}^{l\times n}, e \in \mathbb{R}^l
\tag{6.46}
$$

Naturally, without obtaining the exact optimal solution to a primal problem, the authors instead construct an approximation so as to implement a dual gradient method in an entirely distributed sense.

$$
d(\nu,\omega) = \min_{x\in\mathbb{R}^n} f(x) + \nu(Ax + Ex) - (\nu b + \omega e)
\tag{6.47}
$$

In terms of convergence, contrary to previous work containing only local linearity under the condition of a local error bound, they eventually show that the induced sub-optimality globally and linearly converges using a global error bound property.

The dual gradient (DG) algorithm, which approximates $x = arg\max_{x \in \mathbb{R}^n} L(x, \nu, \omega)$ and updates the parameter projection of $\lambda = \lambda^*(\nu, \omega)$ with weighted step size, is implemented, and sufficient theoretical analysis and mathematical proofs are provided. Their previous work results are presented in [59]. Also, some particular network applications in practice and some basic theories can be found in [60].

It is notable that the above decomposition process must be under two assumptions, which are strong convexity and Lipschitz-continuity of the gradient of the primal objective function, respectively. In subsequent research [61], the authors assert that the assumption of a bounded Hessian matrix of the primal objective function also implies an analogous conclusion. Additionally, they evaluated the proposed fully distributed DG algorithm to indicate that it outperforms the previous centralised dual gradient (CG) algorithm.

6.6 SOME OTHER RESEARCH RELEVANT TO DISTRIBUTED APPLICATION

This section introduces two aspects of parallel computing. The first aspect is to evaluate the basic performance of a couple of existing algorithms in logistic regression. The second one is to parallelize the conjugate gradient optimization.

6.6.1 Evaluation of Parallel Logistic Regression

In [62], several popular and well-performed approaches which aim to overcome the large-scale data shortcoming are discussed and evaluated so as to provide insight and guidelines to accelerate parallel system design. As stated explicitly, the work narrows down its concerns to the LR model.

As is known to all, the LR model prevails in binary because of its ease in implementation and understanding. Admittedly, the LR algorithm is very successful in the traditional linear classification way while its ill-performance in clusters, which contains tens of (moderate), hundreds of or thousands of (large) computing nodes, has attracted great attention. The work [62] not only attempts to implement Sub-Linear Logisitc Regression (SLLR) based on Hadoop and Spark, respectively, but also compares 6 standalone or distributed frameworks for LR using 4 sparse datasets, among which one cannot even fit in a single memory. These methods are as follows: Mahout based on Hadoop [63], LIBLINEAR standalone [64], Sub-Linear Logistic Regression (SLLR) [65], Parallel Sublinear Algorithms on Hadoop, Parallel Sublinear algorithms on Spark and Parallel Gradient Descent in Spark.

After a series of tests, their recommendations are made in accordance with four aspects: precision, runtime, cluster size and fault tolerance. Although LIBLINEAR accomplishes the best results, it requires the entire dataset to be loaded into the memory. Since the parallel SLLR obtains comparable performance with LIBLINEAR, upon which paradigm it is based, Hadoop or Spark depends on how much resource the system has.

6.6.2 Conjugate Gradient Optimization

As stated in numerous publications, the statistical learning framework is regarded as a mainstay in many ML branches. In general, to minimize a collection of risk functions, it is usually involved in a convex optimization assumption. In [68], the LR classification topic is their objective problem, as it is stated that LR is more favourably extendable to multi-class classifiers. It intends to propose a novel algorithm based on the non-linear conjugate gradient (CG) method, which is a commonly used method in optimization domain. It is also a very intriguing result according to their performance being 200 times faster than the iterative reweighted least squares (IRLS) method with either CG [66] or Cholesky decomposition [67]. The work [68] focuses on computational cost. It is obvious that the Newton–Raphson method is suited for quick convergence with respect to appropriate initialization. However, operations, for instance, the inverse of matrix, relevant to Hessian matrix consume large amount of memory resources. To tackle this challenge, the authors directly apply a non-linear conjugate gradient instead of buffering the Hessian matrix.

$$\boldsymbol{\omega}(n+1) = \boldsymbol{\omega}(n) + \alpha(n)\boldsymbol{d}(n) \tag{6.48}$$

Here, $\boldsymbol{\omega}$ is the normal weight vector, $\alpha(n)$ denotes the step size in the nth iteration and $\boldsymbol{d}(n)$ represents the nth direction vector. In addition, the updated step size optimized by Newton methods in the one-dimension condition will utilize the following updated formula to improve the iteration procedure.

$$\alpha(n+1) = \alpha(n) - \frac{\triangledown}{h} \tag{6.49}$$

where $\triangledown$ is the gradient of α, and h denotes the second order derivative. It is convinced that the non-linear CG algorithm avoids computing the Hessian matrix and its inverse. In the meantime, the initial value of ω is determined by employing multiple regression analysis (MRA) to achieve faster convergence, whose results are illustrated in the experiment part. However, only experimental evaluation instead of a rigorous mathematic proof is presented in terms of the step size updated procedure. This might be provided in future work.

6.7 A CASE STUDY

This section provides an overview of a traffic management solution based on classification in smart city. The statements below are merely depicted as a concise case study for readers who are interested in implementations.

Suppose we would like to supervise the traffic bursts on a particular road to predict the road condition and push recommendations to drivers in the next few minutes. Let us assume that we only need the duration of a green light at the first cross, x_1, and the current traffic flow, x_2, based on an isolated road segment. The ith sample vector would be $\boldsymbol{x}^i = (x_1^i, x_2^i)$. Meanwhile, we have some labels $\boldsymbol{y} = (y^1, y^2...y^n)$, "smooth" for 1 and "congested" for -1, corresponding to the samples. The tuple could be denoted as $((x_1^i, x_2^i), y^i)$. In this case, we concentrate on linear binary classification.

Therefore, we would have to set up a "model" (function, $f(x) = \boldsymbol{\omega} \boldsymbol{x}^T = \omega_0 + \omega_1 x_1 + \omega_2 x_2$) as the bound of two classes, where the parameter $\boldsymbol{\omega} = (\omega_0, \omega_1, \omega_2)$ and the first element of $\boldsymbol{x}$ is 1 for the constant. Now, the following risk function or cost function using LR would be

$$R(f(x), y) = \log(1 + e^{-\boldsymbol{\omega} \boldsymbol{x}^T y}) \tag{6.50}$$

At this point, the object is transformed to minimize the risk function which is convex under such circumstance. To gain the optimal solution of $R(f(x), y)$, the previous presented optimization algorithms could be used. As a matter of fact, the volume of data and the size of the sample vector are large and beyond the normal processing capability. Thus, distributed improvements based on large-scale data are required to address this problem.

6.8 CONCLUSIONS

This chapter has presented an overview of several advanced studies for paralleliza-tion of current optimization methods, e.g., SGD, the Newton method and ADMM parallel framework. In addition, it has demonstrated a machine-learning-specific plat-form for parallelism, named Petuum. The chapter has provided a profile of how to adapt the effective sequential single mode methods to distributed applications. Our aim is to give a better understanding of classical ML algorithm parallelism to boost the very promising growth in this area.

Bibliography

[1] Parikh, Neal, and S. P. Boyd. (2014). Proximal Algorithms. *Foundations and Trends in Optimization,* 1.3: 127–239.

[2] D. Xu, Li, W. He, and L. Shancang (2014). Internet of things in industries: A survey. *Foundations and Trends in Optimization Industrial Informatics, IEEE Transactions on,* 10.4: 2233–2243.

[3] Atzori, Luigi, A. Iera, and G. Morabito. (2010).The internet of things: A survey. *Computer Networks,* 54.15: 2787–2805.

[4] J. Gubbi, R. Buyya, S. Marusic, and M. Palaniswami (2013). Internet of Things (IoT): A vision, architectural elements, and future directions. *Future Generation Computer Systems,* 29.7: 1645–1660.

[5] Theodoridis, Evangelos, G. Mylonas, and I. Chatzigiannakis. (2013). Developing an iot smart city framework.*IISA 2013.*

[6] J. Jin, J. Gubbi, S. Marusic, and M. Palaniswami (2014). An information frame-work for creating a smart city through internet of things. *Internet of Things Journal, IEEE,* 1.2: 112–121.

[7] Yuan, Guo-Xun, Chia-Hua Ho, and Chih-Jen Lin. (2012). Recent advances of large-scale linear classification. *Proceedings of the IEEE,* 100.9: 2584–2603.

[8] M. Armbrust, A. Fox, R. Griffith, A. D. Joseph, R. Katz, A. Konwinski, G. Lee, D. Patterson, A. Rabkin, I. Stoica, and M. Zaharia (2010). A view of cloud computing. *Communications of the ACM*, 53.4: 50–58.

[9] I. Foster, Y. Zhao, I. Raicu, and S. Lu (2008). Cloud computing and grid computing 360-degree compared. *Grid Computing Environments Workshop, 2008. GCE'08.*

[10] B. Barney (2010). Introduction to parallel computing. *Lawrence Livermore National Laboratory*, 6.13: 10.

[11] K. Asanovic, R. Bodik, B. C. Catanzaro, J. J. Gebis, P. Husbands, K. Keutzer, D. A. Patterson, W.L. Plishker, J. Shalf, S. W. Williams, and K. A. Yelick (2006). *The landscape of parallel computing research: A view from Berkeley*, Vol. 2. Technical Report UCB/EECS-2006-183, EECS Department, University of California, Berkeley.

[12] R. Riesen, R. Brightwell, and A. B. Maccabe (1998). Differences between distributed and parallel systems. *SAND98-2221, Unlimited Release, Printed October.*

[13] J. Dean and S. Ghemawat (2008). MapReduce: Simplified data processing on large clusters. *Communications of the ACM*, 51.1: 107–113.

[14] K. Shvachko, H. Kuang, S. Radia, and R. Chansler (2010). The hadoop distributed file system. *Mass Storage Systems and Technologies (MSST), 2010 IEEE 26th Symposium on*, IEEE.

[15] M. Zaharia, M. Chowdhury, M. J. Franklin, S. Shenker, and I. Stoica (2010). Spark: Cluster Computing with working sets. *HotCloud*, 10: 10–10.

[16] C. Rudin, D. Waltz, R. N. Anderson, A. Boulanger, A. Salleb-Aouissi, M. Chow, H. Dutta, P. N. Gross, B. Huang, S. Ierome, and D. F. Isaac (2012). Machine learning for the New York City power grid. *IEEE Transactions on Pattern Analysis and Machine Intelligence*, 34(2), 328–345.

[17] C. M. Bishop (2006). Pattern recognition. *Machine Learning*, 128.

[18] V. N. Vapnik (1999). An overview of statistical learning theory. *IEEE Transactions on Neural Networks*, 10(5), 988–999.

[19] L. Bottou (2012). Stochastic gradient descent tricks. *Neural Networks: Tricks of the Trade*, Springer Berlin Heidelberg, 421–436.

[20] M. Zinkevich, M. Weimer, L. Li, and A. J. Smola (2010). Parallelized stochastic gradient descent. *Advances in Neural Information Processing Systems*, 2595–2603.

[21] B. ZRecht, C. Re, S. Wright, and F. Niu (2011). Hogwild: A lock-free approach to parallelizing stochastic gradient descent. *Advances in Neural Information Processing Systems*, 693–701.

[22] S. Y. Zhao, and W. J. Li (2015). Fast asynchronous parallel stochastic gradient descent. *arXiv preprint arXiv:1508.05711*.

[23] J. Keuper, and F. J. Pfreundt (2015). Asynchronous parallel stochastic gradient descent: A numeric core for scalable distributed machine learning algorithms. *Proceedings of the Workshop on Machine Learning in High-Performance Computing Environments, ACM*, 1.

[24] O. Bousquet, and L. Bottou (2008). The tradeoffs of large scale learning. *Advances in Neural Information Processing Systems*, 161–168.

[25] D. E. Rumelhart, G. E. Hinton, and R. J. Williams (1985). Learning internal representations by error propagation. *California Univ. San Diego La Jolla Inst. for Cognitive Science.*

[26] J. Keuper, and F. J. Pfreundt, (2015). Asynchronous parallel stochastic gradient descent: A numeric core for scalable distributed machine learning algorithms. *Proceedings of the Workshop on Machine Learning in High-Performance Computing Environments, ACM*, 1.

[27] R. Mcdonald, M. Mohri, N. Silberman, D. Walker, and G. S. Mann (2009). Efficient large-scale distributed training of conditional maximum entropy models. *Advances in Neural Information Processing Systems*, 1231–1239.

[28] M. Zinkevich, J. Langford, and A. J. Smola (2009). Slow learners are fast. *Advances in Neural Information Processing Systems*, 1231–1239.

[29] J. Langford, A. Smola, and M. Zinkevich (2009). Slow learners are fast. *arXiv preprint arXiv:0911.0491*.

[30] Microsoft Research. V. Wabbit. *https://github.com/JohnLangford/vowpal_wabbit/wiki.*

[31] R. Johnson and T. Zhang (2013). Accelerating stochastic gradient descent using predictive variance reduction. *Advances in Neural Information Processing Systems*, 315–323.

[32] D. Sculley (2010). Web-scale k-means clustering. *Proceedings of the 19th International Conference on World Wide Web, ACM*, 1177–1178.

[33] D. Grünewald and C. Simmendinger (2013). The GASPI API specification and its implementation GPI 2.0. *7th International Conference on PGAS Programming Models*, vol. 243.

[34] Weisstein, Eric W. Newton's Method. From MathWorld—A Wolfram Web Resource. *http://mathworld.wolfram.com/NewtonsMethod.html.*

[35] C. J. Lin, R. C. Weng, and S. S. Keerthi (2008). Trust region Newton method for logistic regression. *Journal of Machine Learning Research,* 9, 627–650.

[36] C. Y. Lin, C. H. Tsai, C. P. Lee, and C. J. Lin (2014). Large-scale logistic regression and linear support vector machines using Spark. *Big Data (Big Data), 2014 IEEE International Conference on,* 519–528.

[37] Y. Zhuang, W. S. Chin, Y. C. Juan, and C. J. Lin (2015, May). Distributed Newton Methods for Regularized Logistic Regression. *Pacific-Asia Conference on Knowledge Discovery and Data Mining,* 690–703.

[38] M. C. Lee, W. L. Chiang, and C. J. Lin (2015). Fast Matrix-vector Multiplications for Large-scale Logistic Regression on Shared-memory Systems. *Data Mining (ICDM), 2015 IEEE International Conference on,* 835–840.

[39] M. Martone (2014). Efficient multithreaded untransposed, transposed or symmetric sparse matrixâĂŞvector multiplication with the recursive sparse blocks format. *Parallel Computing,* 40(7), 251–270.

[40] M. Martone, S. Filippone, S. Tucci, M. Paprzycki, and M. Ganzha (2010). Utilizing Recursive Storage in Sparse Matrix-Vector Multiplication-Preliminary Considerations. *CATA,* 300–305.

[41] M. Martone, S. Filippone, M. Paprzycki, and S. Tucci (2010, October). Assembling Recursively Stored Sparse Matrices. *IMCSIT,* 317–325.

[42] R. Anil, S. Owen, T. Dunning, and E. Friedman (2010). *Mahout in Action.*

[43] X. Meng, J. Bradley, B. Yuvaz, E. Sparks, S. Venkataraman, D. Liu, J. Freeman, D. Tsai, M. Amde, S. Owen, and D. Xin (2016). Mllib: Machine learning in Apache Spark. *JMLR,* 17(34), 1–7.

[44] Y. Low, J. E. Gonzalez, A. Kyrola, D. Bickson, C. E. Guestrin, and J. Hellerstein (2014). Graphlab: A new framework for parallel machine learning. *arXiv preprint arXiv:1408.2041.*

[45] E. P. Xing, Q. Ho, W. Dai, J. K. Kim, J. Wei, S. Lee, X. Zheng, P. Xie, A. Kumar, and Y. Yu (2015). Petuum: A new platform for distributed machine learning on big data. *Transactions on Big Data,* 1(2), 49–67.

[46] S. Lee, J. K. Kim, X. Zheng, Q. Ho, G. A. Gibson, and E. P. Xing (2014). On model parallelization and scheduling strategies for distributed machine learning. *Advances in Neural Information Processing Systems,* 2834–2842.

[47] P. Xie, J. K. Kim, Y. Zhou, Q. Ho, A. Kumar, Y. Yu, and E. Xing (2015). Distributed machine learning via sufficient factor broadcasting. *arXiv preprint arXiv:1511.08486.*

[48] C. J. Hsieh, K. W. Chang, C. J. Lin, S. S. Keerthi, and S. Sundararajan (2008). A dual coordinate descent method for large-scale linear SVM. *Proceedings of the 25th International Conference on Machine Learning,* 408–415.

[49] H. Li, A. Kadav, E. Kruus, and C. Ungureanu (2015). MALT: Distributed data-parallelism for existing ML applications. *Advances in Neural Information Processing Systems,* 629–637.

[50] T. Yang (2013). Trading computation for communication: Distributed stochastic dual coordinate ascent. *Proceedings of the Tenth European Conference on Computer Systems,* 3.

[51] D. Gabay and B. Mercier (1976). A dual algorithm for the solution of nonlinear variational problems via finite element approximation. *Computers & Mathematics with Applications,* 2(1), 17–40.

[52] S. Boyd, N. Parikh, E. Chu, B. Peleato, and J. Eckstein (2011). Distributed optimization and statistical learning via the alternating direction method of multipliers. *Foundations and TrendsÂő in Machine Learning,* 3(1), 1–122.

[53] S. Boyd, L. Xiao, A. Mutapcic, and J. Mattingley (2007). Notes on decomposition methods. *Notes for EE364B, Stanford University,* 1–36.

[54] C. Zhang, H. Lee, and K. G. Shin (2012). Efficient Distributed Linear Classification Algorithms via the Alternating Direction Method of Multipliers. *AISTATS,* 1398–1406.

[55] I. Necoara and V. Nedelcu (2015). On linear convergence of a distributed dual gradient algorithm for linearly constrained separable convex problems. *Automatica,* 55, 209–216.

[56] I. Necoara, V. Nedelcu, and I. Dumitrache (2011). Parallel and distributed optimization methods for estimation and control in networks. *Journal of Process Control,* 21(5), 756–766.

[57] A. Beck, A. Nedic, A. Ozdaglar, and M. Teboulle (2014). Optimal distributed gradient methods for network resource allocation problems. *IEEE Transactions on Control of Network Systems,* 1(1), 64–74.

[58] A. G. Bakirtzis and P. N. Biskas (2003). A decentralized solution to the DC-OPF of interconnected power systems. *IEEE Transactions on Power Systems,* 18(3), 1007–1013.

[59] I. Necoara and V. Nedelcu (2013). On linear convergence of a distributed dual gradient algorithm for linearly constrained separable convex problems. Technical Report. University Politehnica Bucharest. *arXiv preprint arXiv:1406.3720v2.*

[60] I. Necoara and D. Clipici (2013). Parallel coordinate descent methods for composite minimization: Convergence analysis and error bounds. *arXiv preprint arXiv:1312.5302.*

[61] I. Necoara and A. Nedich (2015). A fully distributed dual gradient method with linear convergence for large-scale separable convex problems. *Control Conference (ECC), 2015 European*, 304–309.

[62] H. Peng, D. Liang, and C. Choi (2013). Evaluating parallel logistic regression models. *Big Data, 2013 IEEE International Conference on*, 119–126.

[63] Apache Mahout. (2012). Scalable machine-learning and data-mining library.

[64] R. E. Fan, K. W. Chang, C. J. Hsieh, X. R. Wang, and C. J. Lin (2008). LIB-LINEAR: A library for large linear classification. *Journal of Machine Learning Research*, 9, 1871–1874.

[65] H. Peng, Z. Wang, E. Y. Chang, S. Zhou, and Z. Zhang (2012). Sublinear algorithms for penalized logistic regression in massive datasets. *Joint European Conference on Machine Learning and Knowledge Discovery in Databases*, 553–568.

[66] P. Komarek and A. W. Moore (2003). Fast Robust Logistic Regression for Large Sparse Datasets with Binary Outputs. *AISTATS*.

[67] P. Komarek and A. W. Moore (2005). Making logistic regression a core data mining tool with TR-IRLS. *Fifth IEEE International Conference on Data Mining (ICDM'05)*, 4.

[68] K. Watanabe, T. Kobayashi, and N. Otsu (2011). Efficient Optimization of Logistic Regression by Direct Use of Conjugate Gradient. *Machine Learning and Applications and Workshops (ICMLA), 2011 10th International Conference on*, 1, 496–500.

[69] N. Rosenfeld, M. B. Elowitz, and U. Alon (2002). Negative auto-regulation speeds the response time of transcription networks. *J. Mol. Biol.*, 323: 785–793.

[70] M. A. Savageau (1976). *Biochemical Systems Analysis: A Study of Function and Design in Molecular Biology*. Addison-Wesley. Chap. 16.

[71] M. A. Savageau (1974). Comparison of classical and auto-genous systems of regulation in inducible operons. *Nature*, 252: 546–549.

CHAPTER 7

Security in Smart Grids

Julia Sánchez

La Salle - Ramon Llull University

Agustín Zaballos

La Salle - Ramon Llull University

Ramon Martin de Pozuelo

La Salle - Ramon Llull University

Guiomar Corral

La Salle - Ramon Llull University

Alan Briones

La Salle - Ramon Llull University

CONTENTS

Over the last few years the great advance in technology, the need for greener and more sustainable power sources and energy laws promoted by the governments allowed the **Smart Grid** trend to become a reality. Due the heterogeneous nature of the new electrical grids that integrate very diverse power sources, such as renewable energies, and due to the large amount of sensors and actuators that enable advanced functionalities of the Smart Grid, there is a need for **a highly flexible, scalable and easy-to-manage Internet-like network**. Moreover, new services are offered to users, which are exposed to new **vulnerabilities and threats**, requiring novel security and high-availability mechanisms to guarantee the best quality of service.

At the same time, new networking and computing concepts such as **Service Composition** and **Software-Defined Networks/Anything (SDN/SDx)** have arisen as powerful tools and methodologies for managing the future networks' architectures that propose **modularity, adaptability and centralized management of the communication system**. The characteristics provided by these approaches match perfectly with the requirements stated by the Smart Grid and more concretely against cybersecurity. In the case of the Smart Grid and more concretely in the electricity distribution network, a huge amount of data collected is processed continuously. Nowadays it is treated usually by dedicated and highly expensive devices. Relying on the expertise and experience of the partners of the FINESCE project, we advocate for a **Software Defined Utility (SDU)** concept for managing the Smart Grid and its security, where many of the functions that those dedicated devices perform will rely on programmable commodity hardware, low-cost sensors and high-speed and reliable IP-based communications underneath.

7.1 INTRODUCTION TO CYBERSECURITY

As networks become an integral part of corporations and everyone's lies, advanced network security technologies are being developed to protect data and preserve privacy. Network security testing is necessary to identify and report vulnerabilities, and also to assure enterprise security requirements. Security analysis is necessary to recognize malicious data, unauthorized traffic, detected vulnerabilities, intrusion data patterns and also to extract conclusions from the information gathered in security tests. Then, where is the problem? There is no open-source standard for security testing, there is no integral framework that follows an open-source methodology for security testing (information gathered after a security test includes large data sets), there is not an exact and objective pattern of behavior among network devices or, furthermore, among data networks and, finally, there are too many potential vulnerabilities.

The security of communication and data networks was not considered a priority a few years ago, perhaps because nobody was able to predict such an impressive growth of the use of data networks and their significant importance worldwide. The design of protocols, devices and networks was more focused on their operational function rather than providing systems that fulfilled security requirements. However, this trend has radically changed now. Nowadays this is a very prolific line of research with much effort dedicated to security.

Security is a feature of any system that indicates whether it is free from danger, harm or risk. Unfortunately, total security is a utopia, so there is always the likelihood of threats or dangers and new vulnerabilities that arise every day.

7.1.1 Key Security Aspects for Any System

When the security of a system has to be designed, it is important to take into account that the level of security provided can be defined with and depends on the strength of the following factors:

- Security

- Functionality

- Usability

In any implementation of security controls, these three factors have to be considered carefully, researched for the balanced trade-off for all stakeholders. The final balance will depend on the corporation's needs and it is important to not forget that simply focusing on any one individual factor will severely impair the others [8] (Figure 7.1).

To define the level of security in a system, first it is needed to understand what we are protecting. The main asset of any organization is Information. *Information Security* [17] refers to all measures that seek to protect the information against any irregularities while it may be stored both in a computer or any other medium. Thus, it is imperative that the system through information travels and where is stored, counts

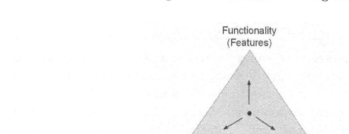

Figure 7.1 Functionality-Security-Usability trade-off

on with a set of policies, rules, standards, methods and protocols used to protect system infrastructure and all information contained or managed by it. Information must be protected from possible destruction, modification, disclosure or misuse. Not only should pay attention to intentional attacks, but also to possible software or hardware failures that threaten security.

As shown in Figure 7.2, Information Security is based on the following fundamental principles [17]:

- **Confidentiality (C)**. It ensures that the information is accessed only by authorized users or processes.

- **Integrity (I)**. It seeks to ensure that data remain unchanged by processes or unauthorized users.

- **Availability (A)**. It is the feature of finding information available at the time required by users or processes.

In an organization it is vitally important to establish standards, policies and security protocols aimed at the preservation of each of these principles, commonly known as the CIA triad.

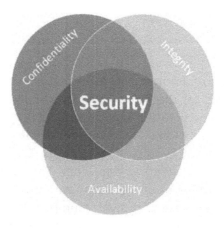

Figure 7.2 CIA triad

Security must be achieved through a continuous process; it must be conceived from the start of each project and not after consolidation, taking into account the potential risks, the probability of occurrence and the impact they may have. Any vulnerability in a system can affect the three main aspects, CIA. Then, you can make wise decisions based on knowledge obtained.

7.1.2 Network Vulnerability Assessment

Data networks are used not only to communicate devices but also to transport sensitive information, critical business applications and even to store mission-critical data. The increasing dependence on networks and the heterogeneous amount of devices interconnected in Smart Grids' systems leads to more security risks and hence more potential vulnerabilities that could be deliberately exploited.

A vulnerability is a condition of a missing or ineffectively administered safeguard or control that allows a threat to occur with a greater impact or frequency (or both) [7] and which could be exploited directly by an attacker or indirectly through automated attacks [71]. Vulnerability assessment must be an important part of a security audit to identify and quantify vulnerabilities in a system (a computer, a communications infrastructure or a whole data network).

A security audit is a systematic, measurable technical assessment of how the organization's security policy is employed at a specific site and provides a measurable way to examine how secure a site really is. Information is gathered through personal interviews, vulnerability scans, examination of operating system settings and network baselines. Usually, a security audit follows a three-phase methodology [7]:

- **Phase one—Planning**. Defines the scope of the effort, the network architecture, and the existing security solutions.

- **Phase two—Testing tools and activities/tests**. Verifies system security settings and identifies system vulnerabilities, in order to view the system from two separate perspectives, that of the external attacker and that of a malicious insider.

- **Phase three—Reporting**. Defines a structured analysis of the collected data and a reporting approach.

7.1.2.1 Security Vulnerabilities

Security vulnerabilities have become a main concern as they are constantly discovered and exploited in computer systems and network devices. Several tools or services exist to analyze them from the point of view of vulnerability definition, modeling and characterization. SANS defines @RISK, the Consensus Security Vulnerability Alert, which reports the new security vulnerabilities discovered during the past week and the actions that other organizations are doing to protect themselves. Also, NVD provides

The SANS Institute. http://www.sans.org
National Vulnerability Database (NVD). http://nvd.nist.org

a security vulnerability database that integrates publicly available U.S. Government vulnerability resources and is synchronized with the CVE vulnerability list. CVE list does not give a measurement that reflects the criticism of vulnerabilities. There exists a public initiative for scoring and quantifying the impact of software vulnerabilities. This score is called the Common Vulnerability Scoring System (CVSS). It represents the actual risk a given vulnerability possesses, helping analysts prioritize remediation efforts [26]. Others like OVAL and VulnXML by OWASP promote open and publicly available security content defining languages for representing system configuration information and transferring results.

7.1.2.2 Security Policies and Standards

Every device connected to a network is exposed to many different threats. While information system and network security professionals struggle to move forward with constantly evolving threats, hackers use technology that conventional security tools and services cannot cope with. Nevertheless, it is important to consider not only deliberated actions but also unintentional human activities, problems with systems or external problems that could derive in negative results like the exposure, modification or even the deletion of information or also the disruption of services.

There exists a serious problem of lack of knowledge about the solutions to establish in order to improve security in a corporation. This fact leads to an insufficient security level in many corporate networks. Cybersecurity standards are generally applicable to all corporations regardless of their size or the industry and sector in which they operate. The most relevant security standards of any cybersecurity strategy are the following [27, 36, 37]:

- **PAS 555**. It is a Publicly Available Specification that supplies a holistic framework for effective governance and management of cybersecurity risk. It not only considers the technical aspects, but also the related physical, cultural and behavioural measures of an organisation's approach to addressing cyber threats for businesses of all sizes.

- **ISO/IEC 27000:2016—Information security management systems— Overview and vocabulary**. This standard provides an overview of Information Security Management Systems (ISMS) and terms and definitions commonly used in the ISMS family of standards available as part of the ISO/IEC 27000 series. It is applicable to all types and sizes of organizations (e.g., commercial enterprises, government agencies, not-for-profit organizations).

- **ISO/IEC 27001:2013—Information security management systems – Requirements**. This standard provides requirements for establishing, implementing, maintaining and continually improving an ISMS. It can help businesses of all sizes in any sector keep information assets secure. It is the only generally recognized certification standard for information and cybersecurity.

Common Vulnerabilities and Exposures (CVE). http://www.cve.mitre.org
Open Vulnerability and Assessment Language (OVAL). http://oval.mitre.org
OWASP Project. http://www.owasp.org

- **ISO/IEC 27002:2013—Code of practice for information security controls.** This standard is designed for use as a reference when selecting controls while implementing an ISMS based on ISO/IEC 27001. However, it can be used for guidance when looking for information on commonly accepted information security controls. It can also be used as a control source when developing industry or organization-specific information security management guidelines. Although strongly associated with ISO/IEC 27001, it can be used independently.

- **ISO/IEC 27005:2011—Information security risk management.** This standard provides guidelines to businesses of all types and sizes which intend to manage risks that could compromise the organization's information security. It supports the general concepts specified in ISO/IEC 27001 and is designed to help implement information security based on a risk management approach. Knowledge of the concepts, models, processes and terminologies described in ISO/IEC 27001 and ISO/IEC 27002 is important for a complete understanding of ISO/IEC 27005:2011.

- **ISO 15408—Evaluation criteria for IT security.** The main goal of this standard is to provide evaluation criteria for IT security in order to compare different models. Functional and assurance requirements can be evaluated when applying this standard.

- **Cloud Controls Matrix (CCM).** It was developed by the Cloud Security Alliance (CSA) in order to offer organisations a set of guidelines that would enable them to maximize the security of their information taking advantage of Cloud technologies and without relying solely on the Cloud provider's assurances.

Other interesting ISO standards are ISO/IEC 22301:2012, ISO/IEC 27003:2010, ISO/IEC 27004:2009, ISO/IEC 27006:2015, ISO/IEC TR 27008:2011, ISO/IEC 27031:2011, ISO/IEC 27032:2012, ISO/IEC 27033, ISO/IEC 27035:2011, ISO/IEC 27036, ISO/IEC 27037:2012 and ISO/IEC 28000 series. Others, like NERC (North American Electric Reliability Corporation) and NIST (National Institute of Standards and Technology), have defined cybersecurity standards. NERC has created many standards, CIP-002-3 through CIP-009-3 (CIP=Critical Infrastructure Protection), to secure bulk electric systems which provide network security administration while still supporting best-practice industry processes. NIST has defined the Special Publications (SP) 800 series to publish computer/cyber/information security guidelines, recommendations and reference materials and SP 1800 series to address cybersecurity challenges in the public and private sectors, practical, user-friendly guides to facilitate adoption of standards-based approaches to cybersecurity. Furthermore, ISA (International Society of Automation) created the ISA99 committee which develops the ISA/IEC 62443 series of standards (technical reports, and related information that define procedures for implementing electronically secure Industrial Automation

---

Cloud Security Alliance (CSA). https://cloudsecurityalliance.org

and Control Systems (IACS)). Finally, it is interesting to mention IASME (Information Assurance for Small and Medium Enterprises), a UK-based standard for information assurance at small-to-medium enterprises (SMEs). It provides criteria and certification for SMEs cybersecurity readiness enabling them to achieve an accreditation similar to ISO 27001 but with reduced complexity, cost and administrative overhead.

7.1.2.3 Security Methodologies and Procedures

Different methodologies for security management and system development have arisen since the growth of networks to support typical and new domains of application. These methodologies are mainly focused on design, implementation and monitoring aspects. Complex systems and networks need adequate methods to formalize and validate their management and development. The main security methodologies are the following:

- **OCTAVE** (Operationally Critical Threats, Assets and Vulnerability Evaluation). It is an open risk-based strategic assessment and planning technique for security [9].

- **MEHARI** (Méthode Harmonisée d'Analyse des Risques). It is a method for risk analysis and risk management developed by CLUSIF (Club de la Sécurité des Systèmes d'Information Francais). It summarizes MARION and MELISA methods.

- **O-ISM3** (Open-Information Security Management Maturity Model). It is an information security management method for ISMS created by The Open Group.

- **EBIOS** (Expression des Besoins et Identification des Objectifs de Sécurité). It is used to assess and treat risks relating to information systems security. It has been designed by DCSSI (Central Information Systems Security Division - Direction Centrale de la Sécurité des Systèmes d'information), department of the French government.

- **OSSTMM** (Open Source Security Testing Methodology Manual). It is an open source methodology proposed by ISECOM [55]. It is a peer-reviewed methodology for performing security tests and metrics. This is the first open source methodology in this environment.

The Open Group global consortium. http://www.opengroup.org

7.1.2.4 Security Assessments

Constantly evolving threats force security professionals to do permanent and exhaustive controls to protect networks and detect weaknesses.

A correct design of a security system needs the understanding of the threats and attacks and how these may manifest themselves in audit data [3]. Consequently, testing is a basic tool to assure that enterprise security requirements are met. The security level can be also evaluated by analysing test results and discovering how attackers can penetrate into systems and networks [16, 71].

Security assessments pursue three main goals. The first goal is to discover design and implementation flaws, as well as to identify the operations that may violate the security policies implemented in an organization. Secondly, it has to ensure that security policy reflects accurately the organization's needs. Finally, it has to evaluate the consistency between the system's documentation and how they are implemented. In fact, a complete test should include the system, communication, physical, personal, operational and also the administrative security [71].

Different security assessments can be performed. The difference between these security assessments is focused on the goal to analyze, cost and time spent in order to obtain results (see Figure 7.3, [55]). The most common security tests are the following: Vulnerability Scanning, Security Scanning, Penetration Testing [22], Risk Assessment, Security Audit, Ethical Hacking, Posture Assessment or Security Testing [71].

7.1.2.5 Network Security Testing Tools

Different vulnerability assessment tools [72] exist in the market. Some of them are free scanners, like Nessus, OpenVAS, Microsoft Baseline Security Analyzer (MBSA), Qualys FreeScan and Secunia PSI. Other commercial tools are GFI LANguard, Retina, Rapid7 Nexpose, Core Impact, QualysGuard, Nipper and SAINT. The latest version of Nessus is closed source, but it is still free without the latest plugins and for

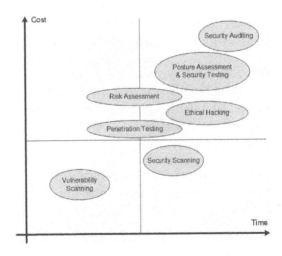

Figure 7.3 Types of system and network security assessments as based on time and cost

home users. The most popular and free web vulnerability scanners are Nikto, Grabber, Vega, Wapiti and ZAP whereas WebInspect, AppScan, Sentinel and N-Stealth are commercial web vulnerability scanners [54]. Not only is cost an important matter in commercial solutions, but also proprietary methodologies that hide the internal testing process and vulnerability assessment.

The first goal of a network vulnerability assessment is to test everything possible [7]. Different improvement proposals to achieve this objective have been found [30, 42, 46, 52, 63, 74]. Most of these systems only take into account servers or computer systems and they do not consider other network devices like firewalls or routers. Furthermore wireless networks need special tests as their properties differ from wired networks. Protecting a network is about more than just securing servers and computer systems. The network should be also protected against attacks that target firewalls and routers. Routers provide services that are essential to the correct operation of the networks. Compromise of a router can lead to various security problems on the network served by that router, or even other networks which that router communicates with. An auditing procedure for a network secured with a firewall is needed. For example, by using Nessus and Nmap tools it is possible to audit manually a firewall.

Moreover, the increasing dependency of wireless networks has opened organizations up to new security threats. Wireless LANs (WLANs) present unique security challenges because they are vulnerable to specialized and particular attacks. Some attacks may exploit technology weaknesses, whereas others may be focused on configuration weaknesses. Compromising only one node or introducing a malicious node may affect the viability of the entire network [16]. Currently, there exist wireless vulnerability scanners, like Aircrack, Kismet, NetStumbler or InSSIDer [72].

7.1.2.6 Summary

- Information is one of the main assets of any organization. The Information System that manages all organization's data, about processes and clients, between devices for daily operations, or whatever, has to ensure a *minimal level of security*. This goal must be accomplished by taking into account the following principles: *Confidentiality and Integrity of data and Availability of services.*

- Many security standards exist to validate the security of an Information Security Management System (ISMS). Nevertheless, there is not a mandatory rule or process to implement a security system. The *most relevant standards* which guide an organization's security design are *ISO/IEC 27000 series*, specifically ISO/IEC 27000:1 and ISO/IEC 27000:2.

- Many testing tools exist to validate the security design of the organization's ISMS. It is important to select those that appropriately test the characteristics of the system and clearly define if it is compliant with the security policy defined.

- The significant need of security systems that integrate *malware and vulnerability detection* in large-scale operational communication networks has become paramount in current-day society. The *automation of the processes related to a security assessment* following a *well-known and established open-source methodology* for security testing may be a good challenge to perform more efficient security analysis.

- As new vulnerabilities and shortcomings need to be solved, control mechanisms are required to *detect existing vulnerabilities periodically,* in order to be able to handle them and apply the corrective measures.

7.2 AUTOMATED SECURITY ASSESSMENT

The need for vulnerability assessment is not just to confirm the security of a network and its devices; it also stems from the concern that a network might not be adequately protected from the exponential number of threats.

The design of a security monitoring surveillance system needs the understanding of how threats and attacks can be performed against systems and how these threats may manifest themselves in audit data [3]. Moreover, a network security analysis must coordinate different sources of information to support effective security models [16]. These sources of information should be distributed through all the networks in order to monitor and manage the maximum number of communications. A correct protection of a network needs to perform periodically security tests so as to control devices and services, and also identify possible vulnerabilities. Testing should be recurring throughout the year and security experts should not rely on just one testing firm. In this sense, network security experts should rely on solutions that provide thorough, flexible, quantifiable, repeatable and consistent network security tests.

Although there exists information about testing networks and about vulnerability or penetration testing, there is no standard that regularizes the involved procedures. The success of a test comes when having a solid methodical plan. Thus a plan that details exactly what, when and how to do is necessary in any vulnerability test. Time and budget constraints can have a negative influence on the depth and extent of a security assessment. The more time and resources needed to plan, learn about the testing tools, install them, configure them with the right parameters, wait for the results, process multiple results files from the different sources, identify the meaningful data and obtain valuable security conclusions, the less profitable that security assessment becomes. Security experts should focus their efforts on the critical issues, being constantly in touch with new threats and exposures, and they should not spend the most part of their resources on routine and mechanical tasks. This is why the automation of these tasks becomes an advantage to alleviate daily work and, consequently, enrich security assessment results, due to the fact that less time, resources, budget and specific knowledge about every single tool will be needed.

7.2.1 Global Architecture of an Automated Security Assessment

An automated security assessment must be based on a framework which provides a platform that improves the processes related to security assessments. It can be defined as a vulnerability detection system with a distributed architecture, useful to simplify security test executions using methodologies like OSSTMM. This solution can reduce the time and resources spent during a manual security test. The automation will minimize the amount of time spent to perform a methodological security test and, simultaneously, it will provide the same amount of active risk prevention in real-time that a manual test would provide. Thus it is needed to automate not only the execution of the different security tools but also the subsequent data processing and storage.

Several issues need to be addressed when designing a new security assessment system that intends to accomplish the different requirements of security experts. This system must be modular, scalable, adaptable to different networks, manageable, compliant with some security methodology and, obviously, secure. The requirements that determine the design goals of this automated system are defined in [11] as part of the *Consensus Model*.

An automated proposal for a vulnerability assessment must be as modular as possible in order to incorporate new technologies, tools and updates whenever they come out. The requirement of modularity can lead to design the automated system as a system composed of eight different modules [12–15]. These modules can be classified in two main groups: an administrative group that includes the *Management, Database and Analysis modules* and, on the other hand, an operational group that comprises the *Internet, Intranet, DeMilitarized Zone (DMZ) and Wireless modules*, which are the probes or sensors that perform the vulnerability assessments. The administrative group can be centralized in the same device or can be distributed and they will be the control modules used by security auditors. The operational group will be used depending on the network technology used to test and on the scope of the vulnerability assessment. It can be seen that only seven modules have been already mentioned. The last module is the Base System. In fact, it is not a separated module as it provides a common and customized platform for the rest of the modules.

Figure 7.4 shows the global architecture of the automated system proposal, the relationship between its modules and the architecture of the different modules. In [11] is presented a complete description of each module.

7.2.1.1 Base System Module

It will provide a secure framework for the whole system, not only supplying a secure physical access but also a secure remote access when collecting data regarding security aspects of a network, so a maximum level of security is needed to assure confidentiality, integrity and authenticity. Every single action is controlled by this module.

7.2.1.2 Management Module

This module is the core acting as an orchestrator for handling the whole system. Its design, the defined communications protocol and its graphical interface allows the interaction between this module and the different testing probes. It also allows the global administration performed by the security expert. The purpose of the Man-

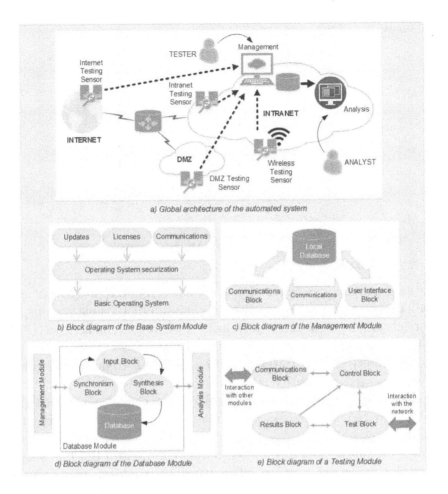

Figure 7.4 Global architecture of the automated system

agement Module is to ensure standardized updates and configuration changes from a single platform. It is centrally aware of all subsystems and tests as they occur to provide a real-time overview, regardless the location of the different testing probes or sensors. The module includes a small local database which contains information related to the granted users that can access to the system, the configured sensors to be activated when testing and the different programmed tests and their logs are stored in this database.

7.2.1.3 Analysis Module

The main goal of the Analysis Module is to show the security test results that the different testing probes have collected. It has to correlate all the data from all the sensors and provide reports for the security analysts. This module is directly related to the last phase of any security audit, which defines a structured analysis of the collected data and a reporting approach [7].

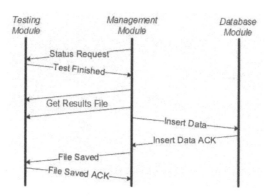

Figure 7.5 Message exchange

7.2.1.4 Testing Modules

The testing modules can be the following ones [12, 14, 15]: (1) Internet Testing Module will provide a thorough security test of all Internet-visible systems, (2) DMZ Testing Module is more focused on the security of the perimetral network, (3) Intranet Testing Module provides tests to verify internal network security, information leaks and network stability and (4) Wireless Testing Module is to provide automatic tests to verify wireless network security. All of them are slaves and act in accordance with the Management Module rules. Their main differences consist in their location in the network, the type of tests they can perform and the open-source security tools that they include.

A Testing Module is composed of four parts. The Control Block manages all the parameters and supervises how the different tests are performed. The probes also incorporate a testing engine in the Test Block, which is the main one responsible to execute the security test. The output of the different testing tools is processed in the Results Block. These results will be sent to the Control Block and the Management Module will be able to collect it to store all the results. The Communications Block defines a communication interface to interact with the Management module.

7.2.2 The Communications Protocol

It must be a specific protocol to perform communication because communication between modules is needed and available protocols [1] only exchange messages between devices in a peer to peer model. It is based on the 3-way handshake mechanism used by the TCP protocol [58]. The messages exchanged between the Management Module and the other modules should permit verifying whether the testing probes are disabled, waiting for instructions, testing or have finished a test and have results ready to be processed. The communications follow a master and a slave pattern. The Management Module is always responsible for starting any dialog with the rest of the modules (see Figure 7.5). The Management Module will poll the other modules to get information about the sensor status, the testing phase, the results file and also it will order the insertion, deletion or reading of testing results in the database.

7.2.3 Other Considerations

As an automated test could be performed by different connection technologies and for different devices, it is needed to take into account the following considerations [11]:

- **Wired security requirements**. These affect Intranet, Internet and DMZ modules. As organizations move more of their business functions to the public network, they need to take precautions to ensure that attackers do not compromise their data or that their data do not end up being seen by the wrong people. Security threats may come from inside, outside or from transit points of the network. Today's networks need security that extends from servers to all its end points, whether they are inside or outside the corporate perimeter. It will be required to find the reachable systems to be tested, to identify the available services and applications, to scan the possible vulnerabilities of these services, to test the routers, firewalls and Intrusion Detection Systems (IDS), to test the containment measures that handle malicious programs, to validate password strength of important systems and also to deploy denial of service (DoS) tests when necessary.

- **Wireless security requirements**. Threats to wireless networks include passive monitoring, unauthorized access to applications and even operation disruption of the network. Most wireless devices are wireless network-ready. End users or network managers often do not change the default settings or implement only weak standard security protocols. Thus network managers need to provide end users with freedom and mobility without offering intruders access to the network or the information sent and received on the wireless network. A security test of a wireless network also needs to follow some methodology in order to correctly perform the test. When performing a security test over a IEEE 802.11 wireless network, it is necessary to be located in its coverage area in order to obtain the information that travels through the air. Then, if a corporation has different wireless networks to cover all its space, the tester will have to move around to detect the different areas and much information will be obtained. In this situation, a previous analysis of the strategical points to locate the testing probes would be necessary to improve the whole process.

- **Router and firewall security requirements**. Routers direct and control much of the data flowing across computer networks. The border router is often the first line of defense. Routers provide many services that can have severe security implications if improperly configured. Some of these services are enabled by default whereas other services are enabled by administrators. As another option, a firewall is a security device between two or more networks. Firewalls provide this security by filtering unused ports and opening ports to allowed hosts. Often, firewalls are employed in conjunction with filtering routers. Then overall perimeter security of a corporation benefits when the configurations of the firewall and router are complementary. Security testing provides a means of verifying that security functions are compatible with system operations and that firewalls and routers are configured in a secure manner. Filtering is a very important responsibility for routers as it allows them to protect network devices from illegitimate traffic. Consequently, forwarding and filtering capabilities of routers should be tested to detect possible vulnerabilities and weaknesses in both functions. As regards forwarding capabilities, compromise of a router's routing table can end in less performance, denial of communication services, and exposure of sensitive data. Moreover, compromise of a router's access control can end in exposure of network configuration or denial of service, and also can ease attacks against other network devices. Related to filtering capabilities, a weak router filtering configuration can decrease the overall security of a corporation network, expose internal devices to attacks and facilitate hackers to avoid detection. Security assessments should test against several services and features on the target routers in order to identify the router type and the features implemented. Firewalls are usually placed at the boundary of the network to block unwanted traffic from or to external networks, as well as to regulate traffic flow between domains inside the same network [77]. Some firewalls can check addresses and ports and look inside the packet header to verify that it is an acceptable packet. Security tests should check the proper implementation and configuration of the firewall.

7.3 SECURITY CONCERNS, TRENDS AND REQUIREMENTS IN SMART GRIDS

Although the Smart Grid has gained relevance in the last few years, not all the parts involved are equally deployed [4]. Actually, the Smart Grid is a system of systems that includes not only the power system itself but also heterogeneous Information and Communication Technologies (ICTs) that represent a fundamental building block. However, many times they have not been integrated together in previous systems [76]. Partial solutions targeted only to specific aspects of the power system are no longer valid given the many services to be provided and the high cost of deployment of many specific systems [61].

The Smart Grid will only be possible with the massive deployment of ICT alongside of power installations and intelligent electronic devices (IED). Although it will provide a great benefit to society, this will not come without important challenges yet to overcome and enormous risks in terms of cybersecurity [73] that have to be urgently tackled. In this respect, there is a great consensus about the enormous cybersecurity threat and the very important damage that security breaches can cause [49, 50].

Smart Grid functions such as smart metering have several potential problems in terms of security threats [43], privacy [6] and fraud risk [44]. Since the network used in smart metering is connected to the whole Smart Grid, its security is something to be taken into account very seriously. Moreover in the near future, new utility functions have to be considered. The management network of the Smart Grid, aside from the grid itself and traditional power plants, should also allow utilities to deploy and control the virtual power plant concept [48] and sustain the rise of electrical vehicles with new features like electricity roaming [64]. It is indeed a very complex network demand. A security breach in the Smart Grid could be used to directly interfere with the electrical infrastructure of a whole territory by damaging it, as has been already proved by the infamous Stuxnet [40] and other similar worms (e.g., DuQu [69]). These kinds of networks also need to deploy systems with redundancy in order to use it to offer high availability to the most critical services against Denial of Service (DoS) attacks [23,32]. It is necessary to use different communication protocols, with the objective of managing this redundancy and obtaining different recovery times according to the requirements of the protected service.

Besides all the benefits provided by the underlying ICT infrastructure of a Smart Grid, the evolution on the remote control of the electrical distribution grids could give back undesirable vulnerabilities if the systems are not correctly secured. Smart Grid network control and monitoring are very important features in order to provide distributed energy generation and storage, quality of service (QoS) and security. Smart Grids link many distinct types of devices–also referred to as Intelligent Electronic Devices (IEDs) - demanding very different QoS levels over different physical media. Indeed, this kind of data network is not exempt from the growing needs of cybersecurity. In addition, availability and secured communications are also crucial for proper network operation [43,73], driving practitioners to consider Active Network Management (ANM) techniques to coordinate the whole communication network.

In addition to this, the Smart Grid relies on sensors, actuators and a management network, usually controlled by Supervisory Control and Data Acquisition systems (SCADA), which are used to control and supervise industrial processes from a computer. That is to say, SCADA systems control items in the physical world through computer systems. This is one of the points in which the main security concern of Smart Grids relies. Some recent cases have demonstrated the critical relevance of it.

- Stuxnet [40], a very complex worm and Trojan discovered in June 2010 that attacked the Iranian nuclear enrichment program. Its code used 7 different mechanisms to expand itself, mainly exploiting 0-day vulnerabilities. It achieved the destruction of about a thousand nuclear centrifuges by changing the behaviour of the actuators while telling the sensors that everything was good.

- A year later, in September 2011, a new Trojan called DuQu was discovered presenting a very similar behavior to Stuxnet and so it is believed that the two worms were related [69].

- Stuxnet was purportedly used again in 2012 against a nuclear power plant in southern Iran but the damage could have been avoided by taking timely measures with the cooperation of skilled hackers.

- In 2013, Iran hacked US energy companies (oil, gas and power) and was able to gain access to control-system software and was also accused of launching DDoS (Distributed Denial of Service) to US banks.

Cyberspace is defined as "an operational domain whose distinctive and unique character is framed by the use of electronics and the electromagnetic spectrum to create, store, modify, exchange and exploit information via interconnected information communication technology (ICT) based systems and their associated infrastructures" [57]. Thus, cyberwarfare is the kind of war that happens in that space in contrast with the traditional kinetic warfare where physical weapons are used. Smart Grids have become a clear potential objective of cyberwarfare considering that nowadays almost everything runs on electrical power and therefore potentially causing outages or, even worse, causing damage especially in some kinds of power plants (e.g., hydroelectric, nuclear, etc.). As brought up in this section, these kinds of attacks are becoming a reality and, recently, the information leakage of the US government that has been brought to light by several initiatives like PRISM [5] has also revealed that the US has drawn up an overseas target list for cyberattacks.

Regarding cybersecurity standards, many of the existing ones are to be taken into account in the Smart Grid as is highlighted in NISTIR 7628 [41], where they are listed and commented upon. A relevant one among them is IEC62351-6 [2] because it is the cybersecurity standard of reference for IEC 61850 and, thus, for the Smart Grid. NISTIR 7628 gives guidelines for cybersecurity implementation in the Smart Grid and provides a logical security architecture of a general nature. Significantly, it contains interesting considerations regarding the use of authentication certificates and secret keys management.

On the other hand, telecommunication networks are becoming suboptimum for the new and stringent requirements imposed by modern necessities of the Smart Grids. Not only is the network being migrated towards IPv6 due to the explosion of addressable devices [70] but also new models of networks are appearing such as Software Defined Networks/Anything (SDN/SDx) [31] in conjunction with Service Composition [24,39]. These kinds of paradigms can make the whole networking ecosystem more sustainable, adaptable, scalable and reusable making it more intelligent and, if used adequately, more secure. Some research has been done about this topic in European projects INTEGRIS [35] and FINESCE [20]. Several threats were detected in the security analysis on these projects and the usage of SDNs and Service Composition technologies were proposed to assist the management and protection of the whole Smart Grid infrastructure. Experiments have been undertaken in specific use cases for providing secure smart metering and a Smart Grid cloud service in a so-called Software Defined Utility (SDU) (see Section 7.6).

7.3.1 Requirements of the Smart Grid

Power network technology has been exceptionally stable for a long time, which is in contrast with the fast evolution of ICT systems. Nevertheless, the Smart Grid is a new concept that arose around the electrical grid. Its main novelty is the addition of a telecommunication network to the electrical infrastructure in order to transport information such as the state of the grid, real-time power consumption, service fault locations, etc. In other terms, the objective of the Smart Grid is to ensure that the grid is economically efficient, sustainable and provides higher standards of power

Table 7.1 Functional classes and requirements [35, 61]

Function	Latency	Reliability	Integrity	Confidentiality
Active Protection Functions	<20 ms	Very High (99,999%)	High	Low
Command and Regulations	<2 s	High (99,99%)	High	Low
Monitoring and Analysis	<2 s	High (99,99%)	High	Low
Advanced Meter and Supply	<5 min <10 s	Low (99%)	High	High
Demand Response	<5 min <5 s	Medium (99,9%)	High	Low

quality thanks to a lower level of losses and enhanced power management and security [18]. This evolution presents many operational problems that cannot be solved by current systems and technologies especially if they are used isolated. Fortunately, the evolution and current maturity of ICT systems makes it possible to cope with the mentioned problems, especially over the distribution grid where today ICT systems are scarcely deployed.

We want to underline the low latency and very high reliability needed for Smart Grid Active Protection Functions (APF) (shown in Table 7.1). This is always difficult to achieve with current technologies, but also the high reliability needed for Commands and Monitoring that is not easy to achieve in practice in a distribution grid environment. Regarding cybersecurity, the parameters in Table 7.1 are Integrity and Confidentiality but also Reliability and Latency. Very high reliability required by some functions implies that special care has to be taken to minimize DoS attacks to a minimum [61].

7.3.2 Smart Grid Security Requirements Definition

The state of the art regarding security in the Smart Grid is in fact defined in IEC62351-6 standard which basically applies security at the transport layer (TLS1.0 [34] with some restrictions) and upper layer communication protocols. It could be argued that protecting the transport layer could be enough since this may provide confidentiality, integrity and device authentication for user data and because many commercial systems rely on protecting systems just like this. However, protecting the Smart Grid only at the transport layer leaves the network and its links open to cybersecurity attacks that may produce DoS, eavesdropping of network management messages and, thus, prohibiting the users from accessing the service. This fact is not aligned with the high reliability feature that is required in the Smart Grid [23]. For this reason, the Smart Grid really urges multilevel security, even above the transport layer [47, 61]. To face this challenge in [20], the Smart Grid is secured by designing a security system in a way that it is really deployable and operative and that (1) balances the many and sometimes conflicting security goals of the different actors and subsystems and (2) accommodates a large and dynamic set of security mechanisms. Table 7.2 was developed jointly with Smart Grid experts from industry and academia, presenting a table with a set of the most important security issues that can affect the proposed infrastructure for the FINESCE's Smart Grid [20]. The main goal is to establish an order of implementation priorities regarding the security aspects. Authors gathered this information from several utilities in order to establish these priorities by numbering them with numbers from 1 to 8 (1 being the highest priority and 8 the lowest). Utilities have to provide the impact level for every problem if the system is crashed down. In the Reason column is presented a brief explanation of the rationale behind the order and decisions of which aspects are more critical than others.

From the requirements mentioned it is possible to draw the following conclusions:

- The stringent requirements in terms of latency and reliability defined in Table 7.1 mean that security decisions are to be taken as close as possible to the affected devices. Therefore, security needs to be as decentralized as possible in the Smart Grid.

- The high integrity needed means that all packets are to be protected by strong enough integrity hashes (at least 64 bits long).

- The high confidentiality needed for meter data implies that it is necessary to encrypt data while being transmitted and when being stored in intermediate systems.

- The need to distribute some of the computations to be performed in the Smart Grid [47,60] means that data have to be also protected to be used by distributed applications.

- Special care has to be taken regarding cryptography keys management given the large amount of keying material to be handled. Special care is required regarding a common policy for the different sets of keys.

- The different context situations of a given Smart Grid domain lead to adapting suitable security policies depending on that context and it requires a context-aware security design.

- The typical long duration of the investments in power network assets (typically 40 years) means that the horizon is about year 2055 and most of encryption techniques used nowadays may be possibly broken at that time; therefore, it is needed that the system has the possibility to upgrade to new coming encryption techniques.

Table 7.2 Smart Grid security issues (Impact: VL, Very Low; L, Low; M, Medium; H, High; VH, Very High)

Security Issue	Problem Description	Priority(P) & Reason & Impact
Data Leakage	Data are stolen and delivered without permission of the proprietary.	**P=5**. If a malicious user can access the system, user stored data could be compromised. This fact could derive in legal problems. *Impact(VH)*
Data Forgery	Data is modified by a malicious user and not detected.	**P=6**. Once the access is accomplished, if notifications of changes are not considered, a malicious user could modify user stored data. *Impact(H)*
Data Lost	Data is erased by a malicious user or a human error.	**P=7**. If a backup system is maintained, this could be an important but not critical problem since data could be restored. *Impact(M)*
Data Transaction	Data is delivered through the network and could be visible to malicious users if it is not encrypted. It depends on the sensibility of the data transmitted that this issue becomes more critical.	**P=1**. It is not necessary to access the system to obtain data under these circumstances. Therefore, it is considered that the most important aspect is that data transactions (data in transit) are encrypted. *Impact(VH)*
Commands execution	Many applications that can reside in IEDs could be sensitive to latency. A DoS attack to the network resources could affect its performance. It affects availability of the services.	**P=8**. Network resources have to be controlled because the access to data stored and applications in IEDs depends on them. It is considered that a network will be designed to detect DoS attacks and avoid latency problems. *Impact(H)*
Authentication	Access to IEDs and data storage has to be controlled and tracked to avoid wrong usage. It affects confidentiality, integrity and availability if a malicious user gets a user with rights granted.	**P=2**. It is very important to maintain control over the users that access data stored in IEDs and track the actions these users perform to avoid problems with data stored and IEDs' functionality. If a wrong usage is detected and users are authenticated, the system can isolate the problematic user to avoid damage. *Impact(VH)*
Authorization	Not all users have the same authorization policies to different zones, resources or stored data. Admin users, privileged users, guest users and third party users must be cataloged with different authorization rules.	**P=3**. It is important to maintain isolated rights to access resources because the system could have third-party users, guests/clients, administrators, etc., and not all should have complete access. The system could be modified by users without complete knowledge or by malicious users if a good authorization policy is not applied. *Impact(H)*
Identity Management (IdM)	The way to maintain a good connection between users and authorization rules is implementing a robust IdM.	**P=4**. Necessary to map users with their respective authorization rules and to maintain control over granted access to the system. *Impact (VH)*

7.4 SECURITY IN A CLOUD INFRASTRUCTURE AND SERVICES FOR SMART GRIDS

The growth of the Internet has fostered the interaction of many heterogeneous technologies under a common environment (i.e., Internet of Things). Smart Grids entail a sound example of this situation where several devices from different vendors, running different protocols and policies, are integrated in order to reach a common goal: bringing together energy delivery and smart services. Moreover, integrating the heterogeneous data generated by every device on the Smart Grid (e.g., wired and wireless sensors, smart meters, distributed generators, electric vehicles and communication network devices) into a single interface has emerged as a hot research topic. To consider this single interface as an efficient solution, an infrastructure capable of storing data from these heterogeneous sources allowing further data analysis is needed. Since data stored and transactions will be made through untrusted networks (Internet) some concern around security and privacy issues arises.

In recent Smart Grid ICT infrastructure proposals, like FINESCE [52], data can reside in both private utility infrastructure and public cloud infrastructure. Data can be moved from one to another, becoming an alternative solution for utilities with a more robust and scalable storage system thanks to the combination of Cloud Computing with their private DC infrastructure.

Cloud Computing is a model that enables on-demand access to a shared set of configurable computing resources that can be rapidly provisioned and released with minimal management effort or interaction by the Cloud Service Provider (CSP) [45]. Cloud Computing solutions offer several benefits [67]. However, the fact that the management of some physical data and machines is implemented by CSPs allowing the customer (utility) a minimum control over virtual machines creates some concern and suspicion. Cloud Computing solutions move some application software and databases of customers to large datacenters where the management and the services do not have the same confidence when housed in an internal infrastructure.

This section aims to gather basic security requirements in deploying a solution based on Cloud Computing. It also exposes attacks and vulnerabilities related to Cloud Computing to be considered for implementing a secure environment. Moreover, the security requirements and the results of a security audit performed over a use case example are presented.

7.4.1 Security Concerns and Requirements in a Cloud Environment

In a Cloud Computing environment there are many security risks depending on how CSPs deliver their services to customers. It is necessary to take into account the characteristics and associated security risks [19, 68] of each cloud deployment model [45] (private, public, hybrid, community) (Table 7.3).

Table 7.3 Cloud Computing deployment models

Deployment models	Basic characteristics
Public	· Clients share same resources of CSP · On demand resource availability · Resources are allocated outside the customer premises · **Security:** (1) It is a more insecure and risky model, more exposed to malicious activity. (2) It involves detailing and analyzing SLAs awareness among customers and CSP
Private	· Clients do not share resources of CSP · Resources can be allocated inside or outside the customer premises · Higher cost · Requires qualified administrators to manage and improve technical safety, control, compliance, fault resistance and transparency · **Security:** (1) Customer data are more secure by providing own infrastructure of CPD. (2) Administered by the customer, security management and responsibilities much easier to carry out and identify
Hybrid	· Combine two or more deployment methods (commonly public and private) · Managed by the organization or by third parties · Resources can be allocated inside or outside the customer premises · Access, through Internet, to multiple well-defined entities but limited · **Security:** More secure than public cloud
Community	· Shared among multiple customers · Operated, managed and commonly secured by customers or by a third party · Resources in outdoor installations, although they may be in the CPD of one of the customers · **Security:** Community members have free access to resources by eliminating risks in public cloud and decreasing costs of private cloud

Above each deployment model, services could be provided through any of the service delivery models (IaaS, PaaS, SaaS) [45]. The security characteristics of each service delivery model are as follows [19, 65]:

- IaaS only provides basic security, including perimeter, such as firewalls, Intrusion Prevention Systems (IPS) and Intrusion Detection Systems (IDS). It also includes load balancing to provide more availability and VMM (Virtual Machine Monitors) to monitor the performance of virtual machines and provide isolation between them.

- PaaS has security problems related to applications developed due to the cloud hosts' SOA (Software Oriented Architecture) environments for hiding the underlying web elements. Thus, and because the attackers are likely to attack

---

Infrastructure-as-a-Service
Platform-as-a-Service
Software-as-a-Service

visible code, a set of metrics are needed to measure quality and security of the encryption in the code written and to prevent the development of applications exposed to attacks.

- SaaS delivers applications via the Internet without installing software. From the customer's perspective, it is difficult to understand if data are secure and if applications are available at all times due to the lack of visibility into how data are stored and applications are deployed. The SOA challenges are focused on how to preserve or enhance the security previously provided by traditional hosting systems.

There is a compromise between system control, data and cost efficiency. The less control by the customer, the lower the costs of implementing business applications. This implies a loss of confidence because security depends largely on the CSP. To define the conditions of service delivery and enhance this confidence, SLAs are established between customers and their suppliers to ensure the quality, availability, reliability and performance of the resources provided.

7.4.1.1 Security Threats

The biggest problem that faces cloud computing is to ensure Confidentiality and Integrity of data and Availability of services. With a comprehensive understanding of the nature of security threats in the cloud it could be possible to manage the risks of any solution proposed based on Cloud Computing technologies.

Table 7.4 describes the risks associated with each threat catalogued with risk models like CIANA, because it conforms to the basic security principles specified in Section 7.1.1, and STRIDE, because it is related to vulnerabilities affecting the cloud. Also, possible countermeasures are provided to minimize these risks [28].

On the other hand, it is important to know the limitations of security problems to which a customer is exposed to minimize risks. Gartner [10] proposes seven specific areas on which customers should collect information before selecting a CSP: (1) What types of users have privileged access and how they are hired, (2) Regulatory compliance and Certifications needed, (3) Preserving privacy requirements regardless of the location of data, (4) Securing data isolation between customers, (5) Availability of data recovery in case of disaster, (6) Support for research and extraction of evidence (due if a crime is incurred), (7) Long-term viability, availability of data regardless of whether the CSP breaks or another company takes over the CSP.

7.4.1.2 Security Issues

The aim is to highlight the problems directly associated with each service delivery model because it is crucial knowing the issues related to IaaS, PaaS and SaaS to develop a secure and robust solution.

CIANA: (C) Confidentiality, (I) Integrity, (A) Availability, (N) Non-repudiation, (A) Authentication.

STRIDE: (S) Spoofing identity, (T) Tampering with data, (R) Repudiation, (I) Information disclosure, (D) Denial of Service, (E) Elevation of Privilege.

Table 7.4 Notorious threats to the cloud and its countermeasures. Principles and characteristics affected are underlined

Threat	Risk Analysis [28]	Countermeasures [68]
Data Breaches	<u>C</u> I A N A S T R <u>I</u> D E	• Isolation of virtual machines and stored data • Full erase data sessions before delivering data to new users to prevent data leakage • Backup data offline
Data Loss	C I <u>A</u> <u>N</u> A S T <u>R</u> I <u>D</u> E	• Use DLP tools (Data Loss Prevention)
Account or Service Traffic Hijacking	<u>C</u> <u>I</u> <u>A</u> <u>N</u> <u>A</u> <u>S</u> <u>T</u> <u>R</u> <u>I</u> D <u>E</u>	• Double authentication techniques • Do not share account credentials among employees • Good definition of SLAs
Insecure Interfaces and APIs	<u>C</u> <u>I</u> A N <u>A</u> S <u>T</u> <u>R</u> <u>I</u> D <u>E</u>	• Evaluate APIs before using it • CSP: Strong access controls, authentication and encrypted transmission
Denial of Service (DoS)	C I <u>A</u> N A S T R I <u>D</u> E	• Use Intrusion Detection and Prevention Systems (IDS, IPS)
Malicious Insiders	N / A <u>S</u> <u>T</u> R <u>I</u> D E	• User access level controls
Abuse of Cloud Services	N / A N / A	• Use registration and validation processes before giving customers access to the cloud • Passive monitoring to ensure that a user does not affect others
Insufficient Due Diligence	N / A <u>S</u> <u>T</u> <u>R</u> <u>I</u> <u>D</u> <u>E</u>	• Security of data, combined with risk transfer in the form of insurance coverage and acceptance of risk taking by CSPs
Shared Technology Vulnerabilities	N / A S T R <u>I</u> D <u>E</u>	• Strong compartmentalization between users • Strong authentication mechanisms • SLAs that include remedy

Table 7.5 Security requirements for service delivery models and different deployment models

Security Requirements	Cloud Deployment Models								
	Public Cloud			Private and Community Clouds			Hybrid Cloud		
	IaaS	PaaS	SaaS	IaaS	PaaS	SaaS	IaaS	PaaS	SaaS
Identification and Authentication	Yes	No	Yes	Yes	No	Yes	No	No	Yes
Authorization	Yes	Yes	Yes	No	No	Yes	No	No	Yes
Confidentiality	No	No	Yes	No	Yes	Yes	No	No	Yes
Integrity	Yes	No	Yes	No	Yes	Yes	Yes	Yes	Yes
Non-Repudiation	No	No	Yes	No	No	Yes	No	No	No
Availability	Yes	Yes	No	Yes	Yes	Yes	No	No	No

First, in the IaaS model there are no security breaches in the virtualization manager. The other important factor is the reliability of the data stored within the hardware vendor. Due to the increasing virtualization of "everything," it becomes an aspect of great interest how the data owner (customer) retains ultimate control over it regardless of location. IaaS is prone to varying degrees of security issues based on the deployment model [19].

Second, in the PaaS model, the CSP can give some control to application developers on top of the platform. But any security is given below the application level, such as IPS at network and host levels. This can concern the CSP, who should pay special attention to offer strong guarantees that data will remain inaccessible between applications [19].

Last but not least, in the SaaS model, the customer depends on the provider who applies appropriate security measures. Security problems in SaaS environments [67] are related to Data security, Network security, *Data location, Data integrity, Data segregation, Data access, Data confidentiality, Authentication and Authorization, Web application security, Data breaches, Virtualization, Availability, Backups and Identity management (IdM).* [59] extends the information about the security problems associated with the SaaS model showing a brief definition of the environment that affects each problem and possible solutions to be applied.

7.4.1.3 Security Requirements

Once there is an analysis of each service delivery model, the underlying possible deployment models, and the threats that arise from each of them, a set of basic security requirements [19] that have to be accomplished can be defined: (1) Identification and Authentication, (2) Authorization, (3) Confidentiality, (4) Integrity, (5) Non-repudiation and (6) Availability. Table 7.5 summarizes these basic security requirements for each service delivery model depending on the underlying deployment model.

7.4.2 Use Case Analysis—FINESCE Cloud for Smart Grid Distribution

7.4.2.1 Use Case Description

FINESCE [35] project's main objective is to contribute to the development of an open infrastructure based on ICT used to develop new solutions and applications in all fields of Future Internet in the energy sector. FINESCE project will incorporate the concept of a virtual substation IEC61850 through FIDEV prototypes, which were initiated in the FP7 European project INTEGRIS [20]. FIDEV is a platform built on commodity hardware, in which different software subsystems provide several communications and data concentrator functionalities. Among these new functionalities, a distributed storage system can be built over different interconnected FIDEVs, acting as a distributed Data Center (DC). These FIDEVs will be interconnected through the FIWARE Lab Cloud using various Generic Enablers (GEs) [21], resulting in a distributed storage system based on a hybrid cloud. They will also provide seamless interaction between this FIDEVs-based private distributed storage system and the FIWARE Lab Cloud. In this sense, the system will be formed by a set of separated FIDEV's testbed devices (physical or virtualized) that will constitute a private cloud, plus public cloud storage capabilities by means of FIWARE Lab. Data can reside in any of the two clouds and be moved from one to another becoming an alternative solution for utilities with a more robust and scalable storage system thanks to the combination of Cloud Computing with their private DC infrastructure.

The project solution requires processing and storing data of different sources like smart meters, electrical vehicles, client data, etc. A hybrid cloud deployment model is selected for the following reasons:

- The private cloud is used to ensure confidentiality of sensitive data stored like critical information about the electrical company.

- The public cloud is used to store non-sensitive data and historical measurement of smart meters, data of electric vehicle charging, etc. when FIDEVs are nearing its storage limit.

- There are applications susceptible to latency. The private cloud avoids vague and uncontrollable latency introduced by the Internet.

As shown in Figure 7.6, this use case interconnects different FIDEVs placed at different sites of ESB in Ireland and FUNITEC–La Salle Lab in Barcelona. These two sites conform a private cloud each one, and the FIWARE Lab performs as the public cloud. The whole cloud infrastructure provides a flexible storage solution for utilities, which can apply different security levels in order to protect their generated data, and select which is the best place to store sensitive or critical data.

In each region, FIDEVs are located and act as virtual substations collecting data

---

FIDEV is defined in FINESCE project as an upgrade of the communication part of IDEV devices defined in INTEGRIS integrating a set of GEs. These GEs will provide a secure interface with the distributed storage system and seamless interfaces to data management for the managers.

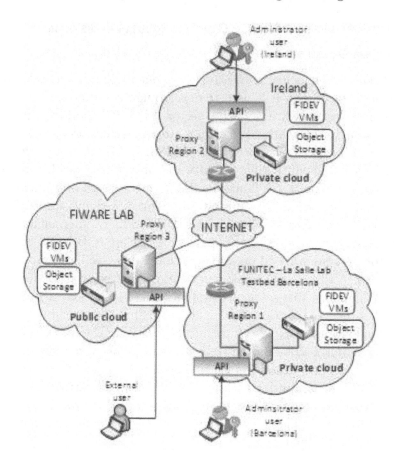

Figure 7.6 Generic overview of the WP5 Stream II Trial topology for Smart Energy use case

on devices connected to the grid. FIDEVs are based on OpenStack Object Storage [51] functionality to provide data storage and the necessary APIs (Application Programming Interface) to interface with them. These APIs are based on FIWARE GE defined in FIWARE project [21]:

- **Object Storage.** It provides robust, scalable object storage functionality based on OpenStack Swift [51]. The OpenStack Swift API provides a standardized mechanism to manipulate both the binary objects that are stored, and the hierarchy of containers (locations that store files) in which they are organized. This RESTful API can be accessed from any client technology that can communicate over HTTP.

- **Identity Management KeyRock.** This API covers a number of aspects involving user's access to networks, services and applications.

In the public cloud, IdM KeyRock and Object Storage are consumed as Software-as-a-Service (SaaS) applications of FIWARE Lab, while in the private cloud (built through the FIDEVs) resources are consumed in Platform-as-a-Service (PaaS) mode.

Object Storage containers in FIDEV proxies are synchronized between them with Rsync to provide a distributed storage system. When FIDEVs are reaching their maximum storage capacity, non-sensitive data are uploaded to FIWARE Lab.

Then, the two main processes performed by the solution are the following:

- *Authentication process.* To ensure that the user can store data in the private cloud, the authentication proxy (FIDEV) will first validate that the user has an Object Storage application membership in FIWARE KeyRock. Then, it will authenticate against Keystone (local IdM that jointly works with KeyRock's public IdM).

- *Storage process.* When a user wants to perform some action, this user must first authenticate. Once the credentials are validated, the user can interact with the proxy node via APIs to perform any action allowed in the distributed storage system (list, upload, download and delete data objects or create new data containers). Files uploaded to the distribution storage system can be encrypted or in plain text. The user is responsible for providing a strong key to encrypt the file that is uploading. The algorithms used are SHA-256, to get a hash of the key provided by the user, and AES-256, to encrypt data with the 256 bit-length encrypting key. The only feasible way of encrypting data is using CDMI(API interface to communicate with public storage nodes).

This paradigm of data being interchanged between public and private clouds faces some security implications due to the lack of control about the infrastructure and applications that manage the information.

7.4.2.2 Security Requirements Analysis

As shown in this section, the basic security requirements for a hybrid cloud must comply with the characteristics of *Identification, Authentication, Authorization, Confidentiality and Integrity for SaaS applications and Integrity for PaaS applications.* All these basic security features are met as follows:

- *Integrity.* Object Storage and Swift are responsible for storing data with integrity.

- *Identification.* Authentication and Authorization. When a user wants to perform operations on data from the private cloud first authenticates against Keystone checking credentials and, if that user is authorized in the storage application requested, the access is permitted.

- *Authorization.* Data transfer operations to FIWARE cloud are made from local FIDEVs, which imply that the user is previously authenticated.

- *Confidentiality.* The data is kept encrypted by a user-defined key.

Cloud Data Management Interface (CDMI) defines the functional interface that applications will use to create, retrieve, update and delete data elements from the Cloud.

However, although a secure system is designed and the result should be a robust system implemented, it is necessary to take into account that security issues can occur and affect the proposed infrastructure for the FINESCE project (considerations and requirements defined in Table 7.2).

7.4.2.3 Security Audit

The main objective of the security audit is to check the vulnerabilities that may have the system identify potential threats and minimize the risk of exposure of data processed and the infrastructure itself. By means of the security audit performance it is possible to verify the minimum security requirements needed as Table 7.5 establishes and the good operation of the code implemented in APIs.

The security analysis is done from the point of view of Ethical Hacking. The process of performing a security audit by means of Ethical Hacking is based in performing some penetration tests (pentesting). To make a good pentesting process there are a few steps to do: *(1) Reconnaissance, (2) Scanning, (3) Exploitation, (4) Maintaining Access and (5) Reporting* [9]. The security audit performed is as follows (details presented in [59]).

Reconnaissance phase is not performed because the information needed about the target system is known. Analysis is performed with Gray Box testing because some information about the system is known. Then, in the ***Scanning phase***, ports opened and services running are discovered. Nmap tool has been used to form a fingerprint of the target. The scanning performed discloses information about opened ports, services running and their versions and operating systems used. After, a *Vulnerabilities analysis step* looks for vulnerabilities associated to components by its technical characteristics and services that are running in the system and discovered in the Scanning phase. It is possible to perform a search on CVE (Common Vulnerabilities and Exposures) databases, to check the vulnerabilities associated to the components used and services discovered. All possible vulnerabilities have been analyzed and the impact to Confidentiality and Integrity of data and Availabilty of services has been valuated. Also, an *Implementation vulnerabilities analysis* is performed to analyze API's code to find vulnerabilities associated to authentication methods, file storage operation and creation of containers to storing data. Some tests have been performed against the API to check its vulnerabilities due to poor code writing or poor application design. In the ***Exploitation phase*** some tests have been performed thanks to information gathered in scanning phase and vulnerabilities analysis, thus checking if the system is robust enough against some feasible attacks and vulnerabilities. Different types of attacks and their impact on Confidentiality and Integrity of data and Availability of services have been evaluated: *Injection attacks* (SQL injection, Cross-Site Request Forgery and Cross-Site Scripting), *SSL related attacks* (using sniffing methods and SSLStrip tool) and *DoS attacks*. Finally, in the ***Reporting phase***, processes, tests and results are reported to conclude the security audit, including solutions or countermeasures to problems found.

The process described is adapted to the needs of the test environment and focused in discovering system vulnerabilities. For this reason, the **Maintaining Access**

phase is not performed and neither backdoors nor rootkits are left on the system to allow future access.

RESULTS

The most important security issues related to the system developed for the Smart Energy use case are the following:

- All interactions with the CDMI are using HTTP making possible that a malicious user could obtain user credentials and the encryption key.

- A malicious user without credentials can use brute force or dictionary attacks to access the system. The Keystone API does not protect the continued attempts.

- Object Storage API has the following operational problems that cause unavailability of the storage service: (1) A user can create an unlimited amount of containers, which could disable the system for other users and (2) The size of the containers is not limited. A single user can upload files until reaching the maximum storage space causing other users inability to upload files.

- Object Storage API does not notify when a user tries to overwrite a file. If the content of the files is not checked, changes in storage are undetectable.

Once the system vulnerabilities are known, solutions can be taken to solve security issues presented in previous sections. Table 7.6 shows some solutions to each security issue described in Table 7.2.

7.4.3 Summary

Cloud Computing solutions provide great flexibility for existing businesses given the wide range of solutions (IaaS, PaaS, SaaS) that can be implemented internally or externally in organizations. Given its cost efficiency by implementing models like SaaS, it has become a growing trend but the fact that more and more users rely on a CSP puts these suppliers in the crosshairs of malicious users. When an organization thinks of moving information or applications to the cloud, it must carefully analyze the threats and risks to which it is exposed as the business model it needs. Above all, it is very important to ensure that the security policy established in the organization extends to the cloud and to perform this will require a level of service by setting a SLA between the CSP and the organization. In this way the organization, operating as a client of cloud services, may have cataloged the risks to which it is exposed.

For the Smart Energy use case, it has been proved the need to apply to the chosen cloud solution various analysis and audits to detect possible threats and risks. Hence, it gives the option to take countermeasures before finally putting the system into production. In addition, the reliance on public cloud of FIWARE demonstrates the importance of knowing how cloud solutions of the CSP provider are implemented to face problems related to Data Location, Data Segregation, Data Violation, Data Availability, Data Access between tenants, Virtualization and Web Applications Security, which are more dependent on agreements with the CSP held by the SLAs and the implementation of physical infrastructure than the proposed system.

Table 7.6 Solutions to security issues in Smart Energy use case

Security Issue	Solution
Data Leakage	Simplest scenario: An attacker has to be authenticated to steal data. Keystone controls user access. Files protected. However, if the attacker sniffs traffic sent to the public cloud and captures valid credentials, data leakage shall not be avoided. Moreover, a file transferred to the public cloud could be intercepted without need to steal credentials. At the moment there is not any solution to this problem. Anyway, at least files are always encrypted from the source.
Data Forgery	A mechanism is needed to notify changes in data stored in the distributed system. At the moment the system has no notification tools. Moreover, if it is taken into account data sent or received from the public cloud, strong hashes used avoid tampering of data.
Data Lost	The system is a distributed system storage that maintains multiple copies of each file using Rsync. Data lost could be restored.
Data Transaction	Data exchanges between FIDEVs inside the private cloud will be encrypted by activating HTTPS and using SSL (only allowing TLS 1.1 or above versions and disabling all weak cypher and all 128 bits encryption mechanism). However, data exchanged with FIWARE Lab (public cloud) through CDMI only can be transmitted with HTTP (insecure). Multiple requests have been launched to FIWARE Lab administrators to activate HTTPS. The infrastructure could be protected against DoS attacks using level 7 firewalls and IPS technology.
Commands execution	Both in the public cloud as in the private cloud, a protection system against DoS attacks has to be implemented. It could be ensured that this premise is true in the private cloud as it depends on organizations but in the public cloud certain SLAs will be established with the CSP.
Authentication	Keystone manages the authentication process but brute force attacks are not controlled.
Authorization	Keystone manages authorization rules.
Identity Management	Keystone manages user accounts.

7.5 THE SMART GRID AS AN IoT

Recent advances on Smart Grids have explored the feasibility of considering the power electrical distribution network as a particular case of the IoT [61, 73]. Certainly, this specific domain poses appealing challenges in terms of integration, since several distinct smart devices or IEDs from different vendors, often using proprietary protocols and running at different layers, must interact to effectively deliver energy and provide a set of enhanced services and features (also referred to as smart functions) to both consumers and producers (prosumers) such as network self-healing, real-time consumption monitoring and asset management [49]. Although the latest developments on the IoT field have definitely contributed to the physical connection of such an overwhelming amount of smart devices [50], several issues have arisen when attempting to provide a common management and monitoring interface for the whole Smart Grid [43, 49].

Indeed, integrating the heterogeneous data generated by every device on the Smart Grid (e.g., wired and wireless sensors, smart meters, distributed generators, dispersed loads, synchrophasors, wind turbines, solar panels and communication network devices) into a single interface has emerged as a hot research topic [6, 73]. So far, some experimental proposals [61] have been presented to face this issue by using the Web of Things (WoT) concept to access a mashup of smart devices and directly retrieve their information using reasonably thin protocols (e.g., HTTP and SOAP) [44]. However, the specific application of these approaches into real-world environments is fairly dubious due to the following reasons: (1) they may open new security breaches [48, 64] (i.e., end-users could gain access to critical equipment), (2) there are no mature electric devices implementing WoT-compliant standards available in the market [49], and (3) industry is averse to include foreign modules (i.e., web servers) on their historically tested and established, but poorly evolved, proprietary systems [40].

A new way to overcome these issues through the European projects INTEGRIS [49] and FINESCE [69] may be explored providing a management interface for the Smart Grid inspired by the WoT. Continuing the work done in these projects, the aim is to implement an ICT infrastructure, based on the IoT paradigm, to handle the Smart Grid storage and communications requirements [32] to manage the whole Smart Grid and link it with end-users using a WoT-based approach, which results in a new bridge between the IoT and WoT. This proposal, which takes the pioneering new form of the WoT, is targeted at providing a context-aware and uniform web-based novel environment to effectively manage, monitor, and configure the whole Smart Grid. Moreover, conducted developments prove the feasibility and reliability of this approach and encourage practitioners to further research in this direction and to envisage new business models [23, 70].

The open IoT-based infrastructure presented in the Web of Energy proposal will provide new tools to manage energy infrastructures at different levels from IoT-based infrastructure enabled machine-to-machine interactions between small and resource-constrained devices on the Smart Grid domain. Thus, the IoT concept has been extended by providing a bidirectional human-to-machine interface, inspired by the WoT, which results in a ubiquitous energy control and management system coined as Web of Energy [4]. This proposal will combine the web-based visualization and tracking tools with the Internet protocols, which enables a uniform access to all devices of the Smart Grid. In order to provide such an effective and reliable management interface to address the heterogeneous nature of devices residing on the grid, we will continue the deployment of an intelligent subsystem devoted to (1) learn from real-world events, (2) predict future situations and (3) assist in the decision-making process.

New security challenges that may affect this ubiquitous energy control and management system involving Internet of Things and, implicitly the Smart Grid arise every day [33, 38]. The success of WoT will depend on an appropiate and controlled security system that takes into account the mechanisms and communications between devices and between humans and devices [38]. In fact, we can define Cybersecurity as an enabler of IoT and a whole cybersecurity strategy is needed [53] to consider new attack surfaces, IoT specific vulnerabilities and firmware problems, among others, to protect stackeholder's interests.

7.6 TOWARDS A SECURE AND SUSTAINABLE SMART GRID MANAGEMENT

7.6.1 A Sustainable Smart Grid Management

As we have seen in previous sections of this chapter, there are many examples of new technologies applied to the Smart Grid. All of them offer new opportunities, and propose new functionalities. However, they also involve the rise of new threats and a higher level of control of the network that increases the complexity in its management. The boost on distributing and virtualizing electrical resources requires new methodologies that simplify the tasks of the administrators of the electrical distribution networks [20]. In this sense, the advances on computer network architectures and management could give a hint about how it could be done. There are two recent computer networks management trends that especially present capabilities completely aligned with the specific needs of the Smart Grid.

On the one hand, Software-Defined Networks/Anything (SDN/SDx) present a way to manage highly distributed resources in a more autonomous and centralized way, programming certain type of resources to be adapted on the fly by themselves (self-configuration) and providing an easier and centralized way to manage the resources remotely. On the other hand, the generalization of the concept of service oriented computing and Service Composition (SC) and its application to Smart Grids could give also many advantages [29, 66, 76]. This methodology that is widely spread in the world of web-services presents a way of modularizing the functionalities as small independent services that could be published, placed, invoked and combined together with other services, to run them remotely and on demand.

Taking into account the critical low latency required in Smart Grid communications and that the resources of the distribution electrical grid such as generators, storage devices or actuators are increasingly becoming more spread, it is considered that the combinations of both technologies could make a great impact in the development of the Smart Grids in the following years. Actually, the necessity of total protection against failures in Smart Grids leads to research different solutions to solve this issue. Moreover, the use of an orchestrator for handling redundancy in different types of networks can efficiently tackle the stringent requirements of recovering services from a failure in the system, even though an overload is produced [47, 60]. The actual performance depends on the status and characteristics of the chosen underlying network behavior.

In fact, the forthcoming Smart Grid transformation towards an intelligent programmable network is possible by translating the philosophy, concepts and technologies from the SDN/SDx and SC in order to implement a Software Defined Utility (SDU).

7.6.2 Software-Defined Network

SDN is a computer network trend which aims to decouple the network control plane and the data plane [25, 31]. Although some research was done in the past with this target, SDN popularity has grown in the last few years fostered by the rise of Open Flow protocol [25]. However, it is necessary to remark that both terms should not be confused. Open Flow is a communication protocol that tries to facilitate the development of SDNs. In fact, Open Flow is a way to implement SDNs by defining the specific messages and message formats exchanged between orchestrator and leaf devices, so in other words, by specifying how the device should react in various situations and how it should respond to commands from the controller. On the other hand, the SDN/SDx concept that is referred to in this section represents a much more generic area that revolves around the programmability of the network for the separation of its control functionalities from the data forwarding.

One of the main advantages that SDN/SDx may contribute is the abstraction of the network to unify and centralize the management of the network. Abstraction could give network administrators a more global vision of the whole network in a cheaper and easier way than nowadays. Therefore, it would allow administrators to achieve also greater scalability and would permit easily integrating different types of middleware, even on demand (i.e., a cognitive system that is able to react and configure the network according to certain parameters [56]). This could make the network act like a living organism that adapts to certain situations making it more versatile and easy to manage and evolve.

7.6.3 Service Composition Paradigm

Another trend that presents very interesting characteristics for creating self-adaptable networks is Service Oriented Computing. Often known as Service-Oriented Architectures (SOA) or Service Composition (SC), it is a type of design pattern based on structured collections of software modules with defined inputs and outputs that can be composed on demand to create more complex functionalities (composite services) that can be represented by workflows of services [24, 39]. This paradigm has been efficiently brought to business services by means of web-services for a long time. However, the generic concept has not been explored in depth yet into the scope of other fields or purposes.

Service composition can bestow the Smart Grid network and applications with great versatility and scalability, especially if used along with some intelligent control middleware that enables and disables the modules depending on the state and context of the network and the requirements of the moment.

7.6.4 The Proof of Concept: A First Approach to Orchestrate Secured Smart Metering

The designed use case is focused on collecting encrypted smart meter and Remote Terminal Units (RTU) data and saving it encrypted in some place (e.g. an IED, a hybrid Cloud environment as shown in Section 7.4, etc.) in a way that the authorized actors can work with the data without having access to user specific data, preserving their customers' anonymity and protecting them from malicious attacks.

As stated before, smart metering represents only a set of the Smart Grid solutions, but it is the part that has already been more regulated, deployed and tested around the world. *Advanced Metering Infrastructure* (AMI) consists of smart meters, data management, communication network and applications. AMI is one of the three main anchors of Smart Grids along with *Distributed Energy Resources* (DER) and *Advanced Distributed Automation* (ADA). Smart metering is usually implemented using automatic meter reading (AMR), a technology that automatically gathers data from energy, gas and water metering devices and transfers it to the central office in order to analyze it for billing or demand side management purposes. Data are read remotely, without the need to physically access the meter. AMR systems are made up of three basic components to be secured: the meter, the Central Office and the communication systems. AMR includes mobile technologies, based on radio frequency, transmission over the electric cables (power line) or telephonic platforms (wired or wireless) [62, 76].

First of all, in order to determine the security requirements of the smart metering function, the work developed by NIST (National Institute of Standards and Technology) called NISTIR 7628 [41] has been of great importance, because of its highly detailed description of requirements and elements that must be taken into account when deploying a Smart Grid. Since this proof-of-concept targets to secure smart metering as a first approach, a limited set of requirements have been selected from among over 200 entries in [41], considering those that affect directly or indirectly smart metering. Once the requirements have been set up, a chart has been developed with those requirements (Table 7.7) on one axis and the technologies that can be used to meet the requirements on the other axis. Technologies and techniques associated to each requirement have been selected based on the authors' experience developed during the INTEGRIS and FINESCE European projects [20, 35] and on some new state-of-the-art techniques such as homomorphism that allows the information to be encrypted at all times, even when having to handle it, in contrast with other traditional techniques that require decryption to be performed before the information is treated and then encrypted again [75]. Based on this knowledge from industrial partners and academia, Table 7.7 specifies which secured-ICT technology can meet more accurately the requirements. Table fields are simplified but a brief description of each requirement is shown in a footnote.

SG.SC-3 Security Function Isolation/ SG.SC-4 Information remnants/ SG.SC-5 DoS Protection/ SG.SC-6 Resource Priority/ SG.SC-7 Boundary Protection/ SG.SC-8 Communication Integrity/ SG.SC-9 Communication confidentiality/ SG.SC-10 Trusted Path/ SG.SC-11 Crypto Key Establishment/ SG.SC-12 Use of Validated Cryptography/ SG.SC-15 PKI certificates/ SG.SC-19 Security Roles/ SG.SC-20 Message Authenticity/ SG.SC-26 Confidentiality at rest/ SG.SC-29 Application Partitioning/ SG.SI-2 Flaw Remediation/ SG.SI-3 Malicious Code and Spam protection/

After the techniques have been selected, use cases that fulfill the security requirements have been defined [20,35]. Some examples of them are the installation process of a smart meter, reading the power consumption, firmware updating, system monitoring, maintenance processes, etc. For providing the basic functionalities needed to cover those use cases, independent modules must be defined taking into account that these modules later can be combined in different workflows. The focus on the SC interoperable standalone modules, which can be invoked or dropped on demand, leads to considerable cheap solutions in the field of Smart Grids and presents a solution that may be integrated incrementally. Moreover, it allows system architects to design and deploy flexible applications that could be modified and evolved according to eventual new needs. Furthermore, the modularization of Smart Grid functionalities and encapsulation into self-contained services facilitates the distribution of the Smart Grid intelligence, approaching the reasoning and decision process and helping to handle its critical constraints of latency on fault reaction.

The one-by-one definition and classification of the modules can guide the SC design process made of composite services and the placement of the different modules in specific physical or logical locations of the Smart Grid. For example, if we consider the description of some security modules such as the AAA module, its execution location could be in a specific segment or end-to-end, while encryption or decryption modules are isolated modules that could be placed in a specific location of the network. This fact can help the reasoning of the administrator person or the specification of automation processes for building and deploying composite services automatically. However, some characteristics are intrinsically related to the functionality offered by the composite service to the end-user, such as the atomic service usage (optional or mandatory) or the order of them inside the workflow (dependent or independent) and can only be completely defined when building the composition. Therefore, aiming at just giving a proof-of-concept demonstration of some of these benefits that SC and SDN could bring into Smart Grid, a basic use case was designed on securing the smart metering in a Software Defined Utility environment [20].

The use case was deployed with a metering operation, carrying out the initialization of the device, reading execution and applying some corrections. The solution provided is based on the following rules:

- To rely as much as possible on proven existing standards, only complementing them when strictly necessary. This comes from the evidence that the first versions of most standards contained serious vulnerabilities.

- From these standards, to choose the right options for the Smart Grid (see Table 7.7).

SG.SI-4 Information System Monitoring/ SG.SI-7 Software and info integrity/ SG.SI-8 Information Input Validation/ SG.AC-3 Account Management/ SG.AC-8 Unsuccessful Login Attempts/ SG.AC-11 Concurrent Session Control/ SG.AC-13 Remote Session Termination/ SG.AC-16 Wireless Access Restrictions/ SG.AC-17 Access control for portable and mobile devices/ SG.AU-X Auditability/ SG.AU-16 Non-repudiation/ SG.CM-x Configuration changes/ SG.IA-5 Device Identification and Auth./ SG.MA-x Remote Maintenance.

Table 7.7 Smart Metering and Smart Grid services and features analysis Totally applies(5), Applies a lot(4), Mostly applies(3), Applies(2), Somewhat applies(1), Does not apply(0)

	PKI	Encryption & Decryption(AES)	NAC	Checksum SHA	DoS Defense System	ACL (Different Layers)	IDS	IPS	NMS	Supervised Cognitive System	Unsupervised Cognitive System	Logging	Segmentation (VLAN,VRF,MPLS)	SSH	QoS	Format Check	Homomorphism
SG.SC-3	4	3	4	0	0	2	1	1	0	0	0	0	5	0	0	0	0
SG.SC-4	5	0	5	0	0	0	4	4	3	0	0	2	1	0	0	0	0
SG.SC-5	0	0	0	0	5	3	4	4	0	0	2	2	0	0	0	0	0
SG.SC-6	0	0	2	0	0	0	0	0	0	0	3	0	1	0	5	0	0
SG.SC-7	5	1	4	0	0	4	3	3	1	0	0	1	5	0	0	0	0
SG.SC-8	5	3	0	5	0	0	0	0	0	0	1	0	0	0	0	5	4
SG.SC-9	5	5	2	0	0	2	2	2	0	0	0	0	0	1	0	0	5
SG.SC-10	5	0	0	0	0	1	1	1	4	1	3	0	0	0	0	0	1
SG.SC-11	5	5	4	0	0	0	1	1	1	0	0	1	0	0	0	0	5
SG.SC-12	5	5	0	0	0	0	0	0	1	0	0	0	0	0	0	0	5
SG.SC-15	5	0	0	0	0	0	0	0	0	0	0	0	0	0	0	0	0
SG.SC-19	5	0	4	0	0	3	1	1	1	0	0	1	2	0	0	0	0
SG.SC-20	5	4	1	4	0	0	0	0	0	0	0	0	0	1	0	5	3
SG.SC-26	0	5	0	3	0	0	0	0	0	0	0	0	5	0	0	0	4
SG.SC-29	5	0	5	0	0	2	1	1	0	0	0	0	3	0	0	0	0
SG.SI-2	0	0	0	0	0	0	0	0	4	5	5	2	0	0	0	0	0
SG.SI-3	0	2	0	0	0	5	5	5	5	0	0	0	3	0	0	0	5
SG.SI-4	0	0	0	0	0	0	4	4	5	2	3	4	0	0	0	0	0
SG.SI-7	5	0	0	5	0	0	0	0	0	0	0	0	0	0	0	0	5
SG.SI-8	5	0	0	0	0	0	0	0	0	0	0	0	0	0	0	5	0
SG.AC-3	5	0	5	0	0	0	0	0	0	0	0	0	0	0	0	0	0
SG.AC-8	5	0	5	0	0	0	3	3	4	2	4	4	0	0	0	0	0
SG.AC-11	0	0	5	0	0	0	1	1	5	2	4	3	0	0	0	0	0
SG.AC-13	5	0	5	0	0	0	0	0	3	2	2	0	0	0	0	0	0
SG.AC-16	5	4	5	0	0	0	0	0	0	0	0	0	0	0	0	0	1
SG.AC-17	5	0	5	0	0	0	1	1	1	0	0	2	0	0	0	0	0
SG.AU-X	2	0	0	0	0	0	0	5	0	0	5	5	0	3	2	0	0
SG.AU-16	5	0	0	0	0	0	0	0	4	0	5	0	0	0	0	0	0
SG.CM-x	5	0	5	4	0	1	1	1	3	3	0	5	0	4	2	5	4
SG.IA-5	5	0	5	0	0	0	3	3	2	0	0	2	0	0	0	0	0
SG.MA-x	5	4	5	1	0	2	2	0	3	3	2	4	1	5	2	1	1

- To place cybersecurity services as close as needed to the sensing and actuation points to improve latency and reliability of applications. In fact, this is done based on SC paradigm by placing them in the cybersecurity server and repository contained in the IEDs.

- To use a common coordinated cybersecurity data repository for all the involved technologies.

- To distribute this repository, either as a whole or partially, in the FIWARE Lab Cloud although having also a central repository located elsewhere. The central cybersecurity repository is replicated so that, in case of disconnection, the system continues to work for some time even allowing the inclusion of new devices and functions.

- To define cybersecurity metrics to feed the context-aware system to enable improved system management.

- To adhere to the principle of the Trusted Computing Group (TCG) of using Trusted Platform Modules (TPM) to protect in-built software and hardware as well as storage of data, including the basic keying material.

- To use, whenever feasible, authentication based on Certificates.

MODULES DEFINITION

The first and most important task that should be accomplished in the definition of the use cases was which functionalities are required to achieve the goal of the use case and which service modules are necessary to cover each of these functionalities. After the in-depth analysis of the security requirements and the technologies that can be used to fulfill them (briefly detailed by means of Table 7.7), one must figure out which functionalities are required and which modules could be useful to accomplish the objective of the final workflow. A correct modularization of the process must present services as loose-coupled as possible. That is important for two main reasons. First, it will help to reuse the services in different use cases avoiding their reimplementation and deployment. Second, it will facilitate the adaptability of the workflow to context changes (e.g. the level of security is reduced and some Smart Grid services could be removed in order to speed up the process, or the opposite, new security requirements are introduced and new services are created and integrated in the workflow chain in order to handle them).

The Smart Grid needs to manage many security schemes that in turn have different native key management schemes and policy enforcement methods which apply to different places or hierarchies in the Smart Grid. It would be unwise to keep those schemes without coordination.

WORKFLOW EXAMPLE

Finally, the modules have to be joined to create the whole workflow. As it has been described before, this process allows the smart meter function to be enrolled into the Smart Grid system. In order to do so, the whole process is being carried out in some

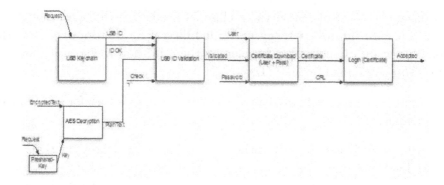

Figure 7.7 Complete workflow approach for "sign up with a non-validated USB" service

steps. For this example, some specific modules (such as USB keychain, AES Decryption, USB validation, etc.) were defined. They are combined as seen in Figure 7.7, building a workflow that executes the complete validation process, and offering different possibilities using a USB authentication token and a pre-shared key.

The importance of this process is not the process itself (that is just a basic proof-of-concept example) but its modularized design and deployment methodology. It focusses on enhancing the flexibility in the operation. It is very different to current-day straightforward deployments by procuring to the system architect a reusable design, a function virtualization and a chance of cloud computing deployment. If this process does not suit the utility's needs it can be easily changed, modules can be quickly swapped for others or even removed to simplify the process. Another interesting characteristic is that it does not depend on how the modules are implemented. As long as the input and output interfaces are well defined the module interoperability is fixed. So it should also help to avoid any vendor lock-in and to foster the interoperability and reusability of the systems.

SC offers a way to modularize complex operations and SDN the technology to orchestrate them around the network and communicate to the different entities in order to invoke and deploy them in a distributive way, while continue managing them from a central point. The greatness comes when those techniques are combined setting up the policies and building workflows autonomously as a SDU orchestrator [25].

7.7 CONCLUSIONS

The Smart Grid is at the same time a part of the Internet of Things and an example of a cyber-physical system where the physical power grid is surrounded by many intelligent communication devices that allow for an enhanced management of it. It is a system of systems that may not bring only great performance benefits to society but also big risks in terms of cybersecurity since it opens the power system to at least the same threats faced by the Internet. In fact, considering the novel, heterogeneous and distributed nature of the Smart Grid, it is reasonable to think that the vulnerabilities will be still larger. Furthermore, cybersecurity in Smart Grids is essential for the survival and feasibility of this electricity concept, thus making the risks still more relevant. To address this issue there is a lot of urgent on-going work worldwide being very relevant the one undertaken by the Standards Developing Organizations such as IEEE, IEC, and NIST.

The whole Smart Grid concept changes radically the way the traditional energy grid works as it becomes dynamic, versatile and autonomous and integrates very different power sources by its nature and size. As said in the introduction, the Smart Grid requires a parallel network that controls and monitors all its capabilities and this network must be as dynamic, versatile and autonomous as the Smart Grid. Software Defined Networks and especially Service Composition techniques can help to fulfill these requirements. It does not only apply for the management of the network but also for its security, one of the most important parts of the grid since it affects all of its stakeholders and can even cause international security problems, as some recent cyberwar attacks demonstrated. To sum up, these new paradigms on network computing architectures present a relevant solution based on the modularization of the required functionalities and their deployment on demand, only when and where required, avoiding redundancies and permitting the reusability of modules.

As presented at the end of this chapter, the characteristics of these techniques suit perfectly with the needs of the Smart Grid field and their applicability could represent a cheaper and more dynamic deployment for the Smart Grids in the following years, anticipating the Software Defined Utilities of the future. The complexity of the problem is really impressive and it is not possible to focus on all of its aspects in a single paper or even a project. The present chapter concentrates on the solutions developed in the context of European research project FINESCE in which the authors focused on the protection of data while being transmitted, stored and used in the context of the distribution Smart Grid, with the objective of proposing a Software Defined Utility solution that meets the data cybersecurity requirements of the Smart Grid. The project tackled issues such as access control, key management and context-aware security design considering the case of the electrical distribution Smart Grid in the cloud, the integration of IoT devices in the Smart Grid and the usage of commodity hardware and high-speed communication networks, all of them pieces that will be present in future Smart Grids but that also trigger a set of new security threats that should be taken into account.

7.8 ACKNOWLEDGEMENTS

The authors would like to thank La Salle–URL (Ramon Llull University) for their encouragement and assistance. This work was partly supported by the EU's seventh framework funding Program FP7 under the FI.ICT-2011 Grant number 604677–FINESCE (Future INtErnet Smart Utility ServiCEs).

Bibliography

[1] Intrusion detection exchange format (IDWG). 59th IETF meeting, 2003.

[2] International Electrotechnical Commission IEC 62351. "Power systems management and associated information exchange - data and communications security-

Part 6: Security for IEC 61850," 2009.

[3] J. P. Anderson. Computer security threat monitoring and surveillance. Technical report, James P. Anderson Company, Fort Washington, Pennsylvania, 1980.

[4] G. W. Arnold. Challenges and opportunities in smart grid: A position article. *Proceedings of the IEEE*, 99(6):922–927, 2011.

[5] B. Barret, "What is PRISM?" http://gizmodo.com/what-is-prism-511875267, 2013.

[6] P. Barbosa. Smart meter privacy: A theoretical framework. 2013.

[7] J. A. Blackley, T. R. Peltier, and J. Peltier. Managing a network vulnerability assessment. *Computing Reviews*, 45(5):261, 2004.

[8] C. Braz, A. Seffah, and D. M'Raihi. Designing a trade-off between usability and security: A metrics based-model. In *IFIP Conference on Human-Computer Interaction*, pages 114–126. Springer, 2007.

[9] J. Broad and A. Bindner. *Hacking with Kali: Practical Penetration Testing Techniques*. Newnes, 2013.

[10] J. Brodkin. Gartner: Seven cloud-computing security risks. *Infoworld*, 2008:1–3, 2008.

[11] G. Corral. Consensus and Analia: New challenges in detection and management of security vulnerabilities in data networks. Research group in internet technologies and storage. La Salle - Ramon Llull University. Link: http://hdl.handle.net/10803/9160.

[12] G. Corral, A. Zaballos, X. Cadenas, and A. Grane. A distributed vulnerability detection system for an intranet. In *Proceedings 39th Annual 2005 International Carnahan Conference on Security Technology*, pages 291–294. IEEE, 2005.

[13] G. Corral, A. Zaballos, X. Cadenas, P. Herzog, and I. Serra. Consensus: Sistema distribuido de seguridad para el testeo automático de vulnerabilidades. *V Jornadas de Ingeniería Telemática Jitel*, 5:351–358, 2005.

[14] G. Corral, A. Zaballos, X. Cadenas, and O. Prunera. Automatización de un sistema de detección de vulnerabilidades desde internet. In *III Teleco-Forum, UPTC*, pages 9–12, 2005.

[15] G. Corral, X. Cadenas, A. Zaballos, and M. T. Cadenas. A distributed vulnerability detection system for WLANS. In *First International Conference on Wireless Internet (WICON'05)*, pages 86–93. IEEE, 2005.

[16] J. Dawkins and J. Hale. A systematic approach to multi-stage network attack analysis. In *Information Assurance Workshop, 2004. Proceedings. Second IEEE International*, pages 48–56. IEEE, 2004.

[17] G. A. T. Implementación de la guía NIST SP800-30 mediante la utilización de OSSTMM. Departamento de Ciencias de la Computación. Universidad Nacional del Comahue.

http://tesis-toth.com.ar/fai/wp-content/uploads/2014/05/Tesis-Toth.pdf.

[18] Y. Ding, S. Pineda, P. Nyeng, Jacob Østergaard, E. Mahler Larsen, and Q. Wu. Real-time market concept architecture for ecogrid EU-A prototype for European smart grids. *IEEE Transactions on Smart Grid*, 4(4):2006–2016, 2013.

[19] D. A. B. Fernandes, L. F. B. Soares, J. V. Gomes, M. M. Freire, and P. R. M. Inácio. Security issues in cloud environments: A survey. *International Journal of Information Security*, 13(2):113–170, 2014.

[20] FINESCE. (2015), European Union's 7th Framework program (Future Internet PPP) Project "Future INtErnet for Smart Utility Services". FI.ICT-2011 grant number 604677 http://www.finesce.eu.

[21] FIWARE. Generic enablers catalog. http://catalogue.fiware.org.

[22] D. Geer and J. Harthorne. Penetration testing: A duet. In *Computer Security Applications Conference, 2002. Proceedings. 18th Annual*, pages 185–195. IEEE, 2002.

[23] I. Ghafoor, I. Jattala, S. Durrani, and C. M. Tahir. Analysis of openssl heartbleed vulnerability for embedded systems. In *Multi-Topic Conference (INMIC), 2014 IEEE 17th International*, pages 314–319. IEEE, 2014.

[24] A. J. Gonzalez, R. Martin De Pozuelo, M. German, J. Alcober, and F. Pinyol. New framework and mechanisms of context-aware service composition in the future internet. *ETRI Journal*, 35(1):7–17, 2013.

[25] P. Goransson and C. Black. *Software Defined Networks: A Comprehensive Approach*. Elsevier, 2014.

[26] T. Grance, R. Kuhn, and S. Landau. Common vulnerability scoring system. *IEEE Security & Privacy*, 2006.

[27] BSI Group. Company website: http://www.bsigroup.com.

[28] Top Threats Working Group et al. The notorious nine: Cloud computing top threats in 2013. *Cloud Security Alliance*, 2013.

[29] V. C. Gungor, D. Sahin, T. Kocak, S. Ergut, C. Buccella, C. Cecati, and G. P. Hancke. A survey on smart grid potential applications and communication requirements. *IEEE Transactions on Industrial Informatics*, 9(1):28–42, 2013.

[30] F. Guo, Y. Yu, and Tzi-cker Chiueh. Automated and safe vulnerability assessment. In *21st Annual Computer Security Applications Conference (ACSAC'05)*, pages 10–pp. IEEE, 2005.

[31] V. K. Gurbani, M. Scharf, T. V. Lakshman, V. Hilt, and E. Marocco. Abstracting network state in software defined networks (SDN) for rendezvous services. In *2012 IEEE International Conference on Communications (ICC)*, pages 6627–6632. IEEE, 2012.

[32] D. He, S. Chan, Y. Zhang, M. Guizani, C. Chen, and J. Bu. An enhanced public key infrastructure to secure smart grid wireless communication networks. *IEEE Network*, 28(1):10–16, 2014.

[33] T. Heer, O. Garcia-Morchon, R. Hummen, S. L. Keoh, S. S. Kumar, and K. Wehrle. Security challenges in the IP-based internet of things. *Wireless Personal Communications*, 61(3):527–542, 2011.

[34] Internet Engineering Task Force (IETF). "RFC2246 - the TLS protocol version 1.0" (1999).

[35] INTEGRIS. (2013) European Union's 7th Framework Program Project "INtelligent Electrical Grid Sensor communicationsâĂİ ICT-Energy-2009 call (number 247938). http://fp7integris.eu.

[36] ISO. Company website: http://www.iso.org.

[37] IT governance. Company website: http://www.itgovernance.co.uk.

[38] Q. Jing, A. V. Vasilakos, J. Wan, J. Lu, and D. Qiu. Security of the internet of things: Perspectives and challenges. *Wireless Networks*, 20(8):2481–2501, 2014.

[39] R. Khondoker, B. Reuther, D. Schwerdel, A. Siddiqui, and P. Müller. Describing and selecting communication services in a service oriented network architecture. In *2010 ITU-T Kaleidoscope: Beyond the Internet?-Innovations for Future Networks and Services*, pages 1–8. IEEE, 2010.

[40] R. Langner. Stuxnet: Dissecting a cyberwarfare weapon. *IEEE Security & Privacy*, 9(3):49–51, 2011.

[41] A. Lee and T. Brewer. "Guidelines for smart grid cyber security: Vol. 1, Smart grid cyber security strategy, architecture, and high-level requirements," 2010.

[42] A. J. Lee, G. A. Koenig, X. Meng, and W. Yurcik. Searching for open windows and unlocked doors: Port scanning in large-scale commodity clusters. In *CC-Grid 2005. IEEE International Symposium on Cluster Computing and the Grid, 2005.*, volume 1, pages 146–151. IEEE, 2005.

[43] X. Li, X. Liang, R. Lu, X. Shen, X. Lin, and H. Zhu. Securing smart grid: Cyber attacks, countermeasures, and challenges. *IEEE Communications Magazine*, 50(8):38–45, 2012.

[44] R. K. Megalingam, A. Krishnan, B. K. Ranjan, and A. K. Nair. Advanced digital smart meter for dynamic billing, tamper detection and consumer awareness. In *Electronics Computer Technology (ICECT), 2011 3rd International Conference on*, volume 4, pages 389–393. IEEE, 2011.

[45] P. Mell and T. Grance. The NIST definition of Cloud computing. 2011.

[46] A. M. B. Mohamed. An effective modified security auditing tool (SAT). In *Information Technology Interfaces, 2001. ITI 2001. Proceedings of the 23rd International Conference on*, pages 37–41. IEEE, 2001.

[47] J. Navarro, A. Zaballos, A. Sancho-Asensio, G. Ravera, and J. E. Armendáriz-Iñigo. The information system of integris: Intelligent electrical grid sensor communications. *IEEE Transactions on Industrial Informatics*, 9(3):1548–1560, 2013.

[48] P. Nezamabadi and G. B. Gharehpetian. Electrical energy management of virtual power plants in distribution networks with renewable energy resources and energy storage systems. In *Electrical Power Distribution Networks (EPDC), 2011 16th Conference on*, pages 1–5. IEEE, 2011.

[49] Department of Energy's Office of Electricity Delivery, Energy Reliability (OE), and the GridWise Alliance (GWA). "Future of the Grid: Evolving to meet America's needs," pp.1-40 (2014).

[50] Official Journal of the European Union. "Commission Recommendation of 9 March 2012 on preparations for the roll-out of smart metering systems" (2012/148/eu).

[51] OpenStack. http://www.openstack.org.

[52] X. Ou, S. Govindavajhala, and A. W. Appel. Mulval: A logic-based network security analyzer. In *USENIX Security*, 2005.

[53] OWASP Internet of Things Project. Project Website: https://www.owasp.org/index.php/OWASP_Internet_of_Things_Project.

[54] OWASP Project. Vulnerability Scanning tools. https://www.owasp.org/index.php/.

[55] 2004) P. Herzog (ISECOM). Open source security testing methodology manual 2.1 (OSSTMM). http://www.isecom.org/osstmm/.

[56] X. Pan, L. N. Teow, K. H. Tan, J. H. B. Ang, and G. W. Ng. A cognitive system for adaptive decision making. In *Information Fusion (FUSION), 2012 15th International Conference on*, pages 1323–1329. IEEE, 2012.

[57] R. C. Parks and D. P. Duggan. Principles of cyberwarfare. *IEEE Security & Privacy Magazine*, 9(5):30–35, 2011.

[58] S. W. Richard. *TCP/IP Illustrated*, vol. 2. Addison-Wesley Publishing Company, 1994.

[59] J. Sánchez, G. Corral, R. M. de Pozuelo, and A. Zaballos. Security issues and threats that may affect the hybrid cloud of FINESCE. *Network Protocols and Algorithms*, 8(1):26–57, 2016.

[60] A. Sancho-Asensio, J. Navarro, I. Arrieta-Salinas, J. E. Armendáriz-Íñigo, V. Jiménez-Ruano, A. Zaballos, and E. Golobardes. Improving data partition schemes in smart grids via clustering data streams. *Expert Systems with Applications*, 41(13):5832–5842, 2014.

[61] J. M. Selga, G. Corral, A. Zaballos, and R. Martín de Pozuelo. Smart grid ICT research lines out of the european project integris. *Network Protocols and Algorithms*, 6(2):93–122, 2014.

[62] J. M. Selga, A. Zaballos, and J. Navarro. Solutions to the computer networking challenges of the distribution smart grid. *IEEE Communications Letters*, 17(3):588–591, 2013.

[63] A. Sharma, J. R. Martin, N. Anand, M. Cukier, and W. H. Sanders. Ferret: A host vulnerability checking tool. In *Dependable Computing, 2004. Proceedings. 10th IEEE Pacific Rim International Symposium on*, pages 389–394. IEEE, 2004.

[64] F. Siddiqui, S. Zeadally, C. Alcaraz, and S. Galvao. Smart grid privacy: Issues and solutions. In *2012 21st International Conference on Computer Communications and Networks (ICCCN)*, pages 1–5. IEEE, 2012.

[65] A. A. Soofi, M. I. Khan, R. Talib, and U. Sarwar. Security issues in SAAS delivery model of cloud computing. *International Journal of Computer Science and Mobile Computing*, pages 15–21, 2014.

[66] E. Spanò, L. Niccolini, S. Di Pascoli, and G. Iannacconeluca. Last-meter smart grid embedded in an internet-of-things platform. *IEEE Transactions on Smart Grid*, 6(1):468–476, 2015.

[67] S. Subashini and V. Kavitha. A survey on security issues in service delivery models of cloud computing. *Journal of Network and Computer Applications*, 34(1):1–11, 2011.

[68] S. Venkata, K. Kumar, and S. Padmapriya. A survey on cloud computing security threats and vulnerabilities. 2004.

[69] N. Virvilis and D. Gritzalis. The big four–what we did wrong in advanced persistent threat detection? In *Availability, Reliability and Security (ARES), 2013 Eighth International Conference on*, pages 248–254. IEEE, 2013.

[70] J. Voas and F. Fieee. Imagineering an internet of "Anything."

[71] J. Wack and M. Tracey. Draft guideline on network security testing: Recommendations of the National Institute of Standards and Technology. *NIST Special Publication*, pages 800–842, 2001.

[72] Sectools.org webpage. http://sectools.org.

[73] Y. Yan, Y. Qian, H. Sharif, and D. Tipper. A survey on cyber security for smart grid communications. *IEEE Communications Surveys & Tutorials*, 14(4):998–1010, 2012.

[74] C. Ying, A. Tsai, and H. Yu. Vulnerability assessment system (VAS). In *Security Technology, 2003. Proceedings. IEEE 37th Annual 2003 International Carnahan Conference on*, pages 414–421. IEEE, 2003.

[75] N. Yukun, T. Xiaobin, C. Shi, W. Haifeng, Y. Kai, and B. Zhiyong. A security privacy protection scheme for data collection of smart meters based on homomorphic encryption. In *EUROCON, 2013 IEEE*, pages 1401–1405. IEEE, 2013.

[76] A. Zaballos, A. Vallejo, and J. M. Selga. Heterogeneous communication architecture for the smart grid. *IEEE Network*, 25(5):30–37, 2011.

[77] E. D. Zwicky, S. Cooper, and D. B. Chapman. *Building Internet Firewalls*. O'Reilly Media, Inc., 2000.

CHAPTER 8

Secret Key Generation under Active Attacks

Wenwen Tu

Department of Electrical and Computer Engineering
University of California, Davis, CA 95616
Email: wwtu@ucdavis.edu

Lifeng Lai

Department of Electrical and Computer Engineering
University of California, Davis, CA 95616
Email: lflai@ucdavis.edu

CONTENTS

I N THIS CHAPTER, we will review the existing results on secret key generation under active attacks. First, we will review the basic information theoretic models for key generation when the adversaries are passive. Then, we will focus on three scenarios where the adversaries are active. In the first scenario, the attacker can modify/falsify the messages exchanged during the public discussion phase. We will review the simulatability condition and the consequence of this condition. We will then discuss how to check this condition efficiently. In the second scenario, the attacker can influence the correlated sources obtained by the key generation parties. We will discuss a key generation scheme under active attack, and the corresponding optimal attack strategy of the attacker. Finally, we will consider the scenario where the attacker attacks the helper. We will discuss how to benefit from the helper if the helper is not under attack, and how to detect the presence of the attack if the helper is under Byzantine attacks.

8.1 INTRODUCTION

Physical layer security, also called information theoretic security, is an emerging field which exploits physical layer properties of communication channels to secure future generations of communication systems. In recent years, the problem of enabling communication users to establish a common secret key via public discussion under both source and channel models from the perspective of physical layer has attracted considerable amount of attention [1, 2, 5–9, 11, 13, 14, 17]. Under the source model, the legitimate users have access to correlated random sources as a prior, and based on this randomness they can generate a secret key which is kept confidential from the eavesdropper, via public discussion, i.e., exchanging messages over a noiseless channel [1, 5, 6, 11]. Under the channel model, the legitimate users typically have no correlated randomness in advance, but they can utilize the channel properties to obtain correlated sequences, from which the users can establish a secret key [7, 9, 17].

The approach to generating the secret key at the physical layer is proposed in the source model by the pioneer works [1, 11]. In the considered model, two users, connected by a public noiseless channel, have access to correlated random sequences that are independently and identically (i.i.d.) generated according to a certain given probability mass function (PMF), respectively. [1, 11] prove that the two users can establish a common secret key via public discussion over the noiseless channel even in the presence of an eavesdropper who can fully observe the public discussion. Even though the physical layer approach of key generation is initially proposed under the source model, it also has its intrinsic application in the channel model or wireless setup. In the case where the channel parameters are known, the legitimate users can obtain correlated sequences by letting one of the users act as the transmitter, and transmit a certain sequence over the channel while the others act as receivers [7, 8]. The transmitted sequence and the received sequences can be used as the correlated sources. On the other hand, in the case where the channel parameters are unknown, the legitimate users can also obtain correlated randomness by letting the legitimate users take turns to transmit training signals so that the others can estimate the channel gains [3, 20]. The estimates of channel gains can be viewed as the correlated sources as well. Once the legitimate users have obtained the correlated randomness, they can follow the physical layer approach as developed in [1, 11] to distill a secret key.

Later on, many researchers realize that the existence of a helper (or say a relay in wireless setting) whose role is to assist the legitimate users in establishing a secret key but not to share the same key with them can significantly increase the rate of the generated key [5, 6, 9]. Typically, the helper can obtain certain randomness during the sources observation phase, which is correlated with the sources the legitimate users have. Thus, for example [5, 6], if the helper releases partial information of the randomness to the public at the least rate such that the legitimate users can decode the helper's randomness correctly, the unreleased information the helper has can be transferred into part of the secret key, which increases the key rate.

However, most of the existing key generation works, e.g., the works mentioned above, are based on the assumption that the adversaries are merely passive eavesdroppers, while few works pay attention to the case when the adversary is active. The term *active* indicates that the adversary is not merely a passive listener; it can intercept, modify and even falsify signals to attack the key generation process so that the generated key rate reduces or the communication users agree on different keys, etc. The assumption of a passive adversary limits its application in many more practical scenarios, e.g., in the wireless setting, the adversary can easily send a contamination signal to interfere with the legitimate users' estimation of the channel gain [23, 24].

The existence of an active attacker complicates the key agreement process. The active attacker may interfere with the key generation process from the following three aspects: (1) The attacker attacks the public discussion, making it unreliable by intercepting, modifying or falsifying messages exchanged over the public channel [15]; (2) The attacker attacks the observation of the sources by sending contamination signals to reduce the correlation of the observed sources and to obtain side information with the observed sources [23]; (3) The active attacker attacks the helper and controls the helper to transmit fake messages with the purpose of making the legitimate users agree on different keys [18]. In this chapter, we will review key generation results under these three different scenarios.

1. *The Attacker Attacks The Public Discussion.* In [15], a scenario where two legitimate users wish to agree on a common secret key in the presence of an active attacker, who may modify the messages exchanged over the public channel, is considered. The two legitimate users along with the attacker observe certain correlated sequences which are i.i.d. generated from a known PMF, and they have full access to a public noiseless channel, over which the two users are allowed of public discussion. However, the public discussion is not authenticated. The attacker can intercept and modify messages exchanged over the public channel according to its attack strategy. Furthermore, the attacker is able to send fake messages. In this model, [15] characterizes the relationship of the generated key rate with that obtained from the case with a passive adversary. It introduces an important concept called *Simulatability Condition* which is ini-

tially defined in [12], and provides a remarkable "all or nothing" result which states that the key capacity equals zero when the simulatability condition holds; otherwise, the key capacity is the same as that obtained from the case with a passive adversary, even though it has not been single-letter characterized [1,11]. Recently, [19] presents a polynomial complexity algorithm to check whether or not the simulatability condition holds for a given PMF. These results will be discussed in Section 8.3.

2. *The Attacker Attacks The Sources Observation.* In [23], the problem of generating a secret key with an active attacker attacking the observation of the sources is considered. In the considered model, two legitimate users are connected with a relay via wireless fading channels. There is no direct wireless channel connecting the two users, while the attacker is assumed to have wireless channels connecting itself with each terminal including the relay. Besides, there is a public noiseless channel which is shared by the four terminals. [23] proposes a key generation scheme in this model. The key generation procedure contains two phases: source observation and key agreement. In the source observation phase, the two users and the relay transmit certain training signals over the fading channels to obtain estimates of the channel gains as the correlated sources. In the key agreement phase, they then use the obtained sources to distill a secret key with the help of the relay over the public channel. Different from [15], the public discussion over the noiseless channel is assumed to be authenticated. However, during the source observation phase, the attacker is allowed to transmit attack signals to contaminate the generation of the sources. [23] characterizes the maximal power of the attack signals, below which one can generate a secret key with a positive rate from the proposed scheme. Furthermore, [23] provides the corresponding optimal attack strategy of the attacker to minimize the generated key rate. This scenario will be discussed in Section 8.4.

3. *The Attacker Attacks The Helper.* The third scenario is considered in [18]. In this model, two legitimate users would like to generate a secret key in the presence of a helper. They observe correlated sources in advance and are allowed of public discussion over a public noiseless channel. Different from the previous two scenarios, the attacker here can take full control of the helper. The helper is honest if it is not under attack; otherwise, it is dishonest and it sends fake messages to the two users. But the legitimate users do not know its identity as a prior, and there is no probability assumption on the identity of the helper. There exist many protocols dealing with the Byzantine helper; the two most simplest ones are: (1) Treating the Byzantine helper as dishonest, and the legitimate users generate their secret key ignoring all signals transmitted by the helper; (2) Take the Byzantine helper as honest, and generate the secret key with the helper using the method proposed in [5]. Obviously, neither approach is optimal. The first approach has a risk of obtaining a secret key with a reduced rate, as [5] has proved that an honest helper actually helps in increasing the key rate. The second approach takes a risk that the legitimate users might be misled by the Byzantine helper and agree on different keys. [18] proposes a scheme such that

the legitimate users can detect the identity of the Byzantine helper correctly without loss of any secret key rate. Thus, the legitimate users can benefit in obtaining a secret key with a larger rate from the helper if it is honest, and they can detect it if the helper is under the control of the attacker.

8.2 BASIC MODELS FOR KEY GENERATION WITH A PASSIVE ADVERSARY

To facilitate understanding, we review some basic information theoretic models for key generation with a passive adversary in this section [1, 5, 6, 11, 20, 22]. These studies mainly fall into two categories: key generation in the source model and key generation in the channel model. We first review the basic ideas of generating a common secret key in different scenarios under the source model, then we discuss how to apply the obtained basic ideas to distill a secret key under the channel model.

8.2.1 Key Generation with Side Information at the Adversary

In the basic setup of key generation via public discussion under the source model [1, 11], two legitimate users Alice and Bob, along with an eavesdropper Eve, have access to three correlated sequences (X^n, Y^n, Z^n), which are i.i.d. generated according to a certain given joint PMF P_{XYZ}:

$$P_{X^n, Y^n, Z^n}(x^n, y^n, z^n) = \prod_{i=1}^{n} P_{XYZ}(x_i, y_i, z_i).$$

Alice and Bob are connected via a public noiseless channel to which Eve has full access. In order to agree on a common secret key, Alice and Bob are allowed to communicate with each other via exchanging messages over the noiseless channel. In particular, at the beginning of public discussion, Alice and Bob can generate two local randomness F_1 and F_2, which are independent of (X^n, Y^n, Z^n). Then, for each round use of the public channel, Alice transmits a message Ψ_i as a deterministic function of (F_1, X^n, Φ^{i-1}), and Bob transmits a message Φ_i as a deterministic function of $(F_2, Y^n, \Psi^{i-1}), i = 1, 2, \cdots$. In the end, after m rounds public discussion, Alice and Bob can compute values for key K and key L as

$$K = K(F_1, X^n, \Psi^m) \text{ and } L = L(F_2, Y^n, \Phi^m),$$

respectively.

Definition 8.1 *A key rate R is said to be achievable if $\forall \epsilon > 0$, there exists a key generation protocol when n is sufficiently large, such that*

$$Pr\{K \neq L\} \leq \epsilon, \tag{8.1}$$

$$\frac{1}{n}I(K; Z^n, \Phi^m, \Psi^m) \leq \epsilon, \tag{8.2}$$

$$\frac{1}{n}H(K) \geq \frac{1}{n}\log|\mathcal{K}| - \epsilon, \tag{8.3}$$

$$\frac{1}{n}H(K) \geq R - \epsilon, \tag{8.4}$$

where $\mathcal{K}$ is the alphabet of K. In addition, define the maximal value of R as the corresponding key capacity C.

Here, (8.1) requires that the keys generated by Alice and Bob should be the same with high probability; (8.2) measures the information Eve has on the generated key and it should be negligible; (8.3) implies that the generated key should be uniformly distributed over the key value alphabet. (8.4) measures the rate of the generated key.

For the simple case of $Z = \emptyset$, we have the following result.

Theorem 8.1 ([1, 11]) *If $Z = \emptyset$, the secret key capacity is*

$$C = I(X; Y). \tag{8.5}$$

To achieve a key with rate defined in (8.5), we can apply the Slepian–Wolf coding to let Alice send partial information of X^n at the rate of $H(X|Y)$, to Bob via the public channel such that Bob can decode X^n correctly with high probability. And the unreleased information of rate $H(X) - H(X|Y) = I(X; Y)$ can be transformed as the final key. In particular, Alice will randomly and independently assign each typical X^n sequence into $2^{nH(X|Y)}$ bins using a uniform distribution, with each bin having around $2^{nI(X;Y)}$ X^n sequences. Upon observing a sequence x^n, Alice transmits the index of the bin in which x^n is to the public channel. Then, with the observed y^n as well as the received bin index, Bob can recover x^n correctly with high probability. Both Alice and Bob will set the subbin index within the bin of x^n as the key value. It can be shown that Eve has negligible information about the generated key, as the subbin index and bin index can be shown to be nearly independent. Thus, the generated key is secure from Eve.

For the general case when $Z \neq \emptyset$, to single-letter characterize C is still an open problem, but we know a lower bound as well as an upper bound as follows.

Theorem 8.2 ([1]) *Given P_{XYZ}, the secret key capacity C of X and Y with respect to Z is lower bounded by*

$$C \geq \max_{V-U-X-Y,Z} I(U; Y|V) - I(U; Z|V), \tag{8.6}$$

in which V and U are two auxiliary random variables. Furthermore, the secret key capacity is upper bounded by

$$C \leq I(X; Y|Z). \tag{8.7}$$

To achieve the lower bound defined in (8.6), we can i.i.d. generate $2^{nI(X;V)}$ sequences V^n, and for each generated V^n, i.i.d. generate $2^{nI(X;U|V)}$ sequences U^n according to $P_{U|V}$. Then randomly and independently assign each sequence U^n generated by V^n into $2^{n(I(U;X|V)-I(U;Y|V))}$ bins, and within each bin, uniformly assign each U^n into $2^{nI(U;Z|V)}$ subbins, and set the index within each subbin as the key value. Within each bin, there are around $2^{n(I(U;Y|V)-I(U;Z|V))}$ U^n sequences. Then, upon observing x^n, Alice finds a sequence V^n that is jointly typical with x^n, and then Alice will find a U^n among those U^n sequences generated by V^n, that is jointly typical with

(V^n, x^n). Finally, Alice sends V^n along with the bin index of U^n to Bob. Thus, the total information Alice needs to send is $I(U; X) - I(U; Y)$. With the received messages along with the observed sequence y^n, Bob can correctly recover (U^n, V^n) with high probability. Additionally, with high probability, there exists at least one U^n within each subbin, which is jointly typical with (V^n, Z^n); thus there is no preference for Eve to decide which subbin U^n lies in. Thus, we can show the generated key is secure from Eve. For more details, one may refer to [1].

8.2.2 Key Generation with a Helper

In certain scenarios, there may exist a third trusty party who can assist the generation of the secret key between Alice and Bob. This problem is studied in [5,6]. In the considered model, three terminals Alice, Bob and the helper (called Charlie in the sequel) have access to correlated random sequences (X^n, Y^n, Z^n), respectively, which are i.i.d generated according to a certain given joint PMF P_{XYZ}, while the adversary is assumed to have no correlated side information with (X^n, Y^n, Z^n).

The role of the helper is to help Alice and Bob agree on a common secret key, instead of sharing the secret key with Alice and Bob. They are also allowed to discuss over a public noiseless channel. Denote the collection of all exchanged messages over the public channel by $\mathbf{F}$. The generated key needs not to be secure from the helper, and hence the security condition in (8.2) should be replaced by

$$\frac{1}{n} I(K; \mathbf{F}) \leq \epsilon.$$

For the considered model, we have the following result.

Theorem 8.3 ([6]) *Given any P_{XYZ}, the key capacity C is given by*

$$C = \min\{I(X; YZ), I(Y; XZ)\}. \tag{8.8}$$

To achieve a key rate of $\min\{I(X; YZ), I(Y; XZ)\}$, we can apply the Slepian–Wolf coding technique to let Charlie send partial information of Z^n at the rate of $\max\{H(Z|X), H(Z|Y)\}$ to Alice and Bob so that both Alice and Bob can decode Z^n correctly with high probability. Then Alice sends partial information of X^n at the rate of $H(X|YZ)$ to Bob. More specifically, Charlie randomly and independently assigns each typical Z^n sequence into $2^{n \max\{H(Z|X), H(Z|Y)\}}$ bins, and Alice randomly and independently assigns each typical X^n sequence into $2^{nH(X|YZ)}$ bins. Upon observing a sequence z^n, Charlie sends its bin index to Alice and Bob, so that both Alice and Bob can decode z^n correctly with the local observed sequence x^n and y^n, respectively. Then, Alice sends the bin index of x^n to Bob. Bob can decode x^n correctly with high probability, using (y^n, z^n) along with the received bin index. Finally, set the subbin indices of both x^n and z^n as the key value. Intuitively, the total information of (X^n, Z^n) is $nH(XZ)$, and the information released to the public is $n \max\{H(Z|X), H(Z|Y)\} + nH(X|YZ)$. Thus, the confidential information of (X^n, Z^n) given by

$$nH(XZ) - n \max\{H(Z|X), H(Z|Y)\} - nH(X|YZ) = n \min\{I(X; YZ), I(Y; XZ)\}$$

can be used as the secret key.

8.2.3 Basic Model for Key Generation in Wireless Setting

In this part, we talk about a simple model as a background of key generation in wireless setup. Different from the key generation process in the source model, the legitimate users typically observe no sources in advance, but they can obtain the sources via estimating the channel gain [20, 22].

In the basic wireless key generation model, Eve is assumed to be passive. Two legitimate users Alice and Bob along with Eve are connected via pairwise fading channels. In addition, there is a public noiseless channel shared by the three terminals. The basic idea of generating a secret key for Alice and Bob is to obtain correlated sources first by sending training signals over the fading channel connecting Alice and Bob, and estimating the channel gain, and then to use the noiseless channel to generate a secret key applying the results from the source model as discussed in [1,11].

The wireless channels are assumed to be reciprocal and ergodic block fading, i.e., $h_{AB} = h_{BA}$ (h_{AB} is the channel gain from Alice to Bob, and h_{BA}, h_{AE}, h_{BE}, etc., are defined in a similar manner), and h_{AB} remains the same within each block period and it changes to another random value at the beginning of the next block according to a pre-known distribution.

The wireless channels are modeled by

1. If Alice sends a signal X_A, the signals Bob and Eve receive are

$$Y_B = h_{AB}X_A + N_B,$$
$$Y_E = h_{AE}X_A + N_E,$$

   respectively;

2. If Bob sends a signal X_B, the signals Alice and Eve receive are

$$Y_A = h_{BA}X_B + N_A,$$
$$Y_E = h_{BE}X_B + N_E,$$

   respectively. Here N_A, N_B and N_E are independent additive noises.

Consider a scenario where $h \sim \mathcal{N}(0, \sigma^2)$ with $h \triangleq h_{AB}$, and $N_A, N_B, N_E \sim \mathcal{N}(0, \sigma_0^2)$. Besides, h, h_{AE} and h_{BE} are independent. Suppose the fading block length is L, and within each block, Alice use L_0 symbols to transmit a signal $\mathbf{S}_A$ and Bob uses the remaining $L - L_0$ symbols to transmit a signal $\mathbf{S}_B$. Then we have

$$\mathbf{Y}_B = h\mathbf{S}_A + \mathbf{N}_B,$$
$$\mathbf{Y}_A = h\mathbf{S}_B + \mathbf{N}_A.$$

With the received $\mathbf{Y}_A, \mathbf{Y}_B$, Alice and Bob compute estimates of h by

$$\tilde{h}_A = \frac{\mathbf{S}_B^T}{||\mathbf{S}_B^T||^2}(h\mathbf{S}_B + \mathbf{N}_A) = h + \frac{\mathbf{S}_B^T}{||\mathbf{S}_B^T||^2}\mathbf{N}_A,$$
$$\tilde{h}_B = \frac{\mathbf{S}_A^T}{||\mathbf{S}_A^T||^2}(h\mathbf{S}_A + \mathbf{N}_B) = h + \frac{\mathbf{S}_A^T}{||\mathbf{S}_A^T||^2}\mathbf{N}_B,$$

respectively.

Applying this method in n blocks, Alice and Bob then obtain correlated i.i.d. generated sequences $(\tilde{h}_A^n, \tilde{h}_B^n)$, respectively. During this process, Eve can also obtain two sequences of $(\mathbf{Y}_{E1}^n, \mathbf{Y}_{E2}^n)$ from channels h_{AE} and h_{BE}, respectively. However, $(\mathbf{Y}_{E1}^n, \mathbf{Y}_{E2}^n)$ are independent of $(\tilde{h}_A^n, \tilde{h}_B^n)$ since h, h_{AE} and h_{BE} are independent. Thus, Alice and Bob are able to generate a secret key of rate

$$R = \frac{1}{L} I(\tilde{h}_A; \tilde{h}_B)$$

via public discussion using the noiseless channel.

Observing that $\tilde{h}_A \sim N(0, \sigma^2 + \frac{\sigma_0^2}{||\mathbf{S}_B||^2})$ and $\tilde{h}_B \sim N(0, \sigma^2 + \frac{\sigma_0^2}{||\mathbf{S}_A||^2})$, one can easily obtain that [4]

$$R = \frac{1}{2L} \log \frac{(\sigma_0^2 + \sigma^2 P L_0)(\sigma_0^2 + \sigma^2 (L - L_0)P)}{\sigma_0^4 + \sigma^2 \sigma_0^2 P L}, \tag{8.9}$$

where P is the power constraint on the transmitted signals, i.e., $||\mathbf{S}_A||^2 = L_0 P, ||\mathbf{S}_B||^2 = (L - L_0)P$. And R is maximized at $L_0 = L/2$.

Observe that even though Eve is connected with Alice and Bob, and it can observe the outputs of the fading channels, Eve obtains no correlated side information related to the sources Alice and Bob generated, in the basic wireless model.

8.3 KEY GENERATION WITH PUBLIC DISCUSSION ATTACKED

Having introduced the scenarios where the adversaries are passive, we begin to review the cases when the adversaries are indeed active attackers. In this part, we review the scenario where the attacker Eve attacks the public discussion which indicates the public discussion is not reliable anymore. This model is studied in [15], which characterizes the relationship between the key rate obtained when the adversary is passive and that obtained when the adversary is an active attacker.

8.3.1 Model Modification

Compared with the model considered in Section 8.2.1, the only difference of the model with an active attacker is that the transmitted messages by Alice and Bob are not guaranteed to be correctly received: Eve may intercept the messages, and modify them into different ones according to its attack strategy. What's more, even if Alice or Bob doesn't send any message, Eve may fake a message impersonating Alice or Bob to cheat the other user. To proceed further, we need the following definition.

Definition 8.2 ([15]) *A $(P_{XYZ}, R, \epsilon, \delta)$-protocol is said to be robust if*

- *When Eve acts passively, the probability that both Alice and Bob accept the outcome of the protocol and the key agreement is successful is larger than $1 - \delta$;*

- *When Eve acts actively, the probability that either both Alice and Bob reject the outcome or the key agreement is successful is at least $1 - \delta$*

where the term successful *is equivalent to* Definition 8.1.

Note that Eve acting passively indicates that the attacker acts as a passive listener.

8.3.2 All or Nothing Result

Denote by $S(X;Y||Z)$ the key capacity obtained when Eve is passive, and $S^*(X;Y||Z)$ the corresponding capacity obtained when Eve is an active attacker. It is easy to see that

$$S^*(X;Y||Z) \le S(X;Y||Z).$$

However, compared with $S(X;Y||Z)$, how less $S^*(X;Y||Z)$ should be? The theorem in the sequel provides the answer to this question in a surprising way. Before stating the theorem, we first introduce a concept defined in [12].

Definition 8.3 ([12]) *Given P_{XYZ}, X is said to be simulatable by Z with respect to Y, if $\exists P_{\bar{X}|Z}$ such that $P_{\bar{X}Y} = P_{XY}$ with*

$$P_{\bar{X}Y}(x,y) \triangleq \sum_{z \in \mathcal{Z}} P_{YZ}(y,z)P_{\bar{X}|Z}(x,z). \tag{8.10}$$

And it is denoted it by $Sim_Y(Z \to X)$.

In the same manner, we can also define $Sim_X(Z \to Y)$. The intuition behind $Sim_Y(Z \to X)$ is that, given Z^n, Eve can generate a sequence $\bar{X}^n$ such that $(\bar{X}^n, Y^n)$ has the same joint distribution as that of (X^n, Y^n). Thus, when $Sim_Y(Z \to X)$ holds, if Eve intercepts Alice's messages, and simulates her to exchange messages with Bob according to a certain pre-agreed protocol, Bob can not decide whether the received message is sent by Alice or faked by Eve with a probability larger than ϵ. By selecting $\epsilon < 1 - \delta$, we conclude that no $(P_{XYZ}, R, \epsilon, \delta)$-protocol exists. Thus, we must have $S^*(X;Y||Z) = 0$. On the other hand, if neither $Sim_Y(Z \to X)$ nor $Sim_X(Z \to Y)$ holds, [15] shows that there exists a scheme such that one can transform a key generation protocol for the case with a passive adversary (or say a passive protocol) into a corresponding key generation protocol against the active attacker with negligible key rate loss, as long as n is sufficiently large. Hence, we have the following result.

Theorem 8.4 ([15]) *Given any P_{XYZ}, if $Sim_Y(Z \to X)$ or $Sim_X(Z \to Y)$ holds, we have*

$$S^*(X;Y||Z) = 0.$$

Otherwise, we have

$$S^*(X;Y||Z) = S(X;Y||Z).$$

Theorem 8.4 states that given P_{XYZ}, the key capacity obtained from an active protocol is the same as that from a passive protocol as long as neither $Sim_Y(Z \to X)$ nor $Sim_X(Z \to Y)$ holds; otherwise, no key can be generated in the active protocol. In the following, we provide a high-level idea of how to transform a passive protocol into an active one.

The case when $S(X;Y||Z) = 0$ is trivial. In the sequel, we provide the main idea of the scheme of the case when $S(X;Y||Z) > 0$. Supposing a key of length $N \cdot S(X;Y||Z)$ bits is generated from a certain passive protocol with N length sequences, we analyze how many bits of realizations are needed in the transformed active protocol. The proposed scheme involves three steps:

Step 1: Alice and Bob generate a short key by applying the same passive protocol with each transmitted bit authenticated by a block of sequence realizations.

Take Alice, for instance, when Alice needs to send i-th bit, she sends $X_{2(i-1)\ell+1}^{(2i-1)\ell}$ for 0 or $X_{(2i-1)\ell+1}^{2i\ell}$ for 1. The authentication is guaranteed by checking the typicality of the received signal block with the counterpart realizations of the observed local sequence. Here ℓ is the block length which is selected to make sure the successful attack probability is less than $\delta/3$. To generate a short key with length k, the total length of sequence realizations is of the order

$$O(k/S(X;Y\|Z) + k \cdot r \cdot \log(3kr/\delta)),$$

where r is the number of rounds of public discussion needed for the passive protocol. [15] proves that the passive protocol with r finite exists.

Step 2: Alice and Bob generate a longer key by applying the same passive protocol with each transmitted message authenticated by the $\varepsilon-$ASU hashing (see [21]), using the short key generated in *Step 1*.

According to [21], there exists an $((i+1)/q)$-ASU hashing class of q^{i+2} functions mapping A to B with $|A| = q^{2^i}$, $|B| = q$ such that a successful probability of an impersonation attack is $1/|B|$ and that of a substitution attack is less than $(i+1)/q$. Since the total length of messages transmitted in r rounds is of order $O(rN)$, we apply the hashing to authenticate each transmitted message and set

$$i = \log q, \ \log q^{i+2} = k \text{ and } \log q^{2^i} = O(rN).$$

Then, without much derivation, we conclude that the successful attack probability is upper bounded by

$$2r\frac{i+1}{q} = 2r\frac{\log q}{q} \leq 2r\frac{\delta/3}{2r} = \delta/3,$$

when N is sufficiently large.

Step 3: A final confirmation signal consisting of a block of realizations is transmitted.

The purpose of this signal is to prevent the sender of the final message from erroneously accepting the outcome, and the corresponding successful attack probability upper bounded by $\delta/3$ is guaranteed by setting the length of realizations of this signal block of order $O(\log \frac{3}{\delta})$.

Hence, the total successful attack probability is upper bounded by δ, and the total realizations of sequences used to generate the $N \cdot S(X;Y\|Z)$ length key in the active protocol is

$$O((\log rN)^2 \log \log \frac{3N}{\delta}) + N + O(\log \frac{3}{\delta}).$$

Hence,

$$S^*(X;Y\|Z) = \frac{N \cdot S(X;Y\|Z)}{O((\log rN)^2 \log \log \frac{3N}{\delta}) + N + O(\log \frac{3}{\delta})} \geq S(X;Y\|Z) - \epsilon,$$

when N is sufficiently large.

Finally, we can conclude that such a proposed scheme can transform the passive key generation protocol into a protocol against the active attacker with negligible key rate loss.

8.3.3 Efficiently Checking the Simulatability Condition

As illustrated by Theorem 8.4, given P_{XYZ}, the simulatability condition (SC) defined by (8.10) plays a significant role in deciding whether one can generate a secret key with a positive rate or not.

However, to efficiently check the SC is not a trivial issue. [16] provides an answer in part to how to check the SC. It proposes an algorithm to check it; however, the outcome of the proposed algorithm is merely a necessary, but not a sufficient condition. Thus, we discuss the problem of efficiently checking the SC based on [19], which proposes an efficient algorithm to completely answer this question.

To illustrate the idea, we discuss how to check $\text{Sim}_Y(Z \to X)$. We first rewrite (8.10) in the matrix form

$$C = AQ, \tag{8.11}$$

in which matrix C represents the PMF P_{YX} with $C_{ij} = P_{YX}(i,j)$, matrix A represents P_{YZ} in a similar manner and Q represents the desired $P_{\bar{X}|Z}$. Combining the fact that $P_{\bar{X}|Z}$ is a conditional PMF, we can easily obtain that (8.11) is equivalent to

$$\mathcal{A}\mathbf{q} = \mathbf{c}, \tag{8.12}$$

in which

$$\mathbf{c} \triangleq \left(\begin{array}{c} \text{Vec}(C^T) \\ \mathbf{1}_{|\mathcal{Z}| \times 1} \end{array} \right), \mathcal{A} \triangleq \left(\begin{array}{c} A \otimes \mathbf{I}_{|\mathcal{X}|} \\ \mathbf{I}_{|\mathcal{Z}|} \otimes \mathbf{1}_{1 \times |\mathcal{X}|} \end{array} \right), \mathbf{q} \triangleq \text{Vec}(Q^T),$$

and $\mathbf{I}$ and $\mathbf{1}$ are identity matrix and all 1 vector, respectively. $\text{Vec}(A)$ represents the vectorization of matrix A, and $A \otimes B$ denotes the Kronecker product of matrices A and B. Thus, $\text{Sim}_Y(Z \to X)$ is equivalent to that in which there exists a nonnegative solution to (8.12). Suppose (8.12) is feasible; otherwise, there does not exist such a channel $P_{\bar{X}|Z}$. Solving (8.12), we have

$$\mathbf{q} = \mathcal{A}^g \mathbf{c} + (\mathcal{A}^g \mathcal{A} - \mathbf{I})\mathbf{p},$$

with $\mathbf{p}$ being an arbitrary $1 \times |\mathcal{Z}||\mathcal{X}|$ vector, and $\mathcal{A}^g$ is the general inverse of matrix $\mathcal{A}$. Thus, if there exists a nonnegative $\mathbf{q}$, there must be a $\mathbf{p}$ such that

$$(\mathbf{I} - \mathcal{A}^g \mathcal{A})\mathbf{p} \preceq \mathcal{A}^g \mathbf{c}.$$

Then, according to Farkas's lemma, we have the following theorem.

Theorem 8.5 ([19]) *Suppose h^* is obtained by the following linear programming*

$$h^* = \min_{\mathbf{t}} \mathbf{t}^T \mathcal{A}^g \mathbf{c} \tag{8.13}$$

$$s.t. \quad \mathbf{t} \succeq \mathbf{0}, \ (\mathbf{I} - \mathcal{A}^g \mathcal{A})^T \mathbf{t} = \mathbf{0},$$

then $\text{Sim}_Y(Z \to X)$ holds if and only if $h^* = 0$.

Figure 8.1 Two-way relay model

Since the linear programming has finite input size, we can conclude that the above algorithm can check SC with a polynomial complexity.

8.4 KEY GENERATION WITH CONTAMINATED SOURCES

In this section, we discuss a recent work on key generation in which the attacker can interfere with the generation of the correlated sources that are used to distill a secret key [23]. In the wireless network, the legitimate users typically have no correlated sources as a prior. Thus, in order to generate a secret key, the users need to generate correlated sources first, then to generate a secret key by applying the source model approach discussed in Section 8.2. However, due to the vulnerability of the wireless network [24], the attacker can easily transmit an interference signal to contaminate the source observations. [23] proposes a new algorithm for key generation in a two-way relay channel model. Under the considered model, [23] discusses how much influence the attacker can have on the key generation process.

8.4.1 Two-Way Relay Channel Model

With the basic model in wireless setting in mind, we discuss the model with an active attacker in a two-way relay channel model which is considered in [23]. The influence of an active attacker lies in two aspects: (1) By sending attack signals during the sources obtaining phase, Eve contaminates the generated sources; (2) By controlling the arrived attack signals at the legitimate users, Eve obtains partial information of the generated sources. Both aspects will reduce the key rate.

As illustrated in Figure 8.1, Alice and Bob wish to agree on a common secret key in the presence of an attacker Eve. Alice and Bob are connected by reciprocal fading channels via a relay. But there is no direct wireless link between Alice and Bob; thus, they need the assistance from the relay. What's more, all four terminals are assumed to have full access to a public noiseless channel.

We assume that all channel gains are independent, and denote the channel gains from Alice and Bob to the relay, from Alice, Bob and the relay to Eve by h_1, h_2, h_{AE}, h_{BE} and h_{RE}, respectively. All wireless channels are assumed to be ergodic block fading with a block length T, and $h_1 \sim \mathcal{N}(0, \sigma_1^2), h_2 \sim \mathcal{N}(0, \sigma_2^2)$. In addition, we assume Eve has extra power to control the arrived attack signals at the legitimate users during the three phases when Alice, Bob and the relay take turns to send training signals, and denote the arrived attack signals by Z_1, Z_2 and Z_3. Thus,

- If Alice transmits a signal X_A, the signals received by the relay and Eve are

$$Y_R = h_1 X_A + Z_1 + N_R,$$
$$Y_E = h_{AE} X_A + N_E,$$

  respectively;

- If Bob transmits a signal X_B, the signals received by the relay and Eve are

$$Y_R = h_2 X_B + Z_2 + N_R,$$
$$Y_E = h_{BE} X_B + N_E,$$

  respectively;

- If the relay transmits a signal X_R, the signals received by Alice, Bob and Eve are

$$Y_A = h_1 X_R + Z_3 + N_A,$$
$$Y_B = h_2 X_R + Z_3 + N_B,$$
$$Y_E = h_1 X_R + N_E,$$

  respectively, where N_A, N_B, N_R and N_E are independent additive Gaussian noises with the same zero mean and the same variance σ^2.

In addition, we assume a power constraint P_T on the transmitted signals $\mathbf{X}_A, \mathbf{X}_B$ and $\mathbf{X}_R$ (assume they are of the same length M) as well as a constraint P_E on the attack signals $\mathbf{Z}_1, \mathbf{Z}_2$ and $\mathbf{Z}_3$, as follows

$$\frac{1}{M} \mathbb{E}\{\mathbf{X}_A' \mathbf{X}_A + \mathbf{X}_B' \mathbf{X}_B + \mathbf{X}_R' \mathbf{X}_R\} \leq P_T,$$

$$\frac{1}{M} \mathbb{E}\{\mathbf{Z}_1' \mathbf{Z}_1 + \mathbf{Z}_2' \mathbf{Z}_2 + \mathbf{Z}_3' \mathbf{Z}_3\} \leq P_T.$$

Under this model, [23] proposes an efficient key generation algorithm as below.

Algorithm: Key Generation Scheme

Phase 1: Channel Estimation

1) Alice sends a known signal $\mathbf{S}_A$ with power P_A through the channel h_1 to the relay, and the relay obtains the channel gain estimate $\tilde{h}_{1,R}$ from the received signal $\mathbf{Y}_R^{(1)}$;

2) Bob sends a known signal $\mathbf{S}_B$ with power P_B through the channel h_2 to the relay, and the relay obtains the channel gain estimate $\tilde{h}_{2,R}$ from the received signal $\mathbf{Y}_R^{(2)}$;

3) The relay broadcasts a known signal $\mathbf{S}_R$ with power P_R. Alice estimates h_1 to be $\tilde{h}_{1,A}$ from the received signal $\mathbf{Y}_A$ and Bob estimates h_2 to be $\tilde{h}_{2,B}$ from the received signal $\mathbf{Y}_B$.

Phase 2: Key Agreement

1) With the correlated pair $(\tilde{h}_{1,R}, \tilde{h}_{1,A})$, Alice and the relay agree on a secret key K_1 via public discussion over the noiseless channel;

2) With the correlated pair $(\tilde{h}_{2,R}, \tilde{h}_{2,B})$, Bob and the relay agree on a secret key K_2 via public discussion over the noiseless channel;

3) The relay broadcasts $K_1 \oplus K_2$ via the noiseless channel. Then Alice and Bob obtain (K_1, K_2). They set the key with the smaller length as the final key.

8.4.2 Efficiency of the Key Generation Algorithm

In order to analyze the efficiency of the above mentioned algorithm, we first assume that Eve is passive, i.e., $\mathbf{Z}_i = \emptyset, i = 1, 2, 3$.

As proposed in [23], each fading block is divided into three time slots, each with duration $T_0 = \frac{T}{3}$, which is selected following a similar reason as selecting L_0 in (8.9). Alice, Bob and the relay take turns to use the time slots to transmit the training signals, and the transmitted signals are

$$\mathbf{S}_A = (\sqrt{P_A}, \cdots, \sqrt{P_A}),$$
$$\mathbf{S}_B = (\sqrt{P_B}, \cdots, \sqrt{P_B}),$$
$$\mathbf{S}_R = (\sqrt{P_R}, \cdots, \sqrt{P_R}),$$

with the same size T_0, under the constraint that

$$\frac{1}{3}(P_A + P_B + P_R) \leq P_T.$$

Then, at the end of Step 1, from channel h_1, Alice and the relay receive

$$\mathbf{Y}_A = h_1 \mathbf{S}_R + \mathbf{N}_A,$$
$$\mathbf{Y}_R^{(1)} = h_1 \mathbf{S}_A + \mathbf{N}_R^{(1)},$$

respectively, while Bob and the relay receive the other two sequences from channel h_2

$$\mathbf{Y}_B = h_2 \mathbf{S}_R + \mathbf{N}_B,$$
$$\mathbf{Y}_R^{(2)} = h_2 \mathbf{S}_B + \mathbf{N}_R^{(2)},$$

respectively. Thus, similar as discussed in Section 8.2.3, Alice and the relay obtain

$$\tilde{h}_{1,A} = h_1 + \frac{\mathbf{S}_R'}{||\mathbf{S}_R||^2}\mathbf{N}_A,$$

$$\tilde{h}_{1,R} = h_1 + \frac{\mathbf{S}_A'}{||\mathbf{S}_A||^2}\mathbf{N}_R^{(1)}.$$

And they agree on a secret key K_1 via public discussion, which is confidential from Eve with a rate

$$\frac{1}{T}I(\tilde{h}_{1,A}; \tilde{h}_{1,R}).$$

Alternatively, Bob and the relay obtain

$$\tilde{h}_{2,B} = h_2 + \frac{\mathbf{S}_R'}{||\mathbf{S}_R||^2}\mathbf{N}_B,$$

$$\tilde{h}_{2,R} = h_2 + \frac{\mathbf{S}_B'}{||\mathbf{S}_B||^2}\mathbf{N}_R^{(2)}.$$

And they agree on another secret key K_2 with a rate

$$\frac{1}{T}I(\tilde{h}_{2,B}; \tilde{h}_{2,R}).$$

Then, during the last step of phase 2, Alice and Bob can obtain (K_1, K_2) by letting the relay broadcast $K_1 \oplus K_2$ over the public noiseless channel. And we can easily conclude that the shorter key is kept secure from Eve, even though Eve observes $K_1 \oplus K_2$. Thus, by setting the shorter key as the final common key, we obtain the final key rate

$$R_{co} = \frac{1}{T} \min\{I(\tilde{h}_{1,A}; \tilde{h}_{1,R}), I(\tilde{h}_{2,B}; \tilde{h}_{2,R})\}. \tag{8.14}$$

By computing $I(\tilde{h}_{1,A}; \tilde{h}_{1,R})$ with $\tilde{h}_{1,A} \sim \mathcal{N}(0, \sigma_1^2 + \frac{\sigma^2}{||\mathbf{S}_R||^2})$, $\tilde{h}_{1,R} \sim \mathcal{N}(0, \sigma_1^2 + \frac{\sigma^2}{||\mathbf{S}_A||^2})$ and $\mathrm{cov}(\tilde{h}_{1,A}; \tilde{h}_{1,R}) = \sigma_1^2$ [4], we get

$$I(\tilde{h}_{1,A}; \tilde{h}_{1,R}) = -\frac{1}{2}\log(1 - \rho_1^2), \tag{8.15}$$

in which

$$\rho_1^2 = \frac{\mathrm{cov}^2(\tilde{h}_{1,A}; \tilde{h}_{1,R})}{\mathrm{Var}(\tilde{h}_{1,A})\mathrm{Var}(\tilde{h}_{1,R})} = \frac{1}{\left(1 + \frac{\sigma^2}{\sigma_1^2 T_0 P_A}\right)\left(1 + \frac{\sigma^2}{\sigma_1^2 T_0 P_R}\right)}.$$

Similarly, we can also obtain

$$I(\tilde{h}_{2,B}; \tilde{h}_{2,R}) = -\frac{1}{2}\log(1 - \rho_2^2), \tag{8.16}$$

in which

$$\rho_2^2 = \frac{\mathrm{cov}^2(\tilde{h}_{2,B}; \tilde{h}_{2,R})}{\mathrm{Var}(\tilde{h}_{2,B})\mathrm{Var}(\tilde{h}_{2,R})} = \frac{1}{\left(1 + \frac{\sigma^2}{\sigma_2^2 T_0 P_A}\right)\left(1 + \frac{\sigma^2}{\sigma_2^2 T_0 P_R}\right)}.$$

Hence,

$$R_{co} = \frac{1}{2T} \min \left\{ \log \frac{1}{1 - \rho_1^2}, \log \frac{1}{1 - \rho_2^2} \right\}, \tag{8.17}$$

and the optimal key generation power allocation is

$$(P_A, P_B, P_R) = \arg \max_{P_A, P_B, P_R} \min \left\{ \frac{1}{2T} \log \frac{1}{1 - \rho_1^2}, \frac{1}{2T} \log \frac{1}{1 - \rho_2^2} \right\}$$

$$\text{s.t.} \quad \frac{1}{3}(P_A + P_B + P_R) \leq P_T.$$

From (8.17), we can see that the existence of a passive attacker has no influence on the final generated key, as it can neither obtain a side information with the generated sources that are used to generate the final key nor interfere with the generation of the key. However, the situation is quite different if Eve is active.

8.4.3 Attack Strategy and Power Allocation

In this part, we review the analysis of the case where Eve is active, and discuss the characterization of the optimal attack strategy to the active attacker [23].

8.4.3.1 *Optimal Attack Strategy*

Since Eve is active, the received sequences from channel h_1 are given by

$$\mathbf{Y}_R^{(1)} = h_1 \mathbf{S}_A + \mathbf{Z}_1 + \mathbf{N}_R^{(1)},$$

$$\mathbf{Y}_A = h_1 \mathbf{S}_R + \mathbf{Z}_3 + \mathbf{N}_A.$$

Thus, the corresponding estimates of h_1 are

$$\tilde{h}_{1,R} = h_1 + \frac{\mathbf{S}_A' \mathbf{Z}_1}{||\mathbf{S}_A||^2} + \frac{\mathbf{S}_A' \mathbf{N}_R^{(1)}}{||\mathbf{S}_A||^2} \triangleq h_1 + \Gamma_1 + N_R^{(1)},$$

$$\tilde{h}_{1,A} = h_1 + \frac{\mathbf{S}_R' \mathbf{Z}_3}{||\mathbf{S}_R||^2} + \frac{\mathbf{S}_R' \mathbf{N}_A}{||\mathbf{S}_R||^2} \triangleq h_1 + \Gamma_3 + N_A,$$

where $\Gamma_1 \triangleq \frac{\mathbf{S}_A' \mathbf{Z}_1}{||\mathbf{S}_A||^2}$ and Γ_3 is defined in the same manner. Note that it's easy to check that $N_R^{(1)} \sim \mathcal{N}(0, \frac{\sigma^2}{||\mathbf{S}_A||^2})$ and $N_A \sim \mathcal{N}(0, \frac{\sigma^2}{||\mathbf{S}_R||^2})$.

Since Γ_1 and Γ_3 are controlled by Eve, Eve has side information related to $(\tilde{h}_{1,R}, \tilde{h}_{1,A})$. Thus, according to [10], an achievable rate of K_1 is given by

$$R_{s1} = [I(\tilde{h}_{1,A}; \tilde{h}_{1,R}) - I(\tilde{h}_{1,A}; \Gamma_1, \Gamma_3)]^+. \tag{8.18}$$

Similarly, from channel h_2, the relay and Bob can obtain estimates of h_2 as

$$\tilde{h}_{2,R} = h_2 + \frac{\mathbf{S}_B' \mathbf{Z}_2}{||\mathbf{S}_B||^2} + \frac{\mathbf{S}_B' \mathbf{N}_R^{(2)}}{||\mathbf{S}_B||^2} = h_2 + \Gamma_2 + N_R^{(2)},$$

$$\tilde{h}_{2,B} = h_2 + \frac{\mathbf{S}_R' \mathbf{Z}_3}{||\mathbf{S}_R||^2} + \frac{\mathbf{S}_R' \mathbf{N}_B}{||\mathbf{S}_R||^2} = h_2 + \Gamma_3 + N_B,$$

respectively. Bob and the relay agree on a key K_2 with a rate

$$R_{s2} = [I(\tilde{h}_{2,B}; \tilde{h}_{2,R}) - I(\tilde{h}_{2,B}; \Gamma_2, \Gamma_3)]^+. \tag{8.19}$$

Thus, same as (8.14), the final common key rate is

$$R_s = \frac{1}{T} \min\{R_{s1}, R_{s2}\}. \tag{8.20}$$

Hence, to the attacker, the optimal attack strategy is to find the optimal $(\Gamma_1, \Gamma_2, \Gamma_3)$ with power constraint P_E to minimize (8.20). According to [10], the minimal R_{s1} and R_{s2} are achieved when $\Gamma_1, \Gamma_2, \Gamma_3$ are zero-mean Gaussian random variables with appropriate correlation coefficients of (Γ_1, Γ_3) and (Γ_2, Γ_3). Thus, the attack signals $z_i, i = 1, 2, 3$ should also be zero-mean Gaussian random variables.

For the simplicity of notation, denote the variance of channel gains h_i as $\sigma_{hi}^2, i = 1, 2$, and

$$\sigma_1^2 \triangleq \mathrm{Var}\{\Gamma_1\} = P_{E1}/||\mathbf{S}_A||^2,$$
$$\sigma_2^2 \triangleq \mathrm{Var}\{\Gamma_2\} = P_{E2}/||\mathbf{S}_B||^2,$$
$$\sigma_3^2 \triangleq \mathrm{Var}\{\Gamma_3\} = P_{E3}/||\mathbf{S}_R||^2,$$

where $P_{Ei} \triangleq \mathbb{E}\{z_i^2\}, i = 1, 2, 3$. The power constraint of Eve given by $\frac{1}{3}(P_{E1} + P_{E2} + P_{E3}) \leq P_E$ is equivalent to

$$\frac{\sigma_1^2}{1/||\mathbf{S}_A||^2} + \frac{\sigma_2^2}{1/||\mathbf{S}_B||^2} + \frac{\sigma_3^2}{1/||\mathbf{S}_R||^2} \leq 3P_E. \tag{8.21}$$

And we can obtain that the optimal correlation coefficients of (Γ_1, Γ_3) and (Γ_2, Γ_3) are

$$\rho_1 = \begin{cases} -\frac{\sigma_{h1}^2}{\sigma_1 \sigma_3}, & \text{if } \sigma_{h1}^2 \leq \sigma_1 \sigma_3 \\ -1, & \text{else} \end{cases}, \tag{8.22}$$

$$\rho_2 = \begin{cases} -\frac{\sigma_{h2}^2}{\sigma_2 \sigma_3}, & \text{if } \sigma_{h2}^2 \leq \sigma_2 \sigma_3 \\ -1, & \text{else} \end{cases}, \tag{8.23}$$

respectively. Then, we can solve (8.18) and (8.19) by

$$\begin{aligned} R_{s1} &= \left[-\frac{1}{2}\log(2\pi e \sigma_{e1}^2) + \frac{1}{2}\log\left(2\pi e(\sigma_{h1}^2 + \frac{\sigma^2}{||\mathbf{S}_R||^2})\right) \right]^+ \\ &= \left[\frac{1}{2}\log\left(\frac{\sigma_{h1}^2 + \sigma^2/||\mathbf{S}_R||^2}{\sigma_{e1}^2}\right) \right]^+, \end{aligned} \tag{8.24}$$

$$R_{s2} = \left[\frac{1}{2}\log\left(\frac{\sigma_{h2}^2 + \sigma^2/||\mathbf{S}_R||^2}{\sigma_{e2}^2}\right) \right]^+, \tag{8.25}$$

where

$$\sigma_{e1}^2 \triangleq \left(\sigma_{h1}^2 + \sigma_3^2 + \frac{\sigma^2}{||\mathbf{S}_R||^2}\right) - \frac{(\sigma_{h1}^2 + \rho_1 \sigma_1 \sigma_3)^2}{\sigma_{h1}^2 + \sigma_1^2 + \frac{\sigma^2}{||\mathbf{S}_A||^2}},$$

$$\sigma_{e2}^2 \triangleq \left(\sigma_{h2}^2 + \sigma_3^2 + \frac{\sigma^2}{||\mathbf{S}_R||^2}\right) - \frac{(\sigma_{h2}^2 + \rho_1 \sigma_2 \sigma_3)^2}{\sigma_{h2}^2 + \sigma_2^2 + \frac{\sigma^2}{||\mathbf{S}_A||^2}}.$$

Hence, the optimal attack strategy for the attacker is simplified as

$$\min_{\sigma_1,\sigma_2,\sigma_3 \geq 0} \min\{R_{s1}, R_{s2}\}$$

$$\text{s.t. } \frac{\sigma_1^2}{1/||\mathbf{S}_A||^2} + \frac{\sigma_2^2}{1/||\mathbf{S}_B||^2} + \frac{\sigma_3^2}{1/||\mathbf{S}_R||^2} \leq 3P_E. \tag{8.26}$$

8.4.3.2 Optimal Attack Power Allocation

By solving the optimization problem (8.26), Eve can obtain the corresponding optimal attack power allocation. We first review the case when $R_s = 0$.

A. Attack Power Allocation when $R_s = 0$

To achieve $R_s = 0$, Eve only needs to make either $R_{s1} = 0$ or $R_{s2} = 0$. From (8.24) and (8.25), we can see that σ_2 does not affect R_{s1}. Similarly, σ_1 does not affect R_{s2}. Thus, the least power P_E that makes $R_s = 0$ is achieved at either $\sigma_1 = 0$ or $\sigma_2 = 0$. We can analyze $R_{s1} = 0$ and $R_{s2} = 0$ separately.

With regards to channel h_1, denote

$$\lambda_1 = \frac{1}{2}\left(\frac{\sigma_{h1}^2 + \frac{\sigma^2}{||\mathbf{S}_A||^2}}{||\mathbf{S}_R||^2} + \sqrt{\frac{(\sigma_{h1}^2 + \frac{\sigma^2}{||\mathbf{S}_A||^2})^2}{||\mathbf{S}_R||^4} + \frac{4\sigma_{h1}^4}{||\mathbf{S}_A||^2||\mathbf{S}_R||^2}}\right).$$

$$a_1^2 = \frac{\sigma_{h1}^4}{\lambda_1}, \quad k_1 = \frac{2\sigma_{h1}^2}{\sigma_{h1}^2 + \frac{\sigma^2}{||\mathbf{S}_A||^2}}\frac{||\mathbf{S}_R||}{||\mathbf{S}_A||}.$$

Denote the counterpart (λ_2, a_2, k_2) regarding channel h_2 using similar formulas. Then, we have the following result.

Theorem 8.6 ([23]) $R_s = 0$ *if*

$$P_E \geq \frac{1}{3}\min\{a_1^2, a_2^2\}. \tag{8.27}$$

In particular, if $a_1^2 \leq a_2^2$, the optimal power allocation for the attacker is

$$\sigma_1^2 = \frac{a_1^2}{2||\mathbf{S}_A||^2}\left(1 - \frac{1}{\sqrt{1+k_1^2}}\right), \tag{8.28}$$

$$\sigma_2 = 0, \tag{8.29}$$

$$\sigma_3^2 = \frac{a_1^2}{2||\mathbf{S}_R||^2}\left(1 + \frac{1}{\sqrt{1+k_1^2}}\right). \tag{8.30}$$

Otherwise, the optimal power allocation is

$$\sigma_1 = 0,$$

$$\sigma_2^2 = \frac{a_2^2}{2\|S_B\|^2}\left(1 - \frac{1}{\sqrt{1+k_2^2}}\right),$$

$$\sigma_3^2 = \frac{a_2^2}{2\|S_R\|^2}\left(1 + \frac{1}{\sqrt{1+k_2^2}}\right).$$

When $a_1^2 \le a_2^2$, the allocation of (σ_1^2, σ_3^2) defined by (8.28) and (8.30) requires the least power of P_E to make

$$\sigma_3^2 - \frac{(\sigma_{h1}^2 + \rho_1\sigma_1\sigma_3)^2}{\sigma_{h1}^2 + \sigma_1^2 + \frac{\sigma^2}{\|S_A\|^2}} \ge 0,$$

which results in $R_{s1} = 0$. And the case of $a_1^2 > a_2^2$ follows in a similar way.

B. Power Allocation when $R_s \ne 0$

When $R_s \ne 0$, we must have $P_E < \frac{1}{3}\min\{a_1^2, a_2^2\}$. In this case, in order to solve (8.26), we analyze the minimum values of R_{s1} and R_{s2} separately.

In the following, we only review the analysis of R_{s1}, as the analysis of R_{s2} follows in a similar manner. Obviously, the optimal allocation of $(\sigma_1, \sigma_2, \sigma_3)$ is achieved on the surface

$$\frac{\sigma_1^2}{1/\|S_A\|^2} + \frac{\sigma_2^2}{1/\|S_B\|^2} + \frac{\sigma_3^2}{1/\|S_R\|^2} = 3P_E.$$

Thus, we can define

$$\sigma_1 = a\cos\gamma\sin\theta,$$

$$\sigma_2 = b\sin\gamma\sin\theta,$$

$$\sigma_3 = c\cos\theta,$$

with $a \triangleq \sqrt{3P_E}/\|S_A\|, b \triangleq \sqrt{3P_E}/\|S_B\|, c \triangleq \sqrt{3P_E}/\|S_R\|$ and $0 \le \gamma, \theta \le \pi/2$. We first review the relationship between γ and θ at the optimal point $(\sigma_1, \sigma_2, \sigma_3)$. We have the following result.

Theorem 8.7 ([23]) *For any $\gamma \in [0, \pi/2]$, the optimal θ minimizing R_{s1} is given by*

$$\theta = \begin{cases} \frac{1}{2}\left(\arctan\frac{A}{B} - \arcsin\frac{C}{\sqrt{A^2+B^2}}\right), & B \ge 0 \\ \frac{1}{2}\left(\pi + \arctan\frac{A}{B} - \arcsin\frac{C}{\sqrt{A^2+B^2}}\right), & B < 0 \end{cases} \quad (8.31)$$

where

$$A = ac\sigma_{h1}^2\cos\gamma\left(\sigma_{h1}^2 + \frac{\sigma^2}{\|S_A\|^2}\right) + C,$$

$$B = \frac{a^2}{2}\cos^2\gamma\left[c^2\left(\sigma_{h1}^2 + \frac{\sigma^2}{\|S_A\|^2}\right) - \sigma_{h1}^4\right] + \frac{c^2}{2}\left(\sigma_{h1}^2 + \frac{\sigma^2}{\|S_A\|^2}\right)^2,$$

$$C = \frac{a^3c}{2}\sigma_{h1}^2\cos^3\gamma.$$

Thus, at the optimal point $(\sigma_1, \sigma_2, \sigma_3)$, R_{s1} can be viewed as a function of γ, and we can obtain the optimal solution γ^* by solving

$$\min_{\gamma \in [1:\pi/2]} R_{s1}.$$

Finally, we can obtain the corresponding optimal attack power allocation $(P_{E1}^{(1)}, P_{E2}^{(1)}, P_{E3}^{(1)})$ with regards to R_{s1}.

Following a similar derivation of $(P_{E1}^{(1)}, P_{E2}^{(1)}, P_{E3}^{(1)})$, we can also obtain an optimal attack power allocation $(P_{E1}^{(2)}, P_{E2}^{(2)}, P_{E3}^{(2)})$ with regards to R_{s1}. And the final optimal attack power allocation is the triple related to $\arg\min\{R_{s1}, R_{s2}\}$.

8.5 KEY GENERATION WITH A BYZANTINE HELPER

In this section, we discuss the third scenario where the attacker attacks the helper. The case is studied in [18].

8.5.1 System Model with a Byzantine Helper

As introduced in [18], two legitimate users Alice and Bob wish to establish a common secret key in the presence of a third party Charlie. Alice, Bob and Charlie have access to correlated random sequences (X^n, Y^n, Z^n), respectively, which are i.i.d. generated according to a certain given joint PMF P_{XYZ}, and the PMF of (X^n, Y^n, Z^n) is given by

$$P_{X^n, Y^n, Z^n}(x^n, y^n, z^n) = \prod_{i=1}^{n} P_{XYZ}(x_i, y_i, z_i).$$

In addition, the three terminals are allowed to discuss over a public noiseless channel. However, different from the previous two scenarios, Charlie in this model is under Byzantine attack in the sense that Charlie can either be an honest helper who can help to increase the generated key rate between the two legitimate users [6], or be under attack and controlled by an adversary whose purpose is to reduce the generated key rate or to mislead the legitimate users to agree on different keys. Here, if Charlie is under attack, he can merely fake messages to cheat Alice and Bob according to its attack strategy, and is assumed to have no ability to contaminate the public discussion between Alice and Bob. Furthermore, the legitimate users do not know the identity of Charlie as a prior, and there is not any probability assumption of it.

At the beginning of communication, three terminals are allowed to independently generate local randomness (F_1, F_2, F_3), respectively, which are independent from (X^n, Y^n, Z^n). Denote $\mathbf{F}$ the collection of all messages exchanged over the noiseless channel. Then, at the end of the public discussion, Alice generates a key K_A as a function of $(F_1, \mathbf{F}, X^n)$, Bob generates a key K_B as a function of $(F_2, \mathbf{F}, Y^n)$.

Definition 8.4 *A key rate R is said to be achievable if $\forall \epsilon > 0$, there exists a key generation protocol when n is sufficiently large, such that*

$$Pr\{K_A \neq K_B\} \leq \epsilon, \tag{8.32}$$

$$\epsilon \geq \begin{cases} \frac{1}{n}I(K_A; F_3, \mathbf{F}, Z^n), & \text{If Charlie is under attack} \\ \frac{1}{n}I(K_A; \mathbf{F}), & \text{Otherwise} \end{cases}, \tag{8.33}$$

$$\frac{1}{n}H(K_A) \geq \frac{1}{n}\log|\mathcal{K}| - \epsilon, \tag{8.34}$$

$$\frac{1}{n}H(K_A) \geq R - \epsilon, \tag{8.35}$$

where $\mathcal{K}$ is the alphabet of K_A.

Here, (8.33) implies that if Charlie is under attack, the generated key should be kept secure from him.

Clearly, there are two simple approaches dealing with the Byzantine helper: (1) Treating Charlie as an attacker, Alice and Bob generate a secret key on their own ignoring the messages transmitted by Charlie following the same scheme as discussed in Section 8.2; (2) Treating Charlie as a helper, Alice and Bob generate a secret key with the assistance of Charlie. However, both approaches have drawbacks. Taking the first approach Alice and Bob may lose the opportunity to generate a key with a larger rate if Charlie is indeed an honest helper, while the second approach puts Alice and Bob at risk that they may be cheated by Charlie and agree on two different secret keys if Charlie is under attack. Thus, it's of interest to check whether there exists a scheme such that Alice and Bob are able to potentially utilize the Byzantine helper and generate a secret key with the assistance of Charlie if he is an honest helper while generating the secret key on their own if Charlie is under attack.

8.5.2 Key Generation Scheme against the Byzantine Helper

To potentially utilize the information contained in Z^n, [18] proposes a key generation algorithm against the Byzantine helper.

An attack is said to be *successful* if it is neither detected by Alice nor by Bob, and it results in that Alice agrees with Bob on different keys. Denote the event of a successful attack by $\mathcal{S}$, and the event that Charlie is detected as an attacker while in fact is an honest helper by $\mathcal{D}$.

Definition 8.5 *A $(P_{XYZ}, R_1, R_2, \epsilon)$-protocol is said to be robust, if*

$$Pr\{\mathcal{S}\} + Pr\{\mathcal{D}\} \leq \epsilon, \tag{8.36}$$

and the key rate R_1 is achievable if the helper is under attack, or R_2 is achievable if the helper is honest.

The key generation algorithm proposed in [18] is summarized as follows:

Algorithm: Key Generation against a Byzantine Helper
***
Public Discussion 1: Alice, Bob and Charlie generate local randomness F_1, F_2 and F_3, respectively. And they send messages in order over the public channel according to a certain agreement.

Identity Detection: Alice and Bob use the information at hand to decide Charlie's identity. If Alice or Bob decides that Charlie is under attack, they proceed to step *Public Discussion 2*; otherwise,they follow step *Key Generation* while skipping step *Public Discussion 2*.

Public Discussion 2: Alice and Bob exchange messages on their own, neglecting all messages Charlie sends.

Key Generation: Setting the collection of the public discussion as **F**, Alice generates a secret key via $K_A = K_A(F_1, \mathbf{F}, X^n)$ and Bob generates a secret key via $K_B = K_B(F_2, \mathbf{F}, Y^n)$. If Charlie is determined to be an honest helper, the generated key should be kept secret from the public discussion but not from Charlie; if Charlie is determined to be under attack, the generated key should be kept secret from both the public discussion and Charlie.

Here, if Charlie is an honest helper, he sends his messages as required. On the other hand, if he is under the attacker's control, he sends modified messages to mislead Alice and Bob.

8.5.2.1 A Key Generation Scheme Example

In this part, we use a secret key generation scheme example provided in [18] to illustrate the efficiency of the key generation algorithm against the Byzantine helper. In this scheme, Alice and Bob can potentially utilize the help of Charlie. If Charlie is under attack, this scheme is able to detect it with high probability. Otherwise, if Charlie is honest, this scheme has the ability to utilize Charlie to generate a secret key with a higher rate.

Without loss of generality, we assume $I(X;Y) - I(X;Z) \geq I(X;Y) - I(Y;Z)$, thus $H(Z|X) \geq H(Z|Y)$.

Codebook Construction: *Codebook at Charlie.* Randomly and independently assign an index pair (m_2, f) to each typical Z^n sequence with $m_2(Z^n) \in \mathcal{M}_2 \triangleq [1 : 2^{nR_Z}]$ and $f(Z^n) \in [1 : 2^{nR_0}]$ uniformly distributed. We denote the set of those sequences sharing the same index m_2 by bin $b_Z(m_2)$. And R_Z and R_0 are given by

$$R_Z = H(Z|X) + \epsilon,$$
$$R_0 = I(X;Z) - \epsilon.$$

Codebook at Alice. Randomly and uniformly assign each typical sequence $X^n \in \mathcal{X}^n$ an index $m_1(X^n) \in \mathcal{M}_1 \triangleq \{1, 2, \cdots, 2^{nR_X}\}$. Denote the set of those sequences with the same index m_1 by bin $b_X(m_1)$. Within each bin $b_X(m_1)$, randomly and uniformly assign each sequence two subbin indices $m_0(x^n) \in \mathcal{M}_0 \triangleq \{1, 2, \cdots, 2^{nR_{X_0}}\}$ and $g(x^n) \in [1 : 2^{nR_1}]$ and denote the set of those sequences in bin $b_X(m_1)$ with the same subindex m_0 by subbin $b_X(m_0, m_1)$. Similarly, within each subbin $b_X(m_0, m_1)$ assign each sequence $X^n \in b_X(m_0, m_1)$ a group index $h(x^n) \in [1 : 2^{nR_2}]$. Set R_X, R_{X_0}, R_1 and R_2 as

$$
\begin{aligned}
R_X &= H(X|YZ) + \epsilon, \\
R_{X_0} &= H(X|Y) - H(X|YZ) = I(X; Z|Y), \\
R_1 &= I(X; Y|Z) - \epsilon - \delta, \\
R_2 &= I(X; Y) - I(X; Z) - \epsilon.
\end{aligned}
$$

$$\text{(8.37)}$$
$$\text{(8.38)}$$

Encoding: At Charlie's side, upon observing a sequence $z^n \in b_Z(m_2)$, $m_2 \in \mathcal{M}_2$, Charlie transmits the index m over the noiseless channel; At Alice's side, upon observing a sequence $x^n \in b_X(m_0, m_1)$, $m_0 \in \mathcal{M}_0$, $m_1 \in \mathcal{M}_1$, Alice transmits m_1 to Bob. Clearly, if Charlie is an honest helper, we have the transmitted index $m = m_2$; otherwise, Charlie will set an optimal value but not m_2 to m and transmit it to the public, according to his attack strategy based on the sequence z^n he observes.

Attack Detection: After receiving the index pair (m, m_1), Bob tries to find a unique sequence pair $(\tilde{x}^n \times \tilde{z}^n) \in b_X(m_1) \times b_Z(m) \cap \mathcal{T}_\epsilon^n(XZ|y^n)$, where y^n is the observed sequence. Meanwhile Alice tries to find a unique sequence $\bar{z}^n \in b_Z(m)$ that is jointly typical with x^n. If both of them find such sequences, they declare Charlie to be an honest helper. Otherwise, they declare that Charlie is under attack, and Alice transmits another index m_0 to Bob.

Decoding: If Charlie is declared to be an honest helper, the sequences $(\tilde{x}^n, \tilde{z}^n, \bar{z}^n)$ decoded in the attack detection phase are the final desired sequences. Otherwise, after receiving the further information m_0, Bob tries to find a unique sequence $\tilde{x}^n \in b_X(m_0, m_1)$ that is jointly typical with y^n. If Bob finds it, declare $\tilde{x}^n$ to be the decoded sequence; otherwise, randomly set a decoded sequence in $b_X(m_0, m_1)$.

Key Generation: If Charlie is declared to be an honest helper, Alice and Bob set $K_A = \{g(x^n), f(\bar{z}^n)\}$, $K_B = \{g(\tilde{x}^n), f(\tilde{z}^n)\}$, respectively; otherwise, if they detect Charlie to be under attack, they set $K_A = h(x^n)$, $K_B = h(\tilde{x}^n)$.

[18] shows that, using this scheme, Alice and Bob can detect the identity of the Byzantine helper correctly with a high probability larger than $1 - \epsilon$, and the corresponding generated key is successful.

If Charlie is under attack, Alice and Bob can agree on a common secret key with a rate

$$R = \max\{I(X; Y) - I(X; Z), I(Y; X) - I(Y; Z)\},$$

while if Charlie is an honest helper, the generated secret key rate is given by

$$R = \min\{I(X; YZ), I(Y; XZ)\}.$$

Hence, Alice and Bob can potentially utilize the assistance from Charlie.

Note that, in order to detect the identity of the third communication party, we do not need to expose extra information to the public. Thus, the corresponding generated key rate will not be reduced by adding the identity detection phase.

8.6 CONCLUSION

In this chapter, we have discussed the problem of key generation under active attacks. We have first reviewed the basic scenarios for key generation in the presence of a passive adversary. Then, we have considered the scenario where the attacker attacks the public discussion, reviewed the relationship between the key rate obtained from the case with an active attacker with the corresponding key rate obtained from the passive case and discussed the condition which decides this relationship. Furthermore, we have discussed the scenario where the active attacker interferes with the source observations that are used to generate the secret key, reviewed a key generation scheme against the active attacker and discussed the attacker's optimal attack strategy correspondingly. Finally, we have reviewed the scenario where the attacker attacks the helper, and discussed a key generation scheme that has the ability to utilize the Byzantine helper.

8.7 ACKNOWLEDGEMENT

The work of W. Tu and L. Lai was supported by the National Science Foundation CAREER Award under Grant CCF-1318980 and by the National Science Foundation under Grant ECCS-1408114.

Bibliography

[1] R. Ahlswede and I. Csiszar. Common randomness in information theory and cryptography, part I: Secret sharing. *IEEE Trans. Inform. Theory*, 39(4):1121–1132, Jul. 1993.

[2] J. Barros and M. R. D. Rodrigues. Secrecy capacity of wireless channels. In *Proc. IEEE Intl. Symposium on Inform. Theory*, pages 356–360, Seattle, WA, Jul. 2006.

[3] M. Bloch, J. Barros, M. R. D. Rodrigues, and S. W. McLaughlin. Wireless information-theoretic security. *IEEE Trans. Inform. Theory*, 54(6):2515–2534, May 2008.

[4] T. M. Cover and J. A. Thomas. *Elements of Information Theory*. John Wiley & Sons, New York, 2012.

[5] I. Csiszár and P. Narayan. Common randomness and secret key generation with a helper. *IEEE Trans. Inform. Theory*, 46(2):344–366, Mar. 2000.

[6] I. Csiszár and P. Narayan. Secrecy capacities for multiple terminals. *IEEE Trans. Inform. Theory*, 50(12):3047–3061, Dec. 2004.

[7] I. Csiszár and P. Narayan. Secrecy capacities for multiterminal channel models. *IEEE Trans. Inform. Theory*, 54(6):2437–2452, Jun. 2008.

[8] I. Csiszár and J. Korner. Broadcast channels with confidential messages. *IEEE Trans. Inform. Theory*, 24(3):339–348, May 1978.

[9] L. Lai, Y. Liang, and W. Du. Cooperative key generation in wireless networks. *IEEE Journal on Selected Areas in Communications*, 30(8):1578–1588, Sep. 2012.

[10] L. Lai, Y. Liang, and H. V. Poor. A unified framework for key agreement over wireless fading channels. *IEEE Trans. Inform. Forensics and Security*, 7(2):480–490, Mar. 2012.

[11] U. M. Maurer. Secret key agreement by public discussion from common information. *IEEE Trans. Inform. Theory*, 39(3):733–742, May 1993.

[12] U. M. Maurer. Information-theoretically secure secret-key agreement by not authenticated public discussion. In *Advances in Cryptology–Eurocrypt' 97*, volume 1233, pages 209–225, Berlin, Germany, May 1997. Springer-Verlag.

[13] U. M. Maurer and S. Wolf. Unconditionally secure key agreement and the intrinsic conditional information. *IEEE Trans. Inform. Theory*, 45(2):499–514, Mar. 1999.

[14] U. M. Maurer and S. Wolf. Information-theoretic key agreement: From weak to strong secrecy for free. In *Proc. Advances in Cryptology*, volume 1807, pages 356–373, Bruges (Brugge), Belgium, May 2000.

[15] U. M. Maurer and S. Wolf. Secret key agreement over unauthenticated public channels - Part I: Definitions and a completeness result. *IEEE Trans. Inform. Theory*, 49(4):822–831, Apr. 2003.

[16] U. M. Maurer and S. Wolf. Secret key agreement over unauthenticated public channels - Part II: The simulatability condition. *IEEE Trans. Inform. Theory*, 49(4):832–838, Apr. 2003.

[17] F. Oggier and B. Hassibi. The secrecy capacity of the MIMO wiretap channel. *IEEE Trans. Inform. Theory*, 57(8):4961–4972, Aug. 2011.

[18] W. Tu and L. Lai. Key generation with a Byzantine helper. In *Proc. IEEE Intl. Conf. on Communication Workshop (ICCW)*, pages 429–434, London, UK, Jun. 2015. IEEE.

[19] W. Tu and L. Lai. On the simulatability condition in key generation over a non-authenticated public channel. In *Proc. IEEE Intl. Symposium on Inform. Theory*, pages 720–724, Hong Kong, China, Jun. 2015.

[20] R. Wilson, D. Tse, and R. A. Scholtz. Channel identification: Secret sharing using reciprocity in ultrawideband channels. *IEEE Trans. Inform. Forensics and Security*, 2(3):364–375, Sept. 2007.

[21] S. Wolf. *Information-theoretically and computationally secure key agreement in cryptography*. PhD thesis, Diss. Techn. Wiss. ETH Zürich, Nr. 13138, 1999. Ref.: Ueli Maurer; Korref.: Claude Crépeau.

[22] C. Ye, S. Mathur, A. Reznik, Y. Shah, W. Trappe, and N. B. Mandayam. Information-theoretically secret key generation for fading wireless channels. *IEEE Trans. Inform. Forensics and Security*, 5(2):240–254, Mar. 2010.

[23] H. Zhou, L. M. Huie, and L. Lai. Secret key generation in the two-way relay channel with active attackers. *IEEE Trans. Inform. Forensics and Security*, 9(3):476–488, Mar. 2014.

[24] X. Zhou, B. Maham, and A. Hjorungnes. Pilot contamination for active eavesdropping. *IEEE Trans. Wireless Communications*, 11(3):903–907, Jun. 2012.

III

Towards Smart World from Interfaces to Homes to Cities

CHAPTER 9

Applying Human–Computer Interaction Practices to IoT Prototyping

Salim Haniff
University of Tampere

Markku Turunen
University of Tampere

Roope Raisamo
University of Tampere

CONTENTS

Human–computer interaction (HCI) enables user friendly interfaces into IoT systems, creating a fully closed feedback loop in which the users can control their environment and see the changes take place, including (self)monitoring of people and their activities. The use of HCI in industrial settings has long been explored; however, a recent trend in aiming HCIs towards consumers is an emerging sector. Consumer devices such as smart energy meters, lighting and home automation systems are popular examples of how HCI complements IoT technologies.

In this chapter, we give an overview of some theory and practices that could be employed in developing user interfaces that could be applied for IoT devices. Three case studies will be presented based on previous works in fields similar to the Internet of Things. This chapter is aimed towards developing devices with the consumer in mind.

9.1 INTRODUCTION

While the Internet of Things (IoT) is an established sector in computing, there are many areas from the Computer Science field that can provide valuable insight in developing user interfaces for IoT devices. The area of Human–Computer Interaction (HCI) has provided a solid foundation for today's user interface developers to reference researched knowledge in developing user interfaces. To begin our application of HCI practices to the IoT sector, we will go over our interpretation of IoT and then provide some background knowledge in the area of HCI. This will help us draw some key areas of interest from HCI that can be applied to design of IoT user interfaces.

9.1.1 Internet of Things

In order to gain some insight on the Internet of Things, we will restrict our interpretation of IoT. This will help us to gain a clear focus of how to create user interfaces for IoT systems. The basic premise of IoT is the utilization of *things* to *interact* with each other to reach a common goal [6]. The *things* could be sensors, actuators, mobile devices and desktop computers. The things *interact* with each other utilizing an Internet Protocol, other industrial protocols (e.g., ZigBee), or a bridge device that receives Internet Protocol messages and converts them to vendor specific protocols. A typical scenario, and one we will go into more depth in this chapter, is the use of wireless communication devices in light bulbs. In this scenario, a user may want to turn on the light or program a lighting schedule for the light. The user could use a mobile device to communicate with the light, indicating his intention. Based on the user's intention the light will respond with the user's request and reflect those changes both in the physical light bulb and in the graphical user interface on the mobile device.

Based on the above paragraph, IoT can be decomposed into four fundamental aspects:

- sensors and actuators—or things depending on the context

- data connectivity

- data analysis

- information presentation

In the above scenario, the lighting would represent the *sensors and actuators*, the communication between the mobile device and the light would be represented in the *data connectivity*, interpreting the user's intention would fall into *data analysis* and *information presentation* would be presented in the graphical user interface.

This current scenario is common practice in industrial settings, where machines or sensors attached to programmable logic controllers are typically generating events or data. This event or data are sent to a central server. The central server provides data analysts access to the data where data analysis algorithms are to be performed and visualizations are created to help gain an understanding of that data. The facility operator can then program the machine to perform in a specific way if needed or be alerted if the machine is experiencing problems. Figure 9.1(a) illustrates an example of a web based user interface that was used in a project that showed the number of events generated by machines at a facility. The first graph illustrates the number of total events in a bargraph histogram. The second bargraph, right below the first bargraph, shows a detailed view of the actual number of events colour coded. From the second bargraph, it immediately becomes clear what types of errors are more frequent given a specific time period. The line graph on the bottom shows which features were commonly used in the control room. If the facility operators desired, they could send other data from the sensors to a centralized server and that data can be graphed instead. For example, temperature, humidity data can be sent if the operator wishes to monitor the climate of the facility (shown in Figure 9.1(b)).

One area that has been receiving constant innovation where industrial user interfaces can be used in a consumer setting is home automation and monitoring. Home consumers can replicate such a system to use in their homes to monitor energy consumption, temperature, humidity and also control electrical equipment through the use of electrical relays or network-enabled appliances [28]. Soliman et al. [35] have proposed such a system that could easily be incorporated into home automation and monitoring that fulfils our four fundamental aspects of IoT systems. In their set up, a house has many sensors just like an industrial setting that records data. They also provide a user interface for the user to view data trends. We will discuss the work of Soliman et al. in a later section once we go over some basics of HCI and some user interfaces. The importance of this section was to understand our interpretation of an IoT system and how concepts used in the industrial settings can be re-applied to the consumer settings.

9.1.2 Human–Computer Interaction

The field Human–Computer Interaction (HCI) is an established research field where social and behavioural sciences complement computer and information technologies [10]. There are many research branches within HCI but for the purpose of this chapter we will look at HCI through the computing and information aspect. The origins of HCI date back to the mid-1960s where government funding towards univer-

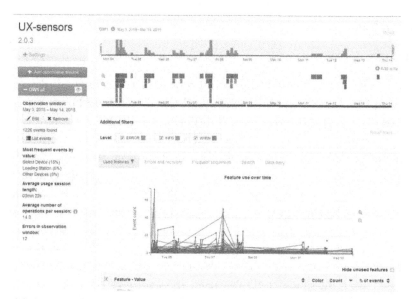

(a) Screen shot of a web interface used to view events from machines at a facility.

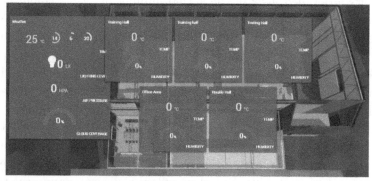

(b) Screen shot of a web interface used to show the values from sensors throughout an industrial facility.

Figure 9.1 These two screen shots depict a HTML application that was used in industrial use-cases to view data coming from machines and sensors located throughout two industrial facilities

sities, government organizations and corporate research labs (e.g., Xerox PARC) allowed scientists and researchers to work freely on developing novel user interfaces [26]. Xerox PARC contributed quite a significant amount of research and products towards the HCI field. Most notably, for this chapter, was the vision of a mobile device by Alan Kay in the 1970s named DynaBook, which resembles today's tablets. Even though there were limitations on implementing DynaBook at the time, Kay and his team went on to develop some key products that would eventually enable the development of DynaBook; these key products were Smalktalk and the Alto GUI [37].

The importance of Smalltalk was that it helped lead the use of object oriented programming that was essential for developing graphical user interfaces [18]. Smalltalk demonstrated how software can be programmed and used in computer systems that can combine various input devices and illustrate the operations of the system in graphical window widgets. The object oriented paradigm nature of Smalltalk influenced the concept of separation of concerns, where a computer system can be broken down into individual modules. The benefit of using individual modules was that the logic is confined to a specific module and didn't impact other modules. The use of modularizing the system to basic components led the way to one of the most widely used frameworks used in graphical user interface programming known as *Model-View-Controller*.

Model-View-Controller was introduced in Smalltalk-80 to satisfy two goals: first was to support the development of highly interactive software and the second was to help programmers create portable interactive graphical applications easily [20]. The *model* provided objects of different classes a way of encapsulating the application domain into easy to manage objects. The *view* provided objects that could be presented visually to the user to show the application's state. Finally, the *controller* provided a mediator between the user's interaction with the *model* and the *view*. Even though the idea of MVC was introduced in the 1970s [33], the concept has been incorporated into many modern day programming languages: for example, C++ utilizing Qt, AngularJS is a JavaScript MVC implementation, Java using Spring MVC and Python using Django. Today, MVC is highly utilized in designing applications on mobile devices such as Android [24] and Apple's iOS [5].

The influence of Smalltalk on the development of graphical user interfaces is one example of how past concepts are still applicable with user interface developments today. Another example that was beneficial to today's style of computer interaction was demonstrated in the 1980s by Richard A. Bolt. The system was called "Put-That-There," which demonstrated how voice and gesture can be combined to produce graphical outputs [7]. To operate the system, the user simply points to the wall-sized screen and gives a verbal command, e.g., "Create a blue square there," and a blue square will appear on the screen where the user was pointing to. Two major technologies were used to accomplish this task. The first technology was a system called the Remote Object Position Attitude Measurement System (ROPAMS) from Polhemus Navigation System used to acquire the space position and orientation. Essentially, the ROPAMS consisted of two cubes that contained three coil antennas in an orthogonal pattern corresponding to x, y and z planes. One cube acts as a transmitter and the other acts as the receiver. The orientation of the transmitter is acquired from the differential signal calculated from the three coils in the receiver. The distance is computed based on the falling signal strength between the transmitter and receiver cubes. The second technology utilised to achieving this task was a DP-100 Connected Speech Recognition System from Nippon Electric Company used for speech recognition. The DP-100 allowed the user to speak small phrases, rather than speaking a single word followed by pauses. The DP-100 is trained by having the user speaking the word once into a microphone. The word is then sampled and saved into the DP-100's active memory. When the user speaks a short phrase, the DP-100 parses the words in the phrase and displays the results. The novelty of this work helped to demonstrate a change from using traditional mouse and keyboard interface to utilizing more seamless interactions with systems.

Research into the field of Automated Speech Recognition (ASR) can be dated as far back as the 1930s by work done by Homer Dudley of Bell Laboratories [19]. From the 1930s to today, a lot of advancements have taken place in the area of ASR, which overlaps the 1980s demo by Richard A. Bolt. The use of statistical models, like Hidden Markov Model (HMM), and the utilization of Machine Learning and Artificial Intelligence has improved the recognition rates of ASR systems. One well known open source project used in ASR research is CMUSphinx (http://cmusphinx.sourceforge.net/). The coming of Sphinx-4 from the CMUSphinx project brings the use of state-of-the-art HMM based speech recognition system [21]. Another open source ASR toolkit is Kaldi (http://kaldi-asr.org/), which was designed to help speech recognition researchers [29]. Kaldi could be coupled with GStreamer, so audio from a device can be sent to a server running Kaldi-GStreamer for remote speech recognition, then the recognized speech could be sent back to the remote device [3].

Gesture based interfaces have evolved over time. Gesture based interface can be broken down into two types: the use of angular and position recording sensors like accelerometers and gyroscopes, and the use of computer vision. Due to the cost of integrated circuits (IC) becoming more affordable and advancements in Very-Large-Scale-Integration (VLSI), not only are ICs becoming more available to researchers and scientists they are also integrating more functionality. Hence, there has been a large growth of off-the-shelf components that can be quickly integrated with microcontrollers to record the sensor data and pass the data onto the system for further processing. Modern game controllers for consoles contain many integrated electronics that help register the orientation, distance and feedback from the user [31]. One great example is the Nintendo Wii Remote. The Nintendo Wii Remote, released in November 2006, looked like a television remote and contained the following electronics: buttons, 3-axis accelerometer, high-resolution high-speed IR camera, speaker, vibration motor and a wireless Bluetooth connectivity component [23]. The price of the controller was US$40 which allowed researchers and scientists access to incorporate the Nintendo Wii Controller in their experiments. There are open source projects only available to allow researchers the opportunity to integrate the Nintendo Wii Remote into their research project. Some examples of these open source projects can be found from the WiiBrew wiki (http://wiibrew.org/wiki/Wiimote/Library) or just entering the term "wiimote SDK" in a search engine.

Computer vision is a field that computationally analyses, modifies or tries to gain a higher level of understanding of images that could be applied to a wide area of application spaces [30]. In our case, computer vision can be applied to developing gesture based user interfaces. OpenCV is a popular open source library used that can be utilized in designing gesture based user interfaces. The application framework for OpenCV usually involves acquiring a frame from a camera, extracting the features from the image, applying the features to a machine learning algorithm and displaying the output in a relevant manner. A comprehensive survey on hand gesture recognition was made by Rautaray and Agrawal [32]. In their survey they broke down each step that other researchers have used in order to detect hand gestures. Essentially, once the image has been acquired the hand needs to be extracted from the image. This can be done by detecting the skin color, searching the image for the shape of the hand, using training classifiers that have been built on data image sets of hands or subtraction of the nonmoving pixels. Once the hand has been detected, then frame-by-frame analysis needs to occur to determine which direction the hand is moving. A simple algorithm that can be used in this stage is to take a series of frames and use the hand detection algorithm used in the previous steps. Locate the center of the hand in the X-Y coordinate space. Those X-Y coordinates can then be used in the machine learning algorithm to determine what gesture was performed. Another popular choice for computer vision gesture recognition is using Microsoft Kinect's API. The Natural User Interface provided from the Kinect API utilizes an infrared camera to detect the user and his joints [25]. The user's joint locations are then mapped onto a skeleton data structure. The developer then utilizes a convenience method to get the location of the joint of interest. By collecting the location of the joints across a series of frames a gesture can be detected.

Tangible user interface is an area that has attracted a lot of attention within the HCI community. The idea behind tangible user interface was published by Hiroshi Ishii and Brygg Ullmer in 1997 [17]. The idea behind the tangible user interface was to create a link between the physical world and cyberspace, essentially moving away from the basic HCI model of humans interacting with the computer through the traditional GUI. By utilizing physical objects, the users were able to create new forms of input to the computer system. The computer would recognize the change of states from these physical objects and render a new view based on the physical objects' state. While this new form of interaction was welcomed, it created some difficulties for some research labs. There would have to be resources allocated from research labs to allocate the electronic components and human resources to assemble these electronic components. Research staff would have to be knowledgeable in integrating the electronics hardware to the computer system. Even if there were kits available, sometimes they didn't have open APIs or expanding functionality made it difficult to integrate into another research product.

A solution to the issues outlined above was addressed by creating products that offered complete modularity in their design and a well-published API. The concept behind Phidgets offers a solution upon which it was easy for developers to focus on the overall project rather than get caught up in the low-level underlines of electronics, in addition to having a product that would allow the average programmer to program and extend the functionalities of their projects [13]. Figure 9.2(a), shows one of the Phidgets I/O board connected to a light sensor. The Phidgets I/O board can then be attached to a computer using the USB cable. An application could then be written to connect to the Phidgets I/O board and read the data coming from the light sensor. If the developer requires a physical rotary interface for their application, they could attach a potentiometer board to the prototype, as shown in Figure 9.2(b), and then into the Phidgets I/O board quickly and update their application code. Phidgets supplies an extensive set of sensors and actuators, in addition to software drivers for many operating systems to communicate with the Phidgets I/O boards, code examples and tutorials on their website (http://www.phidgets.com) that provides

(a) The image shows a Phidgets 1018 I/O board attached to a light sensor

(b) The image shows a Phidgets Rotation sensor, which provides a physical rotary widget that could be used in physical user interfaces. (Permission for image reuse granted by Phidgets Inc.)

Figure 9.2 The image above shows two sensors that could be used by the Phidgets interface board to sense changes in the physical environment

enough knowledge to help developers incorporate physical widgets easily into their applications.

Another innovative device created to help prototype physical user interface is the Arduino. Massimo Banzi wanted an easy-to-use electronics platform that would allow students in interaction design to quickly assemble physical devices to use in prototype testing with users [34]. The Arduino utilizes a microcontroller and an integrated development environment (IDE). The Arduino also comes with an API to assist novice users to write firmware that can interact with various electronic components. The Arduino's hardware has been open sourced so the public can access the schematics and make customizations to the hardware platform. There is also a large community that provides third party components that can make physical prototyping and communication to other devices easier. Sparkfun (http://www.sparkfun.com) and Adafruit (http://www.adafruit.com) are two examples of companies that supply add-on boards and in-depth tutorials on how to use their products with the Arduino.

We started this section looking into the history of the HCI field which allows us to gain a better understanding of how humans can communicate with computers or devices. While GUIs are still dominant in desktop and mobile applications, developing user interfaces for IoT devices should consider different methods, such as those mentioned above. In the earlier days of HCI, the technology was a significant hurdle to overcome. Many ideas were conceived but implementing them was a difficult task due to resource constraints. Given the advances in technology we can seek inspiration from earlier designs and concepts to implement in today's technology. Therefore, a strong link to the past research is highly relevant to developing concepts to solve issues in today's world. GUI programming through MVC and the combination of gesture and speed input for graphical output are just two examples

Figure 9.3 A user interacting with the Nest Thermostat (Permission for image reuse granted by Nest Labs, Inc.)

of user interfaces conceived from early HCI research; there are a few more human computer interaction systems. Virtual reality/augmented reality is one example of a past idea being easily implemented today through a product like Google Cardboard, Oculus Rift, HTC Vive and numerous projector cave implementations. Some early concepts that are very applicable in today's HCI that also relate to IoT are graphical user interfaces, speech and gesture interfaces, and physical user interfaces. The Nest Thermostat (shown in Figure 9.3) is an example where physical user interface and the graphical user interface are combined to help the consumer adjust their room temperature setting to suit their preferences. The consumer can simply turn the face of the thermostat to make a selection or adjust a setting and then press on the dial, confirming the selection or setting.

9.2 HCI METHODOLOGY

In the previous section, we went over some history of the HCI field and tried to bridge the gap to how HCI led to many innovations we are currently using today. Since this chapter is about applying HCI practices to IoT prototyping, previous research will be examined to see how those HCI practices were utilized. The field of IoT is extremely diverse with many services and products to offer to the consumer. There are two approaches that could be taken in developing user interfaces for the consumers. The two approaches are system-centred design and user-centred design. System-centred design focuses on the development of the system without the involvement of the users, whereas user-centred design focuses on creating user profiles and configuring the system to cater to the needs of the profiles [8]. Since the user is the primary focus of this chapter, user-centred design approach provides the best methods to acquire their needs for developing IoT systems. If the IoT systems can be modularized to utilize a MVC framework, this can decouple the developers' dependencies. One group can focus on developing the system and provide software interfaces to the user interface designers. This would allow the user interface designers to quickly change the user interface to meet the needs of the users without requiring the system developers modifying the system code.

User-centered design (UCD) encompasses methods that enables end-users to being more influential on the design of the product or service [1]. During the requirements gathering and usability testing stages, users are solicited to provide input and test the product or service in their environment. This process helps to validate and verify that the correct user requirements have been acquired. Normally, the users are selected based on their expert backgrounds. UCD can also be incorporated into modern software development processes, such as prototyping, spiral, incremental and agile. In a survey amongst UCD practitioners the following methods were commonly used [38]:

- informal usability testing

- user analysis/profiling

- evaluating existing systems

- low-fidelity prototyping

- heuristic evaluation

- task identification

- navigation design

- scenario-based design

Based on previous research studies utilizing case studies for designing user interfaces for industrial applications [36] [15], it was determined that user analysis/profiling, formal/informal usability testing, evaluation of existing systems and scenario-based design were applied with successful outcomes. *User analysis/profiling* are techniques used to isolate who the user of the system will be and their specific needs. Acquiring this data can be achieved from interviewing experts who work in the application domain field, interviewing the users directly and reading previously published use cases in research literature. *Formal/informal usability testing* allows the users to use the system that has been built to see if there are any faults in the system or solicit further requirements the user feels the system should incorporate. *Evaluation of existing systems* can be performed by reading case studies on existing systems and determining new requirements that can be found or implementing an existing system and have the user try it to determine additional requirement needs. *Scenario-based design* allows the development group to describe a fictional scenario of how the user will interact with the overall system; this will enable the development group to generate a persona. The persona could then be used for constructing the system towards that specific user. In the use-cases sections we will apply these HCI practices to help gather user requirements and build systems towards their needs.

9.3 USE CASES

The following use cases are based on the authors' experience in developing full IoT-similar solutions. The first use case was to design a consumer product that would monitor a residential unit, the second use case was to design a smart lighting architecture that could be used by both consumer and retail space owners and the third use case was derived from a research paper which could be applied to the consumers. It is also possible that the first and second use case could be combined with the third use case, in that based on energy consumption patterns the lighting could automatically be adjusted for maximum cost savings by the consumer or retailer through a centralized controller.

9.3.1 Smart Energy Monitoring

One area of growth in providing services in the IoT sector is residential energy management. Smart meters are being equipped to provide the power utility providers access to real-time energy consumption of residential owners. The smart meter is attached to a wireless network that transmits the data from the smart meter back to the utility providers. The consumer can then log onto the power utility providers' web portal to view their usage patterns. In this view, the power utility providers are providing the infrastructure to enable the energy consumption service over a web page; however, we can flip this view where the consumer can be empowered to read their own power consumption which leads to products being created to help the consumer. This could be useful in validating the residential energy consumption with the utility provider's consumption reading; in addition, it allows the user interface to be modified based on the user's requirements.

9.3.1.1 User's Requirements

In this use case, the authors were commissioned in 2013 to develop a full energy-monitoring solution for a client. The energy-monitoring solution was intended to help consumers examine their energy consumption, so they could help determine how to lower their overall energy consumption. The solution needed to collect the data from the smart meter, perform some data analysis and provide the information to the tenant living in the building. The target consumer group were tenants living in apartment buildings that might not have easy access to their smart meters due to the location in which the smart meters had been installed. The client intended to commercialize this solution.

To fully understand the client's need we utilized some of the HCI practices we outlined earlier. First, we performed a **user analysis** to see who all the users would be for the system and their needs. The client had full knowledge of where the prototype testing would occur and the level of technical skills of the tenants living in the building. So we conducted light interviews to gather the types of users for the system. Two main users were identified: the first was the person installing the equipment in the apartments and the second user was the individual apartment owners.

The needs of the installer were as follows:

- equipment must be easy to attach by non-intrusive methods;

- all components must be made from off-the-shelf parts;

- equipment must be expandable from a small apartment size (6 units) to larger complexes (undetermined but more than 12);

- developer API was needed to allow future innovations.

The needs of the individual apartment owners were as follows:

- no personal information about the user can be kept on the system;

- the system must be easily accessible by the apartment owners;

- security measures must be shown to give the user a comfort of security;

- the system must avoid strong technical word usage or imply users have strong technical background;

- the user interface must be clean and show essential data quickly.

9.3.1.2 Hardware Implementation

Following the user analysis, we continued on to **evaluating existing systems** to see if there were lessons we could learn from other systems. There were similar Smart Energy monitoring systems on the market; however, they were either aimed for single user usage or the systems were not able to scale to accommodate a large number of residential dwellings. One exception was the OpenEnergyMonitor project. The OpenEnergyMonitor project is an open source initiative that teaches people how to design software and hardware to monitor their electricity consumption from Smart Meters [22]. The benefit of the OpenEnergyMonitor project was the helpful user forum where users can post problems, solutions, their ideas and educational tutorials.

Based on the information acquired from OpenEnergyMonitor, we consulted the client to get some information on the types of Smart Meters and the access to the wiring of the Smart Meters. To make the installation easy, it was decided that the best approach would be to read the optical output being emitted from the Smart Meter. Some Smart Meters are equipped with a LED light that flashes a pulse once a certain amount of energy is consumed, e.g., 1 impulse per Watt hour (imp/Wh) consumed. Figure 9.4 shows the Smart Meter that was being used at the apartment building of the prototype location. The marker indicates the location of the impulse light being emitted by the meter. To bring this scenario to our interpretation of the IoT, the Smart Meter represents the *thing* or sensors we are acquiring information from.

Figure 9.4 Close-up of the Smart Meter being used at the prototype location site.

After we have received sufficient information of acquiring the data from the Smart Meter, we started a series of **low-fidelity prototyping**. We sketched out some solutions on paper utilizing a computer vision approach that would use a mask to only detect frame changes where the LED lights would be. A quick implementation was made; however, there were some issues since the locations of the Smart Meters undergo variable lighting conditions. In addition, machine learning algorithms would have to be utilized and validated for correctness. This would cause a burden on the installer that would have to find an ideal case for correctly reading the light pulses.

Referring back to the lessons from the OpenEnergyMonitor project, we elected to go with a light-to-voltage converter (TOAS TSL257 [4]) that would be placed over the LED light on the Smart Meter. The TSL257 was wired to an Arduino Due. The Arduino Due was used because it allowed for 54 pins to operate in interrupt mode. So when the TSL257 detects the light pulse, it will create a rising edge that would trigger an interrupt function on the Arduino Due. The interrupt function basically writes the Smart Meter id to the serial port.

The Arduino Due was attached to a Raspberry Pi that performed two main functions. The first function was to store the Smart Meter id and a timestamp of when the light pulse occurred. The second was to upload the stored data to a central server, which was located on campus. At the prototype testing location, a wireless router was installed with a 3G uplink. This represents the first half of the data connectivity portion of the IoT system. A diagram of this hardware set-up can be seen in Figure 9.5.

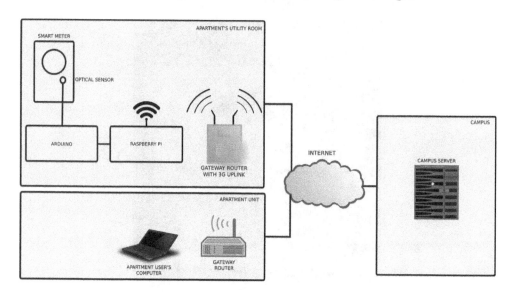

Figure 9.5 This image represents the architecture that was used in the Smart Energy Monitoring use case.

Figure 9.6(a) shows our first prototype that was proposed. The Arduino Due was placed inside a plastic electrical box to prevent debris falling onto the Arduino Due at the test location. Plastic electrical box contained inside a screw terminal block that allowed for quick wire installations for the equipment installer.

The first prototype was installed at the location for early **informal usability testing** on the installer's behalf. Figure 9.7 shows the room with the Smart Meters and the fifth Smart Meter attached to our circuit board. The client recommended that everything should be integrated into a single electrical box to avoid excessive wiring. This led to a second prototype being developed (shown in Figure 9.6(b)).

9.3.1.3 Software Implementation

During the hardware installation aspect of this project, we were simultaneously planning the software aspect of this project. It was decided that a cloud-based architecture would be ideal but a simulated environment would be sufficient for this prototype. There were two areas to consider in the software aspect, transmission of data to the server and reading the data from the server. The Raspberry Pi located at the prototype building location had Rasbian OS installed on it. MongoDB was installed on the Raspberry Pi to provide a cache-style database. This was utilized in the event the 3G uplink was not available for a small duration of time. When the Arduino sends the meter id to the Raspberry Pi through the serial connection, a record is created in MongoDB specifying meter id, a timestamp of the record insertion and the purge column to false. A shell script runs every 5 minutes which checks MongoDB for any rows containing a purge column containing the value false. It then transfers those tuples to the cloud server using a calling URL, which will be explained in the next paragraph, with a timestamp value in the POST data section of the HTTP

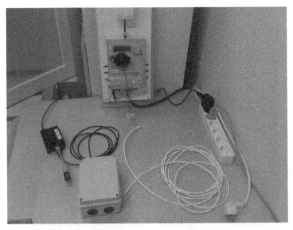

(a) This figure show the first prototype that was developed to be used in the Smart Energy Monitoring system.

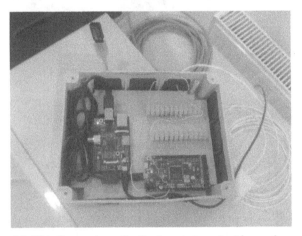

(b) This figure show the final prototype inside an electrical enclousure.

Figure 9.6 This figure shows how we iterated our prototype based on the input of the client.

message. Then it sets the purge value to true. A shell script running every 5 minutes checks for records in MondoDB containing columns where purge is set to true and removes them.

A server on the campus was created to emulate a cloud environment, where API endpoints would be created. This aspect represents our second half of the data connectivity. The server was running Ubuntu Linux 12.04 with Apache, PHP and PostgreSQL. To address security concerns by the residential users, SSL certificates were created and installed on the Raspberry Pis at the installation sites and on the cloud-like server. The REST API endpoints created on the server utilized the MVC methodology. The URL that was used for transmission of data was:

Figure 9.7 This image shows the Smart Meters lined up at the prototype location. The optical sensor was attached from the fifth meter to the Arduino.

https://cloudserver.example.org/iot/test.php/*<site-location-id>*/*<node-id>*/*<meter-id>*

site-location-id: refers to the numerical ID the tenant is living at
node-id: refers to the Raspberry Pi/Arduino node located at the site location
meter-id: refers to the specific tenant's meter id

The *test.php* file contained the logic to parse and validate the correct data coming from the Raspberry Pi nodes at the locations and insert the data into the PostgreSQL database. Our *model* for this IoT system was the timestamps of the LED pulses detected on the Smart Meter.

From the residential user perspective, the *controller* and *view* are handled with the following REST API endpoint:

https://cloudserver.example.org/pmr/scripts/index.php/*<site-location-id>*/*<node-id>*/*<meter-id>*?(chosenDate—week—month)=(yyyy-mm-dd—yyyy-Wmm—yyyy-dd)

Once the user has logged onto the website portal, they were able to use the following URL with a GET call:
https://cloudserver.example.org/pmr/scripts/index.php/1/1/1?chosenDate=2013-06-

26

which would send a JSON response with the following data:

"results": ["hour": 0, "pulses": 11, "hour": 1, "pulses": 10, "hour": 2, "pulses": 11, "hour": 3, "pulses": 10, "hour": 4, "pulses": 11, "hour": 5, "pulses": 10, "hour": 6, "pulses": 11, "hour": 7, "pulses": 10, "hour": 8, "pulses": 11, "hour": 9, "pulses": 10, "hour": 10, "pulses": 11, "hour": 11, "pulses": 10, "hour": 12, "pulses": 11, "hour": 13, "pulses": 10, "hour": 14, "pulses": 11, "hour": 15, "pulses": 10, "hour": 16, "pulses": 11, "hour": 17, "pulses": 10, "hour": 18, "pulses": 11, "hour": 19, "pulses": 10, "hour": 20, "pulses": 11, "hour": 21, "pulses": 10, "hour": 22, "pulses": 11, "hour": 23, "pulses": 11]

The index.php file verifies the user calling the site-location-id, node-id and meter-id data models are authorized. In addition to the standard user id and password log in at the website portal, the index.php file checks to see if the user has the proper SSL certificate to view the data. The query field is then parsed to see what datasets to return. The system was designed to report back three types of datasets with number of pulses in those times windows. The example above used **chosenDate** which shows the number of pulses per hour for the date "2013-06-26" for site-location 1, node-id 1 and meter-id 1. A SQL query is created to collect, group the rows into hours and then perform a summation of the pulses for that time frame. The dataset is then passed on to a *view* where the numeric values are rendered. This represents the data analysis part of our IoT subsystem.

The representation of our information presentation aspect of our IoT system was performed over a web page. The same server that was used for receiving the data from the Raspberry Pi nodes was also used to show the results. The dataset was displayed over a web page. Figure 9.8 shows a web page displaying the results of daily energy for week number 50 in 2013. This dataset would have been created using the following calling URL:

https://cloudserver.example.org/pmr/scripts/index.php/2/1/5?week=2013-512

The importance of Figure 9.8 illustrates how we can take technical data and transform it into a meaningful presentation for the consumer. When the consumer observes this graph, they can gain an understanding of how much energy they have consumed on a given day of the week. Then they will acquire the knowledge to adjust their energy consumption if required.

9.3.1.4 Discussion

The development of an IoT prototype for this project was quite different from the traditional user-centred design approach. When designing prototypes for IoT systems, we can still employ traditional methods but we need to consider the hardware aspects and the software aspects. This requires a strong understanding of the user's needs. In this case study, we identified two major users of the system, which were the installation user and the residential user in a multi-dwelling residential building.

Figure 9.8 A web page showing the week's view of the test user's energy consumption.

Making an overall IoT system required strong insight of the all the users of the system. The client had a lot of knowledge of the users of the system so a series of user-centred design methods were used.

A **user analysis** provided the installer and residential user information that was needed in developing the hardware and software needed for this use-case. An **evaluation of existing systems** was intended to show us what previous problems others have had in this area and how to improve on them; however, at the time of this use case there wasn't a consumer-ready solution for multi-dwelling monitoring. Solutions were either geared towards more industrial applications, soliciting companies for a proprietary solution or solutions were designed for single dwelling purposes. This led us to the OpenEnergyMonitoring project with a lot of advice on how a system could be constructed for our client's needs. An **informal usability testing** was conducted to quickly iterate through what the final prototype would entail. **Low-fidelity prototyping** was briefly used to discuss alternative strategies that could be used but it was evident at a very early stage there were limitations at the install site leading to one solution for acquiring the LED pulses. We conducted a **heuristic evaluation** of the website with the client to see if there were any errors in the web site and how the overall residential users could use the web page; however, with the limited resources these tests could not be completed with the residential users.

The overall IoT system had many subsystems to consider. It was decided in the planning stages that modularizing everything was essential to completing the use case.

Even though we overlapped hardware functionality by using both a Raspberry Pi and an Arduino Due, the purpose was not to set limitations early on in the design stages to avoid a lot of redesigning of the architecture. Today, the cost of 802.11 chipsets and cellular modems are very cost-efficient so that they could be integrated onto an Arduino Due removing the need for the Raspberry Pi and the 3G wireless router. On the other hand, if there is a loss of 3G connectivity then a large buffer would be needed to keep the data during the downtime. From the software side, modularizing helped use decouple a lot of software functionality of the system. We can observe the strong use of the MVC pattern. The PostgreSQL acted as the persistent data store for the pulses from specific Smart Meters. We used the Apache web server as the controller and web pages for the view. With this set-up we can easily make an application for mobile devices to call the URL to request a dataset for a specific time period. Additionally, if there was some aspect of the view the end-users did not like they can create their own application to render a new view. When analysing this use case in terms of applicability to IoT, we can see the four main aspects were utilized. The Smart Meter acted as the sensor, the Raspberry Pi Internet connectivity between the Arduino Due to our campus server provided the data connectivity, the use of SQL queries were utilized for data analysis and the web page provided the information presentation to the users.

We mentioned some other types of user interfaces earlier in the chapter that could be utilized in IoT prototyping. One scenario we should consider are users who want an alternative to going to a web site to view their energy consumption habits. This scenario is quite useful for users that are on time-of-use or tiered pricing rates [11]. Time-of-use enables the user to make more economical decisions concerning their energy consumption. For example, during peak hours of energy consumption on the electrical power grid, users may be charged more for their energy consumption. If the user lowers their energy consumption during high demand times on the electrical power grid, then they lower their costs. Tiered pricing offers the user different rates based on their overall energy consumption within a given time frame. Usually, the first tier pricing for energy consumption is low. When the user crosses over an agreed monthly energy consumption then they are charged at a higher rate.

We can develop a new user interface to take advantage of the data currently being stored on the server. A tangible user interface could be constructed that could be retrofitted over an existing electrical outlet. Figure 9.9 shows a mock-up of the proposed smart outlet which could be installed on a device that consumes a high amount of electricity. The LEDs could be used to indicate to the user what is the current tier zone. The smart socket could query our server to get the total energy consumption for the time frame and display the results through these LEDs. Under normal operations the electrical socket could be disabled if the tier zone is 2 or 3. The user could override the operations of the smart socket by disabling the automatic features which would default the electrical outlet to remain in an on state.

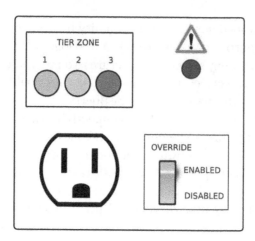

Figure 9.9 This is a mock-up of a smart outlet used to indicate to the user his energy consumption tier zone.

9.3.2 Smart Lighting

Energy conservation has been a major focus of consumers and product providers. As consumers are becoming more conscious of how their energy consumption habits influence many aspects of their lifestyle, they are keen on acquiring products to help adjust and make positive impacts. Many technologies are in the process of being updated to take advantage of modern research and developments. Solid state lighting has now attracted a large audience. Solid state lighting offers more efficiencies compared to compact fluorescent lighting fixtures [16]. In addition, solid state lighting is capable of being controlled by microprocessors, leaving room for researchers, scientists and the home consumer the ability to re-program the behaviour of the light. The addition of the microcontroller also enables these solid state lights to be controlled using a number of wired protocols such as Digital Multiplex (DMX) and Digital Addressable Lighting Interface (DALI). Wired based protocols may be suitable for some locations where the user has access to installation spaces in the ceiling or walls; however, this may be difficult for residential home owners who do not want to renovate their existing place. The advancements of System on a Chip (SoC) with integrated wireless transceivers help offer an easy solution to retrofit existing light bulbs with Smart Light bulbs. These Smart Lights bulbs utilize the electricity coming through the standard electrical bulb socket to power the mircrocontroller and the light bulb. Wireless protocols like Bluetooth, WiFi, or Zigbee can be used to interact with these Smart Light bulbs.

LED lighting is now accessible to the average retail space owner and consumers. While LED lighting can come in multiple form factors, such as light bars and light bulbs, the basic principle in controlling most of them is the same. A popular technique to control a LED light efficiently is through Pulse-Width-Modulation (PWM). PWM can control the intensity of the LED light by adjusting the duty cycle, i.e., the ratio of the time the LED light is on compared to the overall switching period [27]. Since LEDs are solid state devices they can be turned on and off at a quick speed, this allows the LED light to be dimmed. For example, a 100% duty cycle would set the LED light maximum intensity, 50% duty cycle would lower the LEDs' light intensity and 0% would turn the light off. Coloured LED lights are quite often a mixture of red, green and blue LEDs that provide a wide colour spectrum. To control the colour they emit, the intensity of each LED colour is changed.

9.3.2.1 User's Requirements

The origin of this use case started off as a research project in 2014 with certain goals proposed. The first goal was to find a feasible way to install smart lighting into retail spaces to help attract customers. The second goal was to expand the smart lighting system from retail spaces to home consumers, so they can also benefit from the features smart lighting can offer. This use case differs from the Smart Energy Monitoring use case, in that there was no user or clients when the use case started. A system would have to be built first then marketed to prospective retailers and consumers.

Our first step was to perform an **evaluation of existing systems** to see if we could learn lessons based on other automated lighting systems or see if there was a lighting system that could be easily integrated into a smart lighting system. We had a number of solutions to choose based on a combination of previous lighting systems designed in-house [14], consumer available products and a newly designed in-house developer board that was intended for teaching students. With a number of solutions to choose from, the next question we asked was who would be the average user of our system. We needed a strategy to help us identify the needs of the required user to market towards; however, we needed a physical system to show the user in order to solicit further feedback. A **scenario-based design** method was used where a fictional user was created and the events throughout his day was fictionalized. Playing out the scenario helped us gain better knowledge of the type of system we were planning on designing. The knowledge was formalized to create a persona of the intended person that would be using the system. The persona data was handed over to the industry liaison staff member to solicit possible candidates in retail space. The user requirements for retail space owners were as follows:

- lighting equipment must be easy to install

- lighting equipment must be easy to control by tablet or smart phone

- lighting control application must allow individual lights to be grouped

- lighting control application must allow easy-to-program light schedulers

9.3.2.2 Hardware Implementation

Developing the hardware portion of this use case was not as straightforward as the Smart Meter Monitoring use case, mostly due to not having a client who we could interview for some insight. A wide range of hardware devices were analysed to see which would offer the best solution. We narrowed down the decision based on persona user's requirements and easily to source parts. The four options below were considered.

Figure 9.10 This is a DIY kit built in-house to show students how to build a TCP/IP Ethernet LED Light Controller

In the introduction of Smart Lighting, we described how some lights can be controlled by PWM signals. In order to create a PWM signal, microcontrollers are used. Figure 9.10 shows a demo board that was created to show students how to control LED lighting strips. An Arduino Mini Pro is used to create a PWM signal. Due to the voltage differences between the microcontroller being 5 V and the LED strip being 12 V, the PWM pins coming from the Arduino Mini Pro were attached to a NPN-MOSFET transistor. The power supplied to the entire board utilized a 12 V AC-DC wall adapter. There was a switching voltage regulator that efficiently dropped the 12 V to 5 V to drive the microcontroller. This LED demo board was designed to accommodate a common-cathode LED strip. The network connectivity to this device

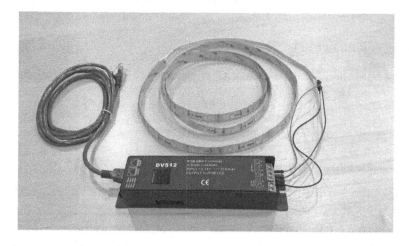

Figure 9.11 This is a DMX LED Lighting controller that can be attached to a DMX-512 network.

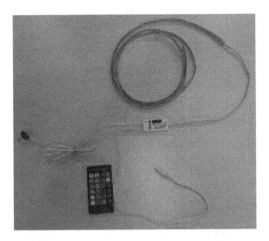

Figure 9.12 This is a store bought LED Lighting controller that can be purchased at any general electronics store.

was handled through an Ethernet NIC; in our case we selected the ENC28J60-H. The concept behind the development of this demo board was to illustrate how a cost-effective DIY kit can be made for consumers that they can assemble and modify for their purposes.

However, while the DIY is of interest for some consumers or retail space owners it does require some knowledge of programming and electronics. An alternative solution for the Smart Lighting system was to look at other consumer options. One product that was reasonably affordable was LED strips powered by IR remote controls. Figure 9.12 shows one product that was considered. This light was adopted from another project where a home automation system was designed; this will be discussed in the next use case. One advantage to this solution was the IR module for controlling the light could be swapped with a different lighting controller, for example a DMX Lighting controller (shown in Figure 9.11). This strategy seemed more adaptable for retail space owners but the cabling required for home consumers may be inconvenient.

Our solution for a Smart Lighting system led to the utilization of the DMX lighting system for retail space owners and the Philips Hue Light bulbs for the consumer users. It could also be possible that the Philips Hue Light bulbs could be utilized in retail space if retailers would be interested.

9.3.2.3 Software Implementation

The design of the software was aimed at accommodating the needs of both the retail space owners and the consumer. So an application needed to be created to perform the needs that were outlined earlier. Unlike the previous use case, the software designed started from the graphical user interface first and then to the server side and hardware abstractions. The focus was trying to find the target user group and letting them get an understanding of the system through the user interface and then amending the hardware side of the project to meet the general needs of the users.

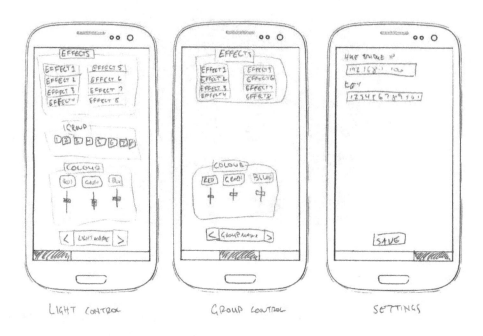

Figure 9.13 An early sketch of the graphical user interface showing the basic functionality.

Before designing the application, we conducted a series of **low fidelity prototyping**. Since the application was targeting mobile devices, we utilized a mobile template upon which to draw our ideas, as shown in Figure 9.13.

Once we felt the graphical user interface encapsulated the user requirements, we proceeded to create a functioning prototype using Apache Cordova [12]. Apache Cordova allowed us to focus on rapidly developing a graphical user interface that allowed easy to change graphical components. Since the graphical user interface was coded in HTML, this made changes to the user interface quick and faster to receive user feedback. Figures 9.14 and 9.15 are a couple of examples of the graphical user interfaces that were the results of the first prototyping. Once the application was created internal staff members went through the application to perform **heuristic evaluation** to see what errors occur. An **informal user testing** was also conducted by internal staff to see how the overall user experience was using the application.

A number of comments were made about the aesthetics of the graphical user interface, most notably the control pane not aligning properly in Figure 9.14. However, the overall functionality was suitable. Two retail space owners were solicited and invited to the lab where they could try out the graphical user interface on a tablet and further discuss their needs. One retail space owner made some remarks about the GUI, in which they would like a pre-defined zone and quick control of the lights in that zone. In addition, they would like to use the lighting automation offered by some of the DMX light fixtures. The second retail space owner was more open to having light fixtures installed in their display cases and have an easy to use controller to control the lighting and program groups of lights.

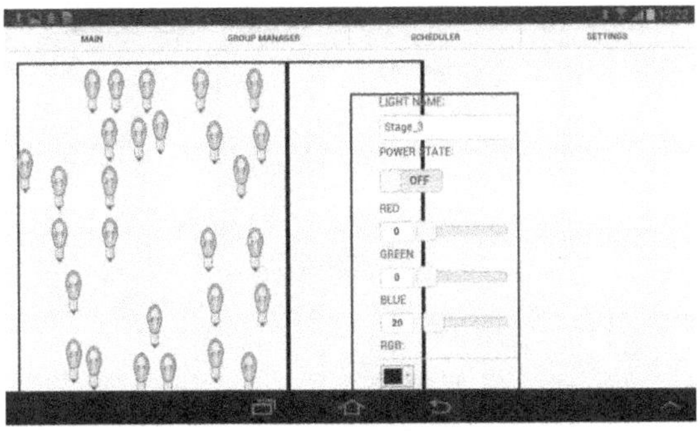

Figure 9.14 The screen shot shows the first working prototype on the tablet. The user can simply tap on the light they wish to adjust.

It was clear that retail space owner 1 wanted a pragmatic approach to the light installation since the DMX lights involved more cost for the infrastructure and elected to go with a simplified user interface. Figure 9.16 illustrates a user interface conceived that would address the needs of retail space owner 1.

Retail space owner 2 was more in line with our consumer persona and the internal staff. The opinions expressed by retail space owner 2 and the internal staff were taken into consideration and implemented into the second to last prototype before being natively coded for the Android device (as shown in Figure 9.17(a) and 9.17(b)). The final **informal user testing** by internal staff seem to favour the new aesthetics GUI. The Haptic feedback allowing them to know that the device has interpreted a "tap," which indicates turn the light on in Figure 9.17(a), or a "tap and hold," which indicates go to a new screen to set light colour in Figure 9.17(b), was also appreciated.

The server side application was split into two different parts. The retail space owner 1 utilized a DMX smart lighting system for their needs. The retail space owner 2 and our vision of the consumer could utilize the Philips Hue lighting system for their needs. We utilized the ready-made DMX Smart Lighting system that was developed in-house [14]. Some modifications were made to the in-house lighting system to

Figure 9.15 Another screen shot of the first working prototype showing the group management screen.

Figure 9.16 This screen shot of the tablet application shows a simple remote control style of interaction to change the lighting in certain zones.

(a) This is a screen shot of the final mobile application used to control the lights.

(b) In this screen shot, after the user has selected the light of interest, a colour wheel shows where the user can adjust the light colour settings.

Figure 9.17 These two screen shots show the final product that was offered to retail space owners and consumers

accommodate JSON data being sent from the Android mobile device to the lighting system. The JSON data contained a sequence of lights to turn on, turn off, adjust the colours and pan and tilt the spotlights. The lighting system contained a defined set of DMX lights in an XML file that was specifically tailored to the needs of retail space owner 1.

For retail space owner 2 and consumers, Philips provides an easy to use API that can be incorporated into mobile applications. When the first implementation of the consumer application was created, we used Apache Cordova to create the user interface. Since the interface was written in HTML, JavaScript was used for the logic handling of the application. The Philips Hue bridge offered a REST API that allowed our application to communicate with the Philips Hue bridge. This helped enforce the MVC framework that has been persistent throughout this chapter. The models are stored on the Philips Hue Bridge and the view is abstracted through the REST API calls. When a call is made to the Philips Hue, a JSON data structure is returned back. The JavaScript is then used to parse the JSON and then render a new view based on the parsed JSON data. For example, the top of our application is dynamically created based on the number of light bulbs the Philip Hue Bridge has detected. If the user adds more lights then that view is automatically updated on the next screen load to reflect the new light being added.

9.3.2.4 Discussion

It can be observed quite easily that designing a product first then marketing it to the consumers (or retail space owners) required a different approach than our first use case. We had to think of what the needs of the user would be first before we could materialize a product for them. To help generate the needs we used two HCI methods, which were **evaluation of existing systems** and **scenario-based design**. The **evaluation of existing systems** allowed us to see what work has already been done by others in the field and what were the needs they were addressing. This helped us to quickly isolate what the hardware aspect would entail but we still needed to find out who the target user would be. Through discussion among the group, we used **scenario-based design** to figure out what a typical day in the life of the user would be and how they would interact with the light. This helped us construct the user's requirements and then develop a couple of solutions to help in the consumer and retail space owner customer acquisition.

Developing prototypes for IoT, from our point of view, is more user oriented than the traditional system oriented design. There are a lot of lighting solutions available for the consumer. The consumer has a lot more choices to select from. To stand out from other products on the market the designer of the IoT products needs to be willing to engage more closely with the user to develop specifically for the user needs. Having an open API like Philips Hue and many other product suppliers can create a wide reach for their products and continue on the development process by utilizing third parties. There are dozens of applications available on the Google Play Store and iTunes that utilize the Philips Hue. Open source hardware products also have their own appeal to the DIY crowd and others that wish to know how electronic consumer products work. In the next use case, we will go over a system that was developed as a base for creating a DIY home automation system or base system for IoT devices.

The popularity of Smart Lighting and open APIs has allowed developers to integrate other user interfaces to help users control their lighting systems. In most scenarios, the user interacts with the lighting system through a mobile application running a graphical user interface, as was illustrated in this use case. Reaching for a mobile device to control the lights may not always be ideal for the user, though. New products available to the consumer in the form of wireless speaker systems (e.g., Amazon Echo and Google Home) are equipped with microphones that are capable of detecting user speech and perform operations based on the speech input. Those products also offer consumers a way of controlling various smart lighting systems. Researchers can replicate a similar set-up in their labs by utilizing Kaldi with the gstreamer plugin. Once Kaldi has been trained to recognize speech then an application on a mobile device can record the user speech and send it to the Kaldi server. The Kaldi server can then convert the spoken speech to text. Based on the text, the Kaldi server could then send data packets to a lighting server (e.g., Philips Hue Bridge or our lighting architecture).

9.3.3 Seamless Home Automation

Home automation has been in existence for over a couple of decades. Many home solutions are available for the consumer to purchase and install in their homes. The X10 system was one of the earliest home automation systems, which was created in the 1970s and marketed towards the consumer. X10 utilised the residential electrical power system as a communications bus to communicate between the X10 controller and the sensors and other X10 equipped devices such as dimmers, outlet receptacles and appliance switches [39]. As hardware has become affordable and more open specifications become available, home automation products are now becoming easier to integrate into the home environment. There has even been a push from wired based technologies to wireless technologies like Bluetooth, WiFi, Z-Wave and ZigBee.

Even though the devices to enable home automation have been around for decades, there have been barriers for general acceptance by the consumers. In research published in 2011 [9], four barriers to mass adoption were outlined based on the user experience of 14 participants. These barriers were cost of ownership, inflexibility, poor manageability and difficulty of achieving security. For the scope of this paper, we will consider the poor manageability barrier. The research paper indicated four issues within the poor manageability barrier that needs to be address.

The first issue was an iteration approach was required in the researched participants' home automation set-ups. The participants usually had to redo their initial installation once they became acquainted with the installed functionality or a change in lifestyle conditions (e.g., another occupant being added to the household). So it was vital that the overall home automation solution must be designed for changes in the environment rather than treated as a one-time install. The second issue raised questions about the system being reliable. Participants indicated that they typically noticed unpredictable behaviour within the system. Rule-based automation systems seem to have created errors which led to difficulties debugging the issue, so the participants either lived with the bugged rule set or disabled the rule. There was also a sense of frustration among the participants when it was not known if the system was responding to their commands. This left a level of uncertainty if the overall system was working. The third issue related to the management and troubleshooting of the home automation system. The participants in the research were divided into either the DIY group or outsourced group. The DIY group took care of their own management and troubleshooting of the home automation system, whereas the outsourced group relied on professional consultants to help resolve issues. The outsourced group discussed some downsides of delegating the task of management and troubleshooting to the consultants. They mentioned being uncomfortable not being able to resolve their own issues, inability to customize various aspects of the system and lack of password management. For this paper, we were interested in the fourth issue brought up in the research study which dealt with complexity of user interfaces in home automation systems. Eight of the participants mentioned that the user interface was one aspect of the home automation system they disliked. They either had a hard time explaining how to use the user interface to others or they had difficulties in learning the user interface themselves.

9.3.3.1 User's Requirements

In this use case, the authors implemented an open source home automation platform in 2012. The goal was to design a system to address the concerns mentioned in the above section pertaining to user interface issues with home automation systems. Researchers could then evaluate new types of user interaction that might alleviate the issues surrounding hard-to-use user interfaces utilizing our open source home automation platform. We derived the consumer's requirements by performing a **user analysis** based on the needs of the users in that research paper. The needs of the users were as follows:

- home automation system must be modular in both hardware and software design

- all components must be made from off-the-shelf parts or sourced easily

- the home automation system must be intuitive for design

- the user should feel empowered that they can use the system or have some guidance where to find help

Those user requirements were used to implement a home automation system where the researchers can focus on resolving user interface issues. To further develop the system we needed a typical scenario a user would encounter when using a home automation system. The **scenario-based design** method was used to come up with the following scenario.

"The user wakes up and sets the lights and television to his preference. As he leaves the house, the house detects he is no longer there and goes into power efficiency mode

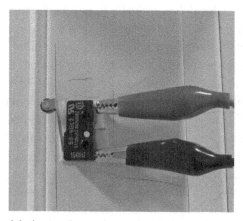

(a) A simple mechanical switch to detect when the door opens.

(b) A servo motor mounted over a light switch, which was used to turn the lights on and off in the room.

Figure 9.18 The switch and servo acted as sensor and actuator that was attached to the Arduino.

by automatically turning off the lights and television. The user returns from work and the system automatically turns on the lights and television. The user will then use their mobile device to adjust the light and change the television channel."

This scenario will serve as a basis to design the hardware and software. By interpreting this scenario we can see that we will need a system that can record all the statuses of the devices in the room. The user will interact with the room through a mobile device; therefore, a GUI will be needed on the mobile device.

9.3.3.2 Hardware Implementation

In an effort to make the home automation system such that it can easily be replicated in any lab, we utilized commercial off-the-shelf parts. We obtained a wireless home router that contained USB ports, a Z-Wave transceiver, WiFi and regular Ethernet jacks. A USB Bluetooth dongle was plugged into the wireless home router to allow communication to Bluetooth devices. The Arduino Duemilanove was utilized since it had a FTDI chipset that was easily detectable by the Linux kernel. An IR transmitter, basic switch, servo motor and an Adafruit RFID/NFC reader [2] were added to the Arduino. The Arduino was attached to the wireless home router by a USB cable.

Attached to the door frame is a basic switch (Figure 9.18(a)). The switch is then plugged into the Arduino on an interrupt pin. When the door opens and presses down on the switch, the circuit closes which creates a rising edge. This rising edge is detected by the Arduino which sends a command to the wireless router to perform a Bluetooth scan of the room for mobile devices.

The IR transmitter on the Arduino was used to control the television and IR RGB light strip. The intention behind this was to develop a subsystem where researchers can work on developing new ways of interacting with home entertainment systems, as

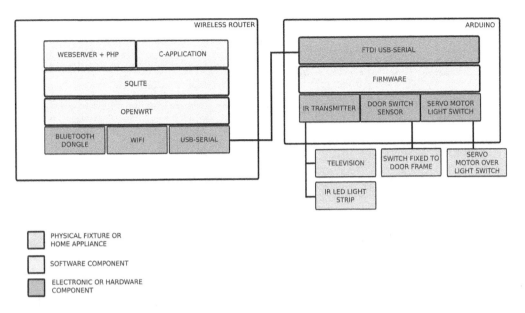

Figure 9.19 This figure represents the architecture that was designed for this use case.

most home entertainment systems utilise an IR based remote controller. We used the servo motor and attached it over the main light switch, as shown in Figure 9.18(b), since we didn't have access to the electrical wiring in the space to control the main lights in the room.

An overview of the architecture can be seen in Figure 9.19. Even though this hardware configuration was relatively simple to put together it does offer tremendous opportunities for user interface researchers to explore and evaluate new user interface interaction techniques.

9.3.3.3 Software Implementation

In most home automation systems, there is a centralized system responsible for the management of devices and sensors. The wireless home router in our set-up served as the centralized system. The wireless home router had OpenWRT Linux distribution installed on it. This provided us an easy way to install packages that would be needed to help develop software for our scenario. There was a C code application created which was used to communicate with the Arduino controller and the USB Bluetooth dongle. The C code application also created a Linux pipe to listen for commands coming through the web server. A web server on OpenWRT with PHP was used to serve interactive web pages to the user (as shown in Figure 9.20 and 9.21). When the user pressed the button on the HTML page, the JavaScript sent the request to the PHP script which opened up the Linux pipe to write the request. The C code application reads the Linux pipe to interpret the code. Based on the code, an associated ASCII code is sent to the Arduino over the USB port. During the process of sending the ASCII code to the Arduino, the command and user id is logged into a

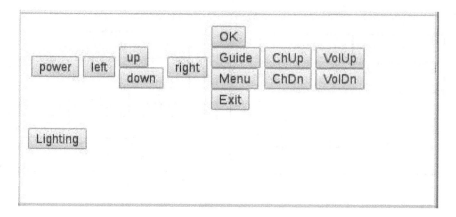

Figure 9.20 A quick HTML mock-up to allow the user to use their mobile phone to control the TV and the RGB lighting strip in the room.

SQLite database. This keeps track of the user's request which can then be used later for pattern analysis.

The Arduino was programmed to detect the door being opened and the interrupt was mapped to a function that would send an ASCII code to the USB port. The Arduino was also programmed to receive ASCII code over the USB port. Each ASCII code was mapped to transmit a specific IR command. For example, if the Arduino received an "0" then it would send the IR command to turn on the television. When the door was open the "z" ASCII code was sent through the USB port to the router.

Going back to our scenario, when the user wakes up and sets the television and lighting to their preference, the user will use the HTML pages served by the WiFi router. This data will get stored in the SQLite database. When the user leaves his apartment, the door triggers the switch which sends a command to the wireless router by the Arduino. This causes the wireless router to initiate a Bluetooth scan which will not detect the user since he has left. Due to no user being found the lights and

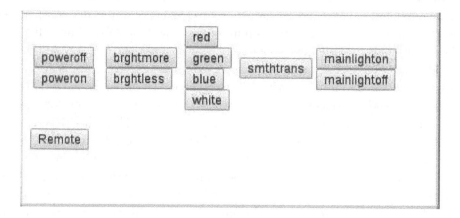

Figure 9.21 Another quick HTML mock-up to allow the user to adjust the light settings on the RGB lighting strip.

television will turn off, assuming the user wants the house to go into power saving mode.

When the user returns home, he will open the door, which triggers the switch. This causes the wireless router to perform a scan of the room and find the user's cell phone and Bluetooth ID. The C application will then search the SQLite database for the last entry of the television and lighting. The C application will then send the last IR codes from the SQLite database to the Arduino. The Arduino will then perform those actions. Essentially, the user will have seamless interaction with his appliances.

Even though the web pages used in this use case are very simple in nature, they serve for demonstrative purposes of how to interact with the system. The views of the web page can always be altered to make them more aesthetically pleasing while all the underlining infrastructure can remain unchanged. If the researcher wants to add a new element into the home he can easily attach it to the existing infrastructure. Most modern IoT devices coming to market are utilizing WiFi, Bluetooth, Zigbee or Z-wave so a developer can quickly expand on the existing open platform.

9.3.3.4 Discussion

This use case tried to develop a platform to develop user interfaces for the home automation sector. The platform was designed to allow user interface research labs to quickly set up their own home automation environment and design more comprehensive and clear user interfaces that would appeal to the consumer. We used a mobile phone and a web page to control the interaction in the room based on the technology available at the time. As the prices have come down considerably and the advancement in mobile hardware has increased tremendously, this has enabled more styles of user interaction to take place. For example, automatic speech recognition APIs are now available that can be installed on mobile device or cloud-based solutions to be utilized to control various devices in the home.

Today, many products are now coming to the market that a consumer can purchase which are ubiquitous in nature. There will be more use cases in which these products will need to work together to form seamless solutions for the user. In our scenario, a combination of easy-to-acquire electronics was combined to detect when a user is present and augment his environment to his last set preference. In the case of IoT, this scenario also fulfils our four fundamental aspects as we have the sensors and actuators. The data connectivity was achieved through Bluetooth and WiFi connections. There wasn't any real particular data analysis but the logged events of the user could be utilized for a predictive system. The information presentation was presented to the user by having the home perform a function the user requested.

At the end of the IoT section (Section 9.1.1) in the introduction, we briefly mentioned a research paper by Soliman et al. [35]. The architecture proposed in the paper is in line with the current REST API solution offered by many IoT devices. Philips Hue and Nest are two products that offer the developers a similar way of interacting with their devices over a REST API. We initially set out to design our home automation platform with the intended location of the router being near the home entertainment console, which is why the approach was to integrate everything into the home router. An alternative approach would be to decouple the sensors and actuators, and provide a REST API to interact with the devices. For example, the wireless router could be hidden from the view of the users and a small microcontroller equipped with a WiFi transceiver and IR transmitter could be placed in the entertainment console. Instead of running cables from the door to the Arduino, a wireless microcontroller combination utilizing WiFi could be installed. To communicate with all the devices a REST API using a JSON structure has become the most commonly used method for device communication. This can be easily reconfigured into our platform since we are already utilizing a web server and a database back-end for keeping persistent data.

The use of a REST API can also allow other user interfaces to communicate with the home router. For example, a web camera, a microphone and a single board computer could be attached to the bezel of any television screen. The web camera could acquire images of the user sitting in front of the television. The images could then be processed by computer vision algorithms to determine what kind of gestures the user has performed. The interpreted gesture could then be encapsulated into a HTTP message and sent to the wireless router. The wireless router would then send the IR command to the television to perform an action. For example, if the user wishes to switch the channel on the television they would either do a swipe gesture to the left or to the right. The microphone attached to the television could also be used for speech recognition, so the single board computer could perform the speech recognition and send a data packet to the wireless router.

Since the other two use cases utilize the Internet for data communication, the wireless home router could be utilized to perform data analysis from the smart energy monitoring server and send a message in the form of visual light through the Smart Lighting network. For example, if the user is about to cross over from the first tier pricing to the more expensive tier pricing, then a signal could be sent in the form of a colour code through the light. The lights could also dim down to utilise less power. The light can now provide information to the user, which is one goal of user interfaces. In a sense, this starts to form a basis of converting our home automation system into an IoT hub.

9.4 CONCLUSION

In our use cases, we tried to demonstrate some HCI practices that could be applied to developing prototypes for the Internet of Things. User-centred design can be applied to the IoT sector when acquiring user requirements from the consumer. With a wide range of products available to consumers, they have much more choices than previously, so a product must stand out to cater to their needs or another product will be selected. This is crucial if developers are competing for the same market share. The Smart Energy Monitoring use case involved interaction with a client to help create a full IoT solution. The user requirements were straightforward to acquire since we utilized user-centred design methods and had direct contact with a knowledge client. We then followed up with a series of prototypes and heuristic testing of the final product.

The Smart Lighting use case was a little bit challenging considering there was no immediate client, which made the solicitation of user requirements difficult. Through an evaluation of existing Smart Lighting systems and creating a scenario of the typical user, we were able to create an initial system that was used to demonstrate our Smart Lighting product. This attracted two retail space owners who provided more input on how they felt the system should operate. This led to two final products that could be implemented in actual retail spaces or used by consumers.

The final use case was derived from previous research publications related to consumers utilising home automation systems in their homes. We showed how we developed a basic platform that could be used to address the complexities involved in the design of user interfaces. This system was designed to be evolving to accommodate newer technologies and test their feasibility in implementing new intuitive user interfaces.

When it came to developing prototypes for these use cases, it was clear an iterative approach was needed. Given all the tools available, we favoured doing early mock-ups of user interfaces utilising low-fidelity means. This allowed us to avoid being caught up on the constraints of software packages. Then we transferred these low-fidelity mock-ups to HTML which allowed us to conduct early informal user testing. There are many tools now available that enable WYSIWYG style of layout offered, so developers can quickly move widgets around to meet the needs of the user. Once the graphical user interface has been agreed on then it is transferred to native mobile applications. One interesting thing to note, while there exist many forms of user interfaces, as mentioned in the HCI section of this chapter, a common default still seems to be a graphical user interface. One reason is that a properly developed graphical user interface can quickly give the user insight into the systems operation. For example, a smart thermostat can quickly show the room's temperature in numerical fonts and the background colour of the numerical fonts could represent the colour of the load demand on the power grid. The user can quickly see if he needs to adjust the temperature in his house to reduce their cost in electricity consumption.

We started this chapter off by looking into the history of HCI and deriving key elements that are still in existence today. The MVC framework is still heavily utilised in developing user interfaces, since it forces an entire system to be modularised. We looked at how past technologies used for gesture and speech recognition were used and how today we have APIs that encapsulate a lot of low level functionalities. This permits researchers an opportunity to quickly incorporate more modalities into their systems. Researchers and scientists should consider looking at past works in HCI as there are many other interesting concepts in research papers.

Bibliography

[1] C. Abras, D. Maloney-Krichmar, and J. Preece. User-centered design. *Bainbridge, W. Encyclopedia of Human-Computer Interaction*. Thousand Oaks: Sage Publications, 37(4):445–456, 2004.

[2] Adafruit. Adafruit pn532 nfc/rfid controller shield for Arduino + extras, 2016. (Access on July 21, 2016.) https://www.adafruit.com/products/789.

[3] T. Alumäe. Full-duplex speech-to-text system for Estonian, 2014.

[4] ams AG. Tsl257 light-to-voltage, 2016. (Access on July 15, 2016.) http://ams.com/eng/Products/Light-Sensors/Light-to-Voltage/TSL257.

[5] Apple Inc. *Model-View-Controller*, 2015. (Access on July 11, 2016.) https://developer.apple.com/library/ios/documentation/General/Conceptual/DevPedia-CocoaCore/MVC.html.

[6] L. Atzori, A. Iera, and G. Morabito. The internet of things: A survey. *Computer Networks*, 54(15):2787–2805, 2010.

[7] R. A. Bolt. "put-that-there": Voice and gesture at the graphics interface. In *Proceedings of the 7th Annual Conference on Computer Graphics and Interactive Techniques*, SIGGRAPH '80, pages 262–270, New York, NY, USA, 1980. ACM.

[8] B. Boussemart and S. Giroux. Tangible user interfaces for cognitive assistance. In *Advanced Information Networking and Applications Workshops, 2007, AINAW '07. 21st International Conference on*, volume 2, pages 852–857, May 2007.

[9] A. J. B. Brush, B. Lee, R. Mahajan, S. Agarwal, S. Saroiu, and C. Dixon. Home automation in the wild: Challenges and opportunities. In *Proceedings of the SIGCHI Conference on Human Factors in Computing Systems*, CHI '11, pages 2115–2124, New York, NY, USA, 2011. ACM.

[10] J. M. Carroll, editor. *HCI Models, Theories, and Frameworks: Toward a Multidisciplinary Science*. Morgan Kaufmann Publishers Inc., San Francisco, CA, USA, 2003.

[11] Toronto Hydro Corporation. Residential electricity rates, 2016. (Access on July 29, 2016.) http://www.torontohydro.com/sites/electricsystem/residential/rates/Pages/resirates.aspx.

[12] The Apache Software Foundation. Apache Cordova, 2016. (Access on July 24, 2016.) https://cordova.apache.org/.

[13] S. Greenberg and C. Fitchett. Phidgets: Easy development of physical interfaces through physical widgets. In *Proceedings of the 14th Annual ACM Symposium on User Interface Software and Technology*, UIST '01, pages 209–218, New York, NY, USA, 2001. ACM.

[14] J. Hakulinen, M. Turunen, and T. Heimonen. Spatial control framework for interactive lighting. In *Proceedings of International Conference on Making Sense of Converging Media*, AcademicMindTrek '13, pages 59:59–59:66, New York, NY, USA, 2013. ACM.

[15] T. Heimonen, J. Hakulinen, S. Sharma, M. Turunen, L. Lehtikunnas, and H. Paunonen. Multimodal interaction in process control rooms: Are we there yet? In *Proceedings of the 5th ACM International Symposium on Pervasive Displays*, PerDis '16, pages 20–32, New York, NY, USA, 2016. ACM.

[16] C. J. Humphreys. Solid-state lighting. *MRS Bulletin*, 33:459–470, 4 2008.

[17] H. Ishii and B. Ullmer. Tangible bits: Towards seamless interfaces between people, bits and atoms. In *Proceedings of the ACM SIGCHI Conference on Human Factors in Computing Systems*, CHI '97, pages 234–241, New York, NY, USA, 1997. ACM.

[18] J. Johnson, T. L. Roberts, W. Verplank, D. C. Smith, C. H. Irby, M. Beard, and K. Mackey. The Xerox star: A retrospective. *Computer*, 22(9):11–26, Sept 1989.

[19] Juang. Automatic speech recognition âĂŞ a brief history of the technology development. 2004. [online <http://www.ece.ucsb.edu/faculty/Rabiner/ece259/Reprints/354_LALI-ASRHistory-final-10-8.pdf> accessed 19. May 2015].

[20] G. E. Krasner and S. T. Pope. A cookbook for using the model-view controller user interface paradigm in smalltalk-80. *J. Object Oriented Program.*, 1(3):26–49, August 1988.

[21] P. Lamere, P. Kwok, W. Walker, E. GouvÅła, Rita Singh, Bhiksha Raj, and Peter Wolf. Design of the cmu sphinx-4 decoder. In *8th European Conf. on Speech Communication and Technology (Eurospeech)*, 2003.

[22] T. Lea, G. Hudson, S. Tagore, C. A. Gabizó, K. Boak, and A. Zayani. Open energy monitor, 2014. (Access on July 15, 2016.) https://openenergymonitor.org/emon/.

[23] J. C. Lee. Hacking the Nintendo Wii remote. *IEEE Pervasive Computing*, 7(3):39–45, July 2008.

[24] K. McCullen. An android application development class. *J. Comput. Sci. Coll.*, 31(6):11–17, June 2016.

[25] Microsoft. Skeletal tracking, 2016. (Access on July 24, 2016.) https://msdn.microsoft.com/en-us/library/hh973074.aspx.

[26] B. A. Myers. A brief history of human-computer interaction technology. *Interactions*, 5(2):44–54, March 1998.

[27] P. Narra and D. S. Zinger. An effective led dimming approach. In *Industry Applications Conference, 2004. 39th IAS Annual Meeting. Conference Record of the 2004 IEEE*, volume 3, pages 1671–1676, vol.3, Oct 2004.

[28] R. Piyare. Internet of things: Ubiquitous home control and monitoring system using android based smart phone. *International Journal of Internet of Things*, 2(1):5–11, 2013.

[29] D. Povey, A. Ghoshal, G. Boulianne, L. Burget, O. Glembek, N. Goel, M. Hannemann, P. Motlicek, Y. Qian, P. Schwarz, J. Silovsky, G. Stemmer, and K. Vesely. The kaldi speech recognition toolkit. In *IEEE 2011 Workshop on Automatic Speech Recognition and Understanding*. IEEE Signal Processing Society, December 2011. IEEE Catalog No.: CFP11SRW-USB.

[30] K. Pulli, A. Baksheev, K. Kornyakov, and V. Eruhimov. Real-time computer vision with opencv. *Commun. ACM*, 55(6):61–69, June 2012.

[31] S. N. Purkayastha, N. Eckenstein, M. D. Byrne, and M. K. O'Malley. Analysis and comparison of low cost gaming controllers for motion analysis. In *2010 IEEE/ASME International Conference on Advanced Intelligent Mechatronics*, pages 353–360, July 2010.

[32] S. S. Rautaray and A. Agrawal. Vision based hand gesture recognition for human computer interaction: A survey. *Artificial Intelligence Review*, 43(1):1–54, 2015.

[33] T. Mikjel H. Reenskaug. The original MVC reports. 1979.

[34] C. Severance. Massimo Banzi: Building Arduino. *Computer*, 47(1):11–12, Jan 2014.

[35] M. Soliman, T. Abiodun, T. Hamouda, J. Zhou, and C. H. Lung. Smart home: Integrating internet of things with web services and cloud computing. In *2013 IEEE 5th International Conference on Cloud Computing Technology and Science*, volume 2, pages 317–320, Dec 2013.

[36] M. Turunen, H. Kuoppala, S. Kangas, J. Hella, T. Miettinen, T. Heimonen, T. Keskinen, J. Hakulinen, and R. Raisamo. Mobile interaction with elevators: Improving people flow in complex buildings. In *Proceedings of International Conference on Making Sense of Converging Media*, AcademicMindTrek '13, pages 43:43–43:50, New York, NY, USA, 2013. ACM.

[37] R. Vertegaal and I. Poupyrev. Organic user interfaces. *Communications of the ACM*, 51(6):26–30, 2008.

[38] K. Vredenburg, J.-Y. Mao, P. W. Smith, and T. Carey. A survey of user-centered design practice. In *Proceedings of the SIGCHI Conference on Human Factors in Computing Systems*, CHI '02, pages 471–478, New York, NY, USA, 2002. ACM.

[39] C. Withanage, R. Ashok, C. Yuen, and K. Otto. A comparison of the popular home automation technologies. In *2014 IEEE Innovative Smart Grid Technologies - Asia (ISGT ASIA)*, pages 600–605, May 2014.

CHAPTER 10

Inclusive Product Interfaces for the Future: Automotive, Aerospace, IoT and Inclusion Design

Patrick M. Langdon

Engineering Design Centre, Engineering Department, University of Cambridge, Cambridge U.K.

Email: pml24@cam.ac.uk

CONTENTS

T He Internet of Things (IoT) is increasingly generating opportunities to connect diverse sensors and on-device processors to provide novel information service to the user. Additionally, the engineering of human machine interfaces in automotive and aerospace domains has, until recently, involved adding new technologies into the sensory context of the human operator. Consequently, improvements and enhancements in available control and display technologies have led to an increase in the requirements for the amount and diversity of information that can be made available to people in Human Machine Interfaces (HMI) in IoT applications/devices, cars and aircrafts. Turn this situation on its head for a moment and consider the design of such interfaces for the older population or those with functionally reduced capability ranges. Here the user of the interface is arguably extraordinary [40, 43, 51]; they may require adaption of their interaction to accommodate the range of their perceptual, cognitive or physical movement capabilities. Simply put, recent developments in research on such interfaces, however from diverse domains, may be usefully applicable to the situational impairments resulting from challenging HMI contexts in IoT.

This chapter addresses the key relevant aspects of inclusive design practices of knowledge interfaces and provides exemplars from four case studies. In Case 1, some of the key issues above are addressed in a design project for a major automotive manufacturer, including the use of inclusive user profiling and multimodal interaction design. In Case 2, Future Aerospace, possible areas of mitigation of the effects of situational impairment are examined in the light of the same techniques. In the third case study, the *Predictive Pointing* technology, which was directly derived from accessibility research, is discussed. It effectively combines gesture tracking and suitable signal processing algorithms in a system that predicts and selects screen items based on tracked free hand movement in 3D. Finally, in Case 4 for mobile applications, we look at how the general principles of inclusive design can be applied in technology to produce a truly personalised interface adapted to the capabilities of the user and making use of the latest technology for interaction: multimodal adaption. Taken together, these domains illustrate the convergence of HMI issues with related considerations, which have originated in design for accessibility and inclusion, and point towards a new transdisciplinary approach for the future, especially with the proliferation of the IoT devices and services.

10.1 THE BACKGROUND TO THE PROBLEMS

In an automotive context, early responses to the timing and physical challenges of driving led to the use of specific strategies to avoid the driver directing attention away from driving task. These include single button radio and single knob selection strategies. Implementation of such novel systems introduces control crowding and is inevitably accompanied by extensive learning requirements. Increasing use of configurable multi-function displays and integrated information displays has mitigated this to some extent. However, this practice has unfortunately led to a proliferation of such non-essential displays resulting in potentially dangerous driver overload. These problems are particularly acute for older or reduced capability drivers, presenting challenges that arise as a result of reaching the limitations of current HMI understanding [35, 37].

10.1.1 The Ubiquity of IoT Technology, and Importance of Inclusive Design

Because of the demands of new, information-heavy HMI and the need for increased convergence of information from context to the human, there is a danger of exceeding the individual's capability to achieve effective Situation Awareness (SA) [25, 26]. This is to say that there may be failures to perceive, comprehend and predict the results of required actions. This is becoming more important as the interconnectivity of engineering elements and context increases. For example, in the future it will be possible for an autonomous car to connect to other cars in its vicinity; to roadside and highway infrastructure; and also to cloud-based processing or information repositories. Applications could include avoidance systems for traffic; the behaviour of automated cars in concert; the provision of real-time parking information and booking of parking, and so on. Indeed, some applications of IoT opportunities are actually created by new markets opened up by connectivity. For example: the provision of vehicles to a user, as and when they are required, on a door-to-door basis; or the use of vehicles for other purposes during traditional "down-time" such as commuter parking [1]. Many of these potential connections are being realised, but this also presents problems of representation of information to the user and of integration of the right information for the current human goal. The thesis outlined in this chapter is that HMI development resulting from the understanding of Situationally Induced Impairment is essentially equivalent to the health impairments resulting from ageing or disease [37, 54]. Inclusive design has successfully addressed the latter and is therefore applicable to the former.

Despite the proliferation of technologies that may be implemented in a modern control situation in automotive or aerospace interactions it should be stated that many, such as eye-trackers, Head-Up Displays (HUDs), augmented reality displays, automated control and dialogue systems, internet of things connectivity, have yet to find effective application in practice and products. One reason for this is the lack of understanding of human acceptance of such novel interactions and their synchronising with human capabilities. Currently, the data are not (easily) available to make design decisions regarding preferred modalities or multimodalities of interaction, e.g., in automotive contexts [62]. Another issue is that general technologies such as IoT connectivity between vehicle, mobile, static, infrastructure and engineering systems have such a huge and combinatorial potential that it is not uncommon to suppose that a plethora of possible opportunities exists or that arbitrary connections are possible [33]. Further, the extrapolation of existing technologies to future systems is unlikely to be successful as many as yet unrealised new technologies cannot be taken into account. For these reasons, it should be noted that the research described in this chapter is intended to illustrate possible future scenarios without being predictive of the details.

10.1.2   What Is the Need for Inclusion?

The field of inclusive design relates the capabilities of the population to the design of products by better characterising the use-product relationship. Inclusion refers to the quantitative relationship between the demand made by design features and the capability ranges of users who may be excluded from use of the product because of those features. By 2020, almost half the adult population in the UK will be over 50, with the over 80s being the most rapidly growing sector (see Figure 10.1). These "inclusive" populations contain a great variation in sensory, cognitive and physical user capabilities, particularly when non-age-related impairments are taken into account. Establishing the requirement of end users is intrinsically linked to the *user centered design* process. In particular, a requirements specification is an important part of defining and planning the variables to be varied and measured and the technology use cases to be addressed during the interactions during user trials.

Inclusive design is a user-centred approach that examines designed product features with particular attention to the functional demands they make on the perceptual, thinking and physical capabilities of diverse users, particularly those with reduced capabilities and ageing. It is known, for example, that cognitive capabilities such as verbal and visuospatial IQ show gradually decreasing performance with ageing. Attending to goal-relevant, task features and inhibiting irrelevant ones is important in interaction and this is known to be affected by ageing. Attentional resources may also be reduced by ageing, such that more mistakes are made during divided attention, dual task situations [43, 47, 53].

Another related design approach that embraces social context is that of universal design. This design approach advocates designing specific products so that they are usable for the widest possible range of capabilities. For example, in an architectural context, buildings should be suitable for all possible end users regardless of functional capabilities, age or social contexts [44]. A further category is that of ordinary and extraordinary design that aims to improve design for older, impaired users of low functionality while at the same time enhancing design for the mainstream and ordinary users in extreme environments [40]. On this basis, design should focus on the extraor-

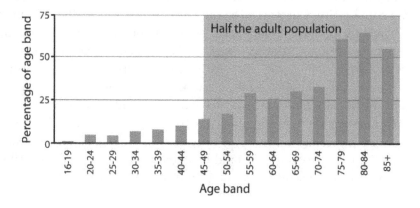

Figure 10.1   Almost half the adult population will be over 50 by 2020.

dinary or impaired first, accommodating mainstream design in the process [40,41,51]. Not all functional disabilities result from ageing. Some common examples of non-age-related impairment include specific conditions such as stroke and head injury, which may affect any or all of perception, memory and movement. Other conditions are generally associated with movement impairment. For example, Parkinson's disease and cerebral palsy involve damage to the brain causing effects such as tremor, spasms, dynamic coordination difficulties and language and speech production impairment. Of course, many other conditions such as Down's syndrome and multiple sclerosis (MS) may affect cognitive capability either directly, through language learning and use or indirectly through its effects on hearing, speech production and writing.

Of all the variations discussed, many differentially affect normal population ranges of capability. They may be rapidly changing and vary in intensity both within and between individuals, leading to a demanding design environment that requires close attention to conflicting user requirements and a better understanding of user capability. Again, this confirms that interaction design for future generations of products must be inclusive.

10.1.3 The Inclusive Design Response

An inclusive design approach offers the opportunity to counteract this trend by making designs accommodate a greater diversity and range of population, by including a wider range of functional capabilities and by better accommodating variation within and between different individuals and groups. Inclusive design aims to make products and services accessible to the widest range of users possible irrespective of impairment, age or capability. To do this, a substantial research effort has been directed towards developing the underlying theory and practice of design analysis in order to develop and provide tools and guidance to designers that they can use to improve the inclusion of a resulting product [23,34,36].

10.1.4 Health Induced and Situationally Induced Impairment

Not all disability arises from ageing or health issues. Sears *et al.* in [54] formulate the concept of Situationally Induced Impairments and Disabilities (SIID) as an example of how widening the boundaries of context in context-aware interaction can impact on the concept of disability, particularly within situations where users interact with ubiquitous, ambient computing environments. They describe how the loss of a limb or impaired hearing can lead to difficulties in using computers but point out that this can also occur as a result of a driving situation or a noisy environment or even interacting whilst walking. Equally, work environments may be disabling, as when interaction is required in a moving ambulance, or in the constrained, physically demanding environment of the construction industry [54]. The applications constrain the task, for example through available modalities, while the environment proscribes physical conditions. The human dimension requires the characterisation of the user, in terms of functional capability, affect and social context. This being the case, designing for disability is quite closely related to designing for situational impairment.

10.2 ADVANCED INTERACTION INTERFACES

Improvements and enhancements in available control and display technologies have led to an increase in the requirements for the amount and diversity of information that can be made available to drivers in cars. A number of approaches, stemming from the recent incursion of methods and theories of cognitive psychology into human factors, are currently in use to address issues of mitigation of information overload and mismatch of displays and controls to task workload or multitasking requirements. It has long been known that complex control tasks combined with situation awareness in tasks such as driving or piloting aircraft place considerable demands on cognitive resources. Under conditions of high control demand, for example, we may select aspects of our environment to attend to, ignoring other important information or we may be distracted by the unconscious focus of our attentional system to unwanted stimuli such as alarms or display clutter [66]. Human attentional capabilities have been likened to that of a searchlight [27]. It is possible to direct it at certain perceptual events at different times in a time-sharing way but the degree of spread of the beam varies and may include other unwanted stimuli whose processing is mandatory. Redundancy gain has been demonstrated where the perceptual detection of a signal is enhanced by the presence of an accompanying redundant signal in a different modality. However, there may be response conflict [39], as well as redundancy gain, as a result of mandatory cognitive processing of two channels that are perceptually close. When complementary responses are required to a display (turn left at letter X; turn R at letter H), performance may be enhanced but when incompatible (turn L at letter X; brake at letter H), performance can be impeded [69]. When the colour of letters does not match semantic targets in a Stroop task (e.g., searching for the ink colour of the word red in blue letters), detection performance is impaired [61].

Multiple scalar variables can sometimes be coded as integral dimensions in displays and the resulting object appears as a single "emergent" percept (e.g., [65]). This idea forms part of the basis of Ecological Interface Design [19]. However, multiple dimensions can alternatively result in distraction if the spatial proximity is low or the variables are poorly related through the task, or if the combined display fails to form a coherent object [67, 70].

The cross-modality redundancy gain discussed above suffers from visual dominance such that visual information is attended to when conflicting with haptic information, where haptic refers to active and passive touch and kinaesthetic body sense. Auditory and haptic channels have been thought to be better suited to warning interruptions when vision is dominant. However, they may be effective for control when vision is not dominant. Furthermore, they may be well suited to analogue-spatial judgements. It is known, for example, that head-up displays can facilitate parallel processing when representing well-learned elements such as runway indicators but inhibit responses when a dangerous situation such as a runway incursion is unexpected [38]. This research has shown that poor choice of psychological dimensions or perceptual primitives can lead to attentional tunnelling problems whereby pilots find it difficult to unlock their attention from display elements to consider other more critical display elements.

More recent research into "cross-modal" interfaces has developed the modal concept experimentally. For example, Ho and Spence [31], in the context of attentional limitations in driving, review recent evidence supporting the Multiple Resource theory [66–68]. They challenge the account of independent resources separated into pathways (e.g., the visual-spatial-manual vs. the auditory-verbal-vocal) instead citing recent evidence that integration of sensory modalities is a norm in human cognition [56,57]. Experimentally, for example, there is evidence that sensory efficiency is greater if different modalities are coming from the same spatial direction. Van Erp *et al.* in [64] tested a tactile torso display effectively conveying directional and distance information simultaneously during a complex control task such as driving.

10.2.1   The State of the Art

Other academic issues that have emerged as implicit in the use of new technologies for integrated and situational displays and interfaces include 1) modelling interface design using advanced cognitive modelling of perception, executive function and working memory (e.g., [66,68,70]), and 2) situation awareness and mental workload in automation of high-workload interfaces and the problems with disengagement of automated systems for human operators during tasks [9,45,58]. Understanding and predicting human performance in complex systems is proving to be a huge and ever-advancing research field. However, the new approaches discussed here focus on the key design elements of multimodality, inclusion and user modelling, rather than on automation, and uses these as a focussing point for the assessment of new technology within this domain.

10.2.2   Solutions and Issues with User Modelling

Cognitive architectures model the uncertainty of human behaviour in detail but they are complex and not easily accessible to interface designers. For example, the ACT-R architecture models the content of a long-term memory in the form of a semantic network. Researchers have already attempted to combine the GOMS family of models and cognitive architectures to develop more usable and accurate models [8]. Numerous other models since have tried combining simple front ends to complex underlying architectures [15,52]. Recently, using the principles of inclusive interaction from product design [36,37] the requirement has been an inclusive user model that can model perception, cognition and motor action in more detail than the GOMS model, while being easier to use than cognitive architecture based models. A successful contribution here has been that of [70] who demonstrate the value of ACT-R and network analysis approaches can be effectively used to predict the usability of multi-branched screens on mobile device interfaces.

Another such approach addresses the design of integrated multimodal display and control technologies for ease of input and task completion. Initially implemented in the domain of better design for elderly and impaired computer and TV users, (GUIDE) [29], this work is directly transferrable to the domain of the Situationally Impaired Interface Disability users (SIID) as proposed by Sears *et al.* [54] and in the form of extraordinary user interfaces [42,51].

User models can also be used to personalise or adapt the design of user interfaces based on the perceptual, cognitive and motor capabilities of specific users and task situation or context. Very little research work has explored this beyond academic studies of user interaction with simple interfaces or content personalization. The SUPPLE project at University of Washington [28] personalises interfaces by changing layout and font size for people with visual and motor impairment for mobile devices, while the AVANTI project [60] provides a multimedia web browser for people with movement impairments and low vision. However, a future interaction system may require a more complete approach to displays using multiple modalities and interaction devices such as direct voice command input, head tracking or eye-gaze detection, and touch screens. Additionally, in a modern control situation the operator can be overloaded with perceptual, cognitive and motor tasks involving monitoring the external situation, listening to speech, decision making and operating the controls of the vehicle. Work in the GUIDE project [29] reviewed recent trends in multimodal data fusion and fission for interactive digital TV and combined multimodality with a user model resulting in a novel, research-based, multimodal adaptive software framework [11, 14, 21, 22, 29]. This software framework allows the design of user profiles based on interface adaptation in terms of both layout modification, such as font size, colour contrast and button spacing, along with appropriate modality selection based on user capabilities and immediate context.

10.3 THE INCLUSIVE ADAPTION APPROACH

The research area particularly addressed here originates in the domain of better design for elderly and impaired computer users. This approach assumes that any human user can be impaired (disabled) in their effectiveness by characteristics of their environment, the task and the design of the user interface they are presented with [54]. Such impairment may take the form of perceptual, cognitive and physical movement functional limitations that then translate into inability. It can arise out of capability limitation or from excessive demands of new technology interfaces. For example, attempting to use a SMS text editor while driving a vehicle in a road environment presents difficulties in perceiving the interfaces for both tasks and also in performing the cognitive and attentional tasks necessary to safely carry out the required separate tasks. Requirements include switching attention between tasks; tracking, monitoring and correcting vehicle movement; and carrying out the correct semantic task. This is especially true if you include the disruption of the physical movement demanded by the designs of the physical control interface (pointing, pressing, steering and braking). Importantly, an Inclusive design approach extends beyond the scope of conventional Usability methods as it must accommodate extremes of capability range that are not normally accommodated by product design. For this reason the approach is well suited to the design of automotive future HMI.

10.4   CASE STUDY 1: FUTURE AUTOMOTIVE

Vehicle technologies have been progressing rapidly, and for some manufacturers, human machine interfaces have incrementally evolved to offer the range of control and displays needed to use some complex technology. Other manufacturers have adopted step changes in HMI technology, such as BMW with the i-drive system, which have not always met with universal customer acceptance due to the long and steep learning curves required to reach reasonable levels of performance [16]. Advanced Driver Assistance Systems (ADAS) functions such as Adaptive Cruise Control (ACC), satellite navigation, lane departure warning system, lane changing control, traffic sign reading and collision avoidance systems all vie for the driver's attention and create opportunities for potentially dangerous errors. This is despite their goal of reducing the burden of the driving task alongside the demands of conventional in-vehicle systems such as climate control and infotainment. Current developments suggest that automation level 3 (Conditional automation) will be commercially available within 5 years, and Level 4 (High automation) by 2020; BASt (German Federal Highways Research Institute, 2013).

The technology innovations currently available within the HMI systems, such as dial controls, soft keys, joysticks, gestures, touchscreens, voice command and so on, have often been incrementally implemented, leading to proliferation in the complexity of the interface itself. This is compounded by the invention of new functions resulting from the introduction of autonomous features. An inclusive design approach offers the opportunity to counteract this trend by making designs accommodate a greater diversity and range of population by including a wider range of functional capabilities and by better accommodating variation within and between different individuals and groups.

Against the backdrop of advancing technological possibility is a dramatically ageing population customer base in the developed world, and a growing younger and more technology accepting customer base within developing countries. The HMI requirements of a younger technophile user are generally easier to satisfy with the march of technology, but the older driver may desire something that is familiar, requiring less learning and sensory ability to use, as well as less distracting to use while driving. The effects on the proportion of the population with less than "full ability" with age can be calculated with exclusion auditing techniques that are capable of evaluating interface demands against capability in the likely user population.

Increasing use of computerised reconfigurable LCD multi-function displays has the benefits of providing reconfigurable graphic, image and textual information on the same physical panel. Early use of such displays often simply replicate existing physical instrument displays. However, this also brings the option of designing displays that integrate different sources of information into single screens and allow a potentially infinite number of display possibilities, potentially on a touch screen control basis.

Many existing in-car display and controls are not unified or harmonised in terms of coordinate systems or representations of vehicle or environment, leaving the driver to perform complex cognitive tasks. The effectiveness of integrated display-controls is highly dependent of how they interact with human cognition particularly in the domain of perception, attention, working (short term) memory and movement control. Also, the existence of better displays has opened up the operational possibility of increasing driver activity requirements and therefore workload and creates the potential for overloads of capability, particularly when tasks are unfamiliar.

A key approach to task overload has been increased automation of tasks (adaptive cruise control) and warning systems. However, although successful in simple implementations this has produced a raft of problems that have been shown to generate errors and affect safety [9, 45, 58]. In particular, operator awareness of system status on failure or deactivation of automated systems is poor and situation awareness in driving can suffer [25, 26]. Current technology development offers new modalities of input, such as HUD display, speech control, eye-tracking and passive brain computer interfaces. Extrapolating further, augmented reality, natural language understanding, multi-touch gestures (2D) and gestural (3D) inputs are now technically feasible along with novel output modalities such as speech, avatar and haptic displays (force touch and vibration). It may be possible to develop new approaches that can utilise the understanding of cognitive science to better facilitate driver situational awareness, especially when multitasking is required [55]. For the purposes of this chapter we will focus mainly on the technologies that are likely to emerge in the near future: in around 5–10 years.

10.4.1 Key Future HMI Design Elements

There are several fundamental axial technology concepts behind potential new designs:

Connectivity: An "anything anywhere", internet-of-things environment where 4/5G (instantaneous) data connections are both possible and normal between individuals, organisations, vehicles, vehicle systems and components, buildings, personal products and the family household. In short, the future usage cases can maintain considerable context information [33].

Integration: The entire HMI system is integrated into one unified computational and display "device". This not only adapts the complexity of the HMI displays and controls but also reduces cognitive load and bounds attentional parameters [66].

Legacy Controls: Specific areas of the vehicle's cabin are custom designed to maintain actual "legacy" physical controls such as knobs and gear selector lever. These are mechatronically constructed to deploy in the context of specific surfaces but for some user profiles will be flush with the "reflective" touch-based surface [59].

Adaption: The cars seats are adjustable with many more degrees of freedom than in present designs and pedals and other key controls such as steering wheel are multiadjustable with x, y, and z configurations [14].

Head-Up Displays: The primary head-up display is infinity focussed and aligned with the visual field of the driver. Most of the information displayed on this is the minimal essential driving information. However, wider areas of the visual field are usable, including areas around the dashboard and edges of the windscreen. These areas can display other screens' "objects" as a continuation of the car's display surfaces. The front passenger may experience more but this will be invisible to the driver [30, 50].

User Profiling: User profile data is collected as soon as a new user is detected by the vehicle. Over time the dataset will capture the history of the user, their driving style and preferences, their anthropometry and their abilities, so that the vehicles will be able to better predict their needs and wants using a user model approach and machine learning [11, 14, 46].

Multimodal Interaction: Allows input of commands and outputs in a number of modalities [48]. Likely input and output modalities in future automotive applications are listed in Table 10.1.

Cognitive Load and Attentional Control: The car's integrated display interface maintains an intelligent monitor of user actions in relation to road conditions and other contextual information (road conditions, weather, infotainment status, passenger activity, modes and levels). Alarms will be displayed on a master alarm in the driver's horizontal line of sight and central console display areas which activate and introduce the full relevant object panels when interrogated by the driver. All alarms operate in multiple modalities and are configurable to the individual [36, 49, 67].

Autonomous and Automated Driving: There has been an accelerated speculation regarding development of technology for autonomous driving that is partly due to OEM's airing concepts for commercial competition, and partly due to the deployment of new technologies (radar, lidar, camera-based scene recognition) in road-going vehicles. We recall that current developments suggest that automation level 3 (Conditional automation/Highly automated) will be widely commercially available within 3 years, and Level 4 (High automation/Driverless) within 10 years [63]. Definitions vary but the approximate narrative definitions of these levels are taken from the following National Highway Traffic Safety Administration (NHTSA) document [63]:

- *Level 3—Limited Self-Driving Automation:* Vehicles at this level of automation enable the driver to cede full control of all safety-critical functions under certain traffic or environmental conditions and in those conditions to rely heavily on the vehicle to monitor for changes in those conditions requiring transition back to driver control. The driver is expected to be available for occasional control, but with sufficiently comfortable transition time.

- *Level 4—Full Self-Driving Automation:* The vehicle is designed to perform all safety-critical driving functions and monitor roadway conditions for an entire trip. Such a design anticipates that the driver will provide destination or navigation input, but is not expected to be available for control at any time during the trip. This includes both occupied and unoccupied vehicles. By design, safe operation rests solely on the automated vehicle system.

Inclusive Design: Using a standard multimodal paradigm, all functions may be accessed using any combination of modalities. For example, the user may select a tangible object from the menu by grabbing it using touch pinch, and may manipulate it this way. At the same time or alternatively they may use eye gaze and continuous speech sentences in dialogue with the car to achieve the same end. Outputs will be simultaneously in multiple modalities but this will be configured for workload and context and for individual user's preference profile or impairments. The inclusive

Table 10.1 Likely input and output modalities in future automotive applications

MULTIMODAL INPUTS
Conventional 2D layered and tabbed visual screens with touch controls, multitouch control with a range of multitouch gestures (swipes, pinches, circles, two hand, multi-finger)
3D pointing and gestures including body/head, hands and fingers. These can be localised over displays or more spatially separated.
Speech based. Either command recognition hierarchy or a speech interaction dialogue using speech understanding. The latter can be effectively used as part of an interaction with an artificial personality.
Eye gaze tracking and limb and head tracking. Eye gaze and limb and head tracking is possible for all occupants continuously to allow any individual to interact with the car. The driver has a reserved super-user status. This allows the car to maintain communication from users' body language. The driver has a reserved super-user status.
The key and key fob is available but interaction with users is possible by means of their personal mobile devices.
Tangible interaction. The car seats and most controls are tangible in that in adjustment mode they can be pulled or manipulated for comfort. Mechanical response in instant and silent. The key car display surfaces maintain the available options for the driver and passenger as a "tangible" object, many that are experienced as a set of objects on the central console and right of driver display "regions". These display objects may be selected and moved by surface or 3D gesture to one of the main display areas including the HUD spaces on the windscreen. This is intelligently managed by the car which will prevent dangerous configurations.

MULTIMODAL OUTPUTS
There are 3D stereo screen displays with multilayer outputs and directional bias towards the normal to the viewer. Passenger and driver can, of course, see differing output.
Car speech interface: the car maintains intelligent dialogue with the occupants. This is managed using intelligent speech understanding. Text to speech is transparent if required for all displays including primary instruments. Advanced (3rd party) software converts web pages to auditory modality (audio description) if required. The car maintains a constant intelligent awareness of available actions/menus and these are also visible if required. The car can present itself as an animated artificial person (avatar) and this can be configured for gender, age, cultural style. Avatar profiles of famous personalities are available.
Car sound interface: all car display sounds are surround sound directionally managed. This is configurable to accommodate impairment (see Inclusive Design below).
Haptic: all surfaces and controls generate physically appropriate haptic sensations, such as vibrations or sensations of movement, solidity, inertia etc. Some warnings are haptic but located on the seat, focused controls or area.

system will maintain a profile of perceptual, cognitive and physical capability for each user that will offer the optimal interface for their capability ranges. This will be decided using a user model based on data unobtrusively obtained from the user's interactions during an initialisation or sign-in program for new users. The configuration of controls and interfaces is linked to the system such that different generational

GHF3

Figure 10.2 A future scenario used in design ethnography.

groups may be offered individually configured interfaces based on their generational "cohort" era vehicles. Learning of newer interface components will thereafter be managed on an offering and acceptance basis [37].

10.4.2 Visualisation of Key Concepts

In order to realise the impact of both advanced HMI and human centred Inclusive interaction design such approaches utilise usage cases derived from considering specific personas as potential users. These usage cases often then detail explicit interactions and their associated future HMI concepts. The *personas* are created from marketing information and by use of *design ethnography*; the use of anthropological, ethnographic and qualitative research techniques to establish rich and detailed data on real-life experiences [20,24,35]. Such approaches often originate in inclusive design where the focus is on ageing and combating design exclusion [10]. However, since the technology predicted from the extrapolations has not yet been realised or is in development, visualisation is required. Key outputs therefore include drawings of entire scenario task elements (Figure 10.2) and drawings of specific interaction concepts (Figure 10.3).

10.5 CASE STUDY 2: FUTURE AEROSPACE

Importantly, an Inclusive design approach extends beyond the scope of conventional usability methods as it must accommodate extremes of capability range or situational contexts of task or stress that are not normally accommodated by product design. For this reason, the approach is well suited to the human centred design of military aircraft interfaces. However, a future cockpit control system demands a more complex type of interface adaptation as it is likely to be equipped with multiple modalities of displays and interaction devices such as direct voice input, head

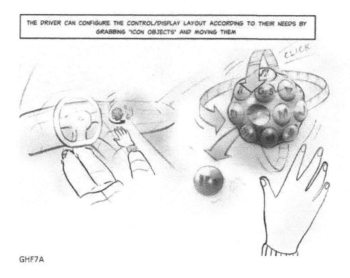

THE DRIVER CAN CONFIGURE THE CONTROL/DISPLAY LAYOUT ACCORDING TO THEIR NEEDS BY GRABBING "ICON OBJECTS" AND MOVING THEM

CLICK

GHF7A

Figure 10.3 A specific interaction concept based on a usage case [35].

tracker, helmet mounted displays, HOTAS (Hands-On Throttle and Stick) controls and so on. This leads to higher workload and the necessity for more careful use of multiple modalities of interactions.

10.5.1 Need for Multimodal Solutions

Individual flight instruments require considerable monitoring and, in order to integrate information across instruments, a substantial and continuous cognitive effort. This has been found to affect pilot performance on primary tasks such as control of the aircraft or physical operation of controls and is particularly relevant during IMC or instrument flying conditions. Functional overloading of physical input devices such as stick and throttle is vulnerable to modality errors as single controls (e.g., mini-joystick or buttons) may be used for very different functions during different mission phases. During conditions of physiological stress and high cognitive workload, attentional tunnelling may occur or perseveration of incorrect hypotheses regarding the display due to confusion or misinterpretation of perception, leading to errors.

Increasing use of computerised reconfigurable "glass" cockpits with LCD multi-function displays has the benefits of providing reconfigurable graphic, image and textual information on the same physical panel. Early use of such displays often simply replicate existing physical instrument displays. However, these cockpits also bring the possibility of designing displays that integrate different sources and timings of information into single representations (weather radar, navigational maps, engine status) and allow a potentially infinite number of display possibilities. This has, however, created a new problems associated with design decisions on the nature of the information to be represented and the graphical, visual spatial and coordinate system choice or representation. Current technology development offers new modalities of input such as helmet-based displays, touch screen, speech command, eye-tracking (see Figure 10.4), brain computer interfaces and data gloves, but these are not be-

Figure 10.4 Hands-free pointing using head-tracking or eye-gaze can be included as a possible interaction modality for aircraft cockpit multi-function displays.

ing exploited to their full potential. Extrapolating further, multi-touch gestures (2D) and gestural (3D) inputs are technically feasible along with novel output modalities such as speech, or natural language, and haptic displays (force, touch and vibration). Remote future projections might include direct brain control or input/output interfaces embedded into the pilot's body. Combinations of these separate modalities, so-called multimodal interfaces, offer the possibility of unique advantages and economies of interface but only if they are acceptable or matched to the human perceptual and cognitive system of the pilot. Currently the development of design patterns and practices for achieving improved performance and ease of use is highly dependent on subtleties of human cognition that are not well understood [31, 66]. A number of approaches, stemming from the recent incursion of methods and theories of cognitive psychology into human factors, are currently in use to address issues of mitigation of information overload and mismatch of displays and controls to task workload or multitasking requirements. Because the proposed studies involve measuring skilled human performance during aerospace tasks the principal research tool requires a method of presenting a task; usually some form of simulator, combined with methods for recording human behaviour and performance. The intention is to capture the difference between performance measures as various elements of the task and simulated environments are manipulated. Measures can be taken; including error at task, deviation from path, reaction times and task timing, workload, stress, usability. Also, ground truth measures record human cognitive and physical parameters as a direct indication of the effects of usage of specific features, displays and controls, as with heart rate variations, Galvanic skin response and other indicators of stress. Research questions generally address the efficacy and salience of various modality combinations.

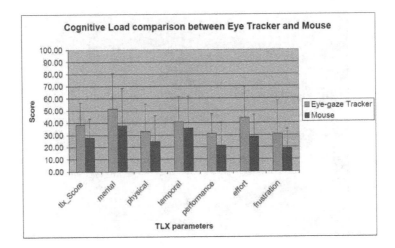

Figure 10.5 Shows workload measures for trials of pointing and selection trials using eye-tracking or traditional mouse.

10.5.2 Multimodal Interface Experiments

In such multimodality experiments direct experimentation uses novel multimodal input and control interfaces, examining novel interaction techniques combining modalities such as eye-gaze with speech. For example, there is a requirement for pilots to point at and select objects on the multi-function displays in the cockpit. Presently this is managed using soft-keys and a mouse but a multi-modal approach introduces the possibility of another modality such as eye-gaze pointing. Standard tests of workload include the NASA TLX subjective scales. In a recent study of multimodal interfaces for Indian agricultural applications, as can be seen from Figure 10.5, there is clear evidence of workload improvement for the eye-gaze over the mouse modality of interaction [14]. Promising results in an aerospace context utilise the same approach.

10.6 CASE STUDY 3: PREDICTIVE POINTING IN AUTOMOTIVE TOUCH SCREENS

Predictive pointing enables realising smart interfaces, which are capable of inferring the user intent, early in the pointing task, and accordingly assisting the on-display target acquisition (pointing and selection). The objective of the predictive pointing system is to minimise the cognitive, visual and physical effort associated with acquiring an interface component when the user input is perturbed due to a situational impairment, for example, to aid drivers in selecting icons on a display in a moving car via free hand pointing gestures.

Interactive displays, such as touchscreens, are becoming an integrated part of the car environment due to the additional design flexibilities they offer (e.g., combined display-interaction-platform-feedback module whose interface can be adapted to the context of use through a reconfigurable Graphical User Interface GUI) and their ability to present large quantities of information associated with In-Vehicle Infotainment

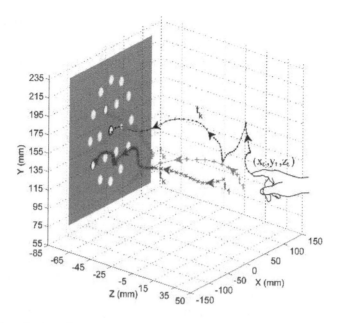

Figure 10.6 Full pointing fingertip trajectories in 3D during three pointing gestures aimed at selecting a GUI item (circles) on the in-vehicle touchscreen surface (blue plane), whilst the car is driven over a harsh terrain with severe perturbations present [4]. Arrows indicate the direction of travel over time, starting at $t_1 < t_k$.

Systems IVIS [17, 18]. Using an in-car display typically entails undertaking a free hand pointing gesture to select an on-screen GUI icon. Whilst this input modality is intuitive, especially for novice users, it requires dedicating a considerable amount of attention (visual, cognitive and physical) that can be otherwise available for driving [32]. Additionally, the user pointing gesture and input on the display can be subject to in-vehicle accelerations and vibrations due to the road and driving conditions (see Figure 10.6), which can lead to erroneous selections [3, 32].

Predictive interactive displays, proposed in [2, 7], utilise a gesture tracker to capture, in real-time, the pointing hand/finger locations in 3D in conjunction with an appropriate probabilistic destination inference algorithm. It can establish the icon the user intends to select on the display, remarkably early in the free hand pointing gesture, and in the presence of perturbations due to road and driving conditions, i.e., SIID.

Predictive displays can notably improve the usability of in-car interactive displays by reducing distractions and workload associated with using them. The Bayesian formulation of the fundamental problem of intent inference, see ([6, 7]) enables the predictive displays to effectively handle varying levels and types of present SIID-originated perturbations and user pointing behaviour, as well as incorporating additional sensory or contextual data when available.

The free hand pointing gesture movements towards an on-screen item in 3D are not deterministic, but are rather governed by a complex motor system subject to numerous physical constraints. The gesture can also be subjected to external motion, jolting, rolling, or acceleration (e.g., in a moving platform). Nonetheless, stochastic models can capture the inherent uncertainty in the pointing finger movements, albeit being driven by intent. This implies that predictions of the pointing object motion are not single deterministic paths, but are rather probabilistic processes, with the pointing finger position at a future time expressed as a probability distribution in space. By adequately incorporating this uncertainty, relatively simple models of the pointing finger motion can be used successfully to track finger movements and evaluate the corresponding observation likelihoods. It is emphasised that the objective of a predictive pointing is not to accurately model the complex human motor (pointing) system. Formulating approximate pointing motion models that enable determining the on-display endpoint (i.e., intent) of a free hand pointing gesture suffices.

Figures 10.7 and 10.8 depict the results of utilising an in-vehicle predictive display under varying levels of SIID due to road and driving conditions when the predictive capability is off and on. In the former case, the experiment becomes a conventional task of interacting with an in-car touchscreen where the user has to physically touch the intended icon on the screen to select it. The benefits of the predictive display are assessed in terms of the system ability to reduce the workload of interacting with the in-car touchscreen and the pointing tasks' durations. Whilst here a Leap Motion controller is employed to produce, in real-time, the locations of the pointing finger in 3D, the predictive display auto-selects the intended on-screen icon once a particular level of prediction certainty is achieved (the user need not touch the display surface to make a selection). This pointing facilitation scheme is dubbed mid-air selection [2]. Pointing finger observations are utilised by a probabilistic intent predictor to calculate the probability of each selectable on-screen icon being the intended destination of the free hand pointing gesture.

Figures 10.7 and 10.8 demonstrate that the predictive display system can reduce the workload of interacting with an in-car display by nearly 50% (reductions in pointing times can be as high as 40%). NASA TLX forms, widely utilised in HCI studies, are used to evaluate the subject workload experienced by the users. The ability of the probabilistic predictor to suppress/eliminate perturbations in the pointing finger motion is addressed in [5], for example utilising a sequential Monte Carlo method (namely a variable rate particle filter) to remove highly non-linear perturbation-related unintentional pointing movements.

10.7   CASE STUDY 4: ADAPTIVE MOBILE APPLICATIONS

We recall that an inclusive design approach extends beyond the scope of conventional usability methods as it must accommodate extremes of capability range or situational contexts of task or stress, that are not normally accommodated by product design. In the case of capability ranges, functional changes of capability in the vision, hearing, touch, thinking and dexterity and strength areas can be used to characterise ageing and specific impairments.

This approach uses cognitive modelling tools that represent perception, cognition and motor action in more detail than the conventional models, and are easier to use than complex cognitive architecture based models. They are intended for use

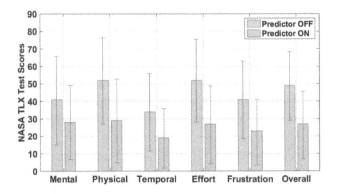

Figure 10.7 NASA TLX scores with and without the predictive functionality whilst driving on a motorway [2], i.e., minimal in-car vibrations.

over a wide range of users' functional capabilities in perceptual, cognitive or physical interactions and are applicable to situational impairment [11,13,36,37]. The approach identifies a set of human factors that can affect human computer interaction and formulates models that relate those factors to interface parameters. The combined modelling system can then predict how a person with objectively measured visual acuity and contrast sensitivity will perceive displays and controls, or how a person with a specific grip strength and range of motion of the wrist will use a pointing device. It is then possible to develop a set of rules relating users' range of capabilities to interface parameters. For example, these rules can be formulated into a web service that can dynamically adjust font size, cursor size, colour contrast, audio volume and spacing between interface elements of any displays. These user models are validated using experimental trials of participants with surveyed ranges of capabilities, for example, from EU studies [12, 13] of more than 120 older citizens, and more recent survey data from 33 Indians sampled across India in the IU-ATC project [11].

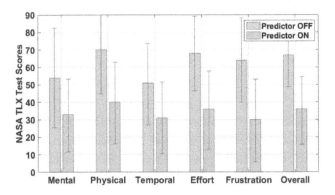

Figure 10.8 NASA TLX scores with and without the predictive functionality whilst driving on a harsh terrain [2], i.e., severe experienced in-car vibrations-accelerations.

10.7.1 The IU-ATC Project

The India-UK Advanced Technology Centre (IU-ATC) was commissioned as a collaborative programme funded by the UK's Engineering and Physical Sciences Research Council (EPSRC), the Government of India's Department of Science and Technology (DST) and industrial partners. The project aims to develop next generation telecommunications networks, leveraging state-of-the-art technologies in sensor network and multimodal interaction, to provide a plethora of services such as e-Governance, networked education, e-Health, social networking and communication in disaster management, for rural and urban populations in both UK and India. The goal is to make this technology accessible for the widest range of end users and applications.

The adaptive interface system has also been integrated to one of the application demonstrators for advanced wireless communication; an ICT based agriculture advisory system that promotes use of technology to increase the agricultural efficiency through urgent forecasting and crop disease diagnosis. This ICT based agriculture advisory system has two components:

- The Pest-Disease Image Upload (PDIU) application used by farmers to upload images of diseased crops, while they are in the field. The uploaded images are automatically sent to remotely located experts, who advise farmers about treatments. The application is designed to run on low-end "legacy" mobile phones or smart phones. It not only makes it easier for farmers who have difficulty in operating a keypad but also accommodates those suffering from poor vision or cognitive impairments.

- A web-based dashboard system that runs on a personal computer and is used by the domain experts to advise farmers. Experts may come from all ages and capability levels so it is therefore important to design a user interface that takes into account the degrees of impairments of different kinds.

Figure 10.9 demonstrates the different rendering of the dashboard and PDIU application for image capture, and, importantly, the sign-up survey dialogue used to collect the match of individuals to different survey modelled user profiles.

10.7.2 Mobile Interfaces

The application demonstrators developed for the IU-ATC also included a system for exploiting the advantages of fast wireless communication in major disasters for early warning, first response and coordination of governance. Natural disasters are becoming more and more prominent in our world today, increasing by over 400% in the last 20 years [1]. For example, in India and the UK, such events have impacted entire cities, as exemplified by the Uttarakhand Flooding of 2013, the Carlisle Flooding of 2005 and the Terrorist Bombings of London in 2005. To address this growing trend, many are working on technological solutions to help deal with such situations. One technology is that of Emergency Warning Systems (EWS). These offer mechanisms to send information from authorities to the populace during times of crisis. Information

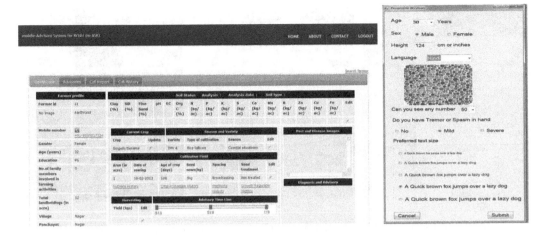

Figure 10.9 (left) The IU-ATC *e-agri* application. The text, colour and layout is modified for each user based on his data from the sign up application in (right).

may include evacuation details, medical advice and precautionary actions to take. In recent years, several countries have started deploying their own nationwide EWS. From a technological perspective, the primary challenge comes from the fact that the country still relies heavily upon GSM networks for cellular connectivity. Hence, some of the newer approaches like the 4G ETWS and the Commercial Mobile Alert System (used in the US) are not feasible. From a social perspective, there are two further challenges. First, India is a linguistically diverse country; secondly, low literacy levels, which are only at around 65%–74%, render traditional text broadcast techniques useless. To handle the above concerns, a user modelling application sign-up has been developed (shown in Figure 10.10) to adapt civilians' EWS user interfaces. It is offered as part of the registration procedure and asks users to perform a small set of tasks that help the system model physical capability and visual acuity. This generates a user capability profile consistent with EU and ITU standards. The profile is then processed by the smartphone application to dynamically render warnings, adapting such things as font size, colour contrast, zooming level, button layout and line spacing. For example, it changes background and foreground colours for colour blind users, turns on a text-to-speech converter for visually impaired users and increases line spacing and default zooming level for users with a tremor or spasm in the hand. An example of this can be seen in Figure 10.10.

10.8 DISCUSSION

As human computer interfaces have advanced there has been a need for representation of increasing amounts of information and consequently the methods to interact with it, especially for IoT applications. This has been fuelled by development of many new innovative technologies for input, control and display that utilise more modalities of human interaction such as speech, sound, 2D and 3D input with multitouch and gestural inputs. There has been a convergence of requirements and solutions in the

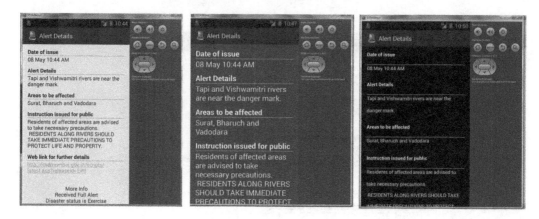

Figure 10.10 Different Mobile Device renderings of a single Emergency Early Warning system (EWS) adapted for users with differing capabilities of vision and movement.

two domains dealt with here: that of computer assistive and adaptive displays and controls for disability and accessibility, and that of advanced interaction support for pilots and drivers, primarily engaged in controlling a vehicle while carrying out other tasks, with a range of difficulties from pointing and selection to keypad entry. Recent engineering solutions to the former have recognised the parallels between designing for extraordinary individuals with health related impairments and for the general population with situational impairments. Increasingly, these solutions are finding their way into the latter world of automotive and aerospace engineering design. Mainly, the solutions lie in the use of provisions of redundant input and control methods based on the human capability to deal simultaneously with perception of, and reaction to, stimuli in different physical modes. However, there are related innovations, such as predictive touch, based on cognitive psychology and modelling. These were originally designed to address capability impairment but are now yielding novel displays and controls in touchable, cross-modal, haptic, visual, auditory and spoken forms. Importantly, the increased interconnectedness of internet enabled systems, IoT, along with the increased speed and volume of information has made possible a whole new raft of applications with concomitant increases in complexity. Taken together, these domains illustrate the convergence of HMI issues arising from similar considerations that have originated in high performance interfaces and design for accessibility and inclusion. It may be that these are pointing to a new transdisciplinary approach for the future. Four case studies have illustrated this convergence for the automotive, aerospace and mobile device applications. This has shown how the use of multimodal adaptive displays can successfully benefit the inclusive population: widening the mainstream design solutions to include those with capability variation due to health or situational impairment. Furthermore, this has also shown how literacy, language variation and context information from sensors can be incorporated into adaptive interactions to further benefit users, in both advanced and developing countries. Although the research communities in HCI, HMI, ergonomics and computer accessibility have been investigating this convergent area for some time, advances have been steady and the level of certainty required for true technology transfer and impact hard to reach. Nevertheless, as the case studies indicate, the multimodal adaptive approach holds great promise in numerous research applications and has already proved itself effective in experiments. It is already clear that the future will include interactions of this sort, including, by design, many people who would have otherwise been unconnected, and improving the effectiveness of those who wish to take advantage of highly demanding technologies. The design of interactive interfaces with diverse technologies through the IoT is undergoing major and revolutionary changes. The designs of the future will be predictive, multimodal, connected, adaptive, context dependent, integrated and intelligent.

Bibliography

[1] Mobility as a service, exploring the opportunity for mobility as a service in the UK. Technical report, Transport Systems Catapult, UK, 2016.

[2] B. I Ahmad, P. M Langdon, S. J Godsill, R. Donkor, R. Wilde, and L. Skrypchuk. You do not have to touch to select: A study on predictive in-car touchscreen with mid-air selection. In *Proc. of the Int. Conf. on Automotive User Interfaces and Interactive Veh. Apps. (AutomotiveUI '16)*, 2016.

[3] B. I Ahmad, P. M Langdon, S. J Godsill, R. Hardy, L. Skrypchuk, and R. Donkor. Touchscreen usability and input performance in vehicles under different road conditions: An evaluative study. In *Proceedings of the 7th International Conference on Automotive User Interfaces and Interactive Vehicular Applications*, pages 47–54. ACM, 2015.

[4] B. I Ahmad, J. Murphy, P. M Langdon, and S. J Godsill. Bayesian target prediction from partial finger tracks: Aiding interactive displays in vehicles. In *2014 17th International Conference on Information Fusion (FUSION)*, pages 1–7. IEEE, 2014.

[5] B. I Ahmad, J. Murphy, P. M Langdon, and S. J Godsill. Filtering perturbed in-vehicle pointing gesture trajectories: Improving the reliability of intent inference. In *2014 IEEE International Workshop on Machine Learning for Signal Processing (MLSP)*, pages 1–6. IEEE, 2014.

[6] B. I Ahmad, J. K Murphy, P. M Langdon, and S. J Godsill. Bayesian intent prediction in object tracking using bridging distributions. *to appear in IEEE Trans. on Cybernetics (arXiv preprint:1508.06115)*, 2016.

[7] B. I Ahmad, J. K Murphy, P. M Langdon, S. J Godsill, R. Hardy, and L. Skrypchuk. Intent inference for hand pointing gesture-based interactions in vehicles. *IEEE Transactions on Cybernetics*, 46(4):878–889, 2016.

[8] JR Anderson. How can the human mind exist in the physical world, 2007.

[9] V. A Banks, N. A Stanton, and C. Harvey. Sub-systems on the road to vehicle automation: Hands and feet free but not ŚmindŠfree driving. *Safety Science*, 62:505–514, 2014.

[10] J.-A. Bichard and R. Gheerawo. The designer as ethnographer: Practical projects from industry in design anthropology: Object culture in the 21st century. 2010.

[11] P. Biswas, P. M. Langdon, J. Umadikar, S. Kittusami, and S. Prashant. How interface adaptation for physical impairment can help able bodied users in situational impairment. In *Langdon, Lazar, Heylighen and Dong, (Eds.) (2014), Inclusive Designing: Joining Usability, Accessibility, and Inclusion*, pages 49–58. Springer, 2014.

[12] P. Biswas, C. Duarte, P. Langdon, L. Almeida, and C. Jung. *Editorial: A Multimodal End-2-End Approach to Accessible Computing*, Springer, ISBN 978-1-4471-5081-7. Springer, 2013.

[13] P. Biswas and P. Langdon. Inclusive user modeling and simulation. In *P. Biswas, C. Duarte, P. Langdon, L. Almeida and C. Jung (Ed.), A Multimodal End-2-End Approach to Accessible Computing HumanŰComputer Interaction Series 2013*, pages 71–89. Springer, 2013.

[14] P. Biswas and P. Langdon. Multimodal target prediction model. In *Proceedings of the extended abstracts of the 32nd annual ACM conference on human factors in computing systems*, pages 1543–1548. ACM, 2014.

[15] A. Blandford, R. Butterworth, and P. Curzon. Models of interactive systems: A case study on programmable user modelling. *International Journal of Human-Computer Studies*, 60(2):149–200, 2004.

[16] BMW. BMW i-drive. Accessed on: 18 May 2014 from http://www.bmw.com/com/en/insights/technology/technology_guide/articles/idrive.html.

[17] G. Burnett, G. Lawson, L. Millen, and C. Pickering. Designing touchpad user-interfaces for vehicles: Which tasks are most suitable? *Behaviour & Information Technology*, 30(3):403–414, 2011.

[18] G. E. Burnett and J. M. Porter. Ubiquitous computing within cars: Designing controls for non-visual use. *International Journal of Human-Computer Studies*, 55(4):521–531, 2001.

[19] C. M. Burns and J. Hajdukiewicz. *Ecological interface design*. CRC Press, 2004.

[20] A. J. Clarke. *Design anthropology: Object culture in the 21st century*. Springer-Verlag, Vienna, 2011.

[21] J. Coelho, C. Duarte, P. Biswas, and P. Langdon. Developing accessible tv applications. In *The proceedings of the 13th international ACM SIGACCESS conference on computers and accessibility*, pages 131–138. ACM, 2011.

[22] J. Coelho, T. Guerreiro, and C. Duarte. Designing tv interaction for the elderly–a case study of the design for all approach. In *A Multimodal End-2-End Approach to Accessible Computing*, pages 49–69. Springer, 2013.

[23] R. Coleman. Designing for our future selves. *Universal Design Handbook* W.F.E. Preiser and E. Ostroff (Eds.) pp. 4.1–4.25, MacGraw-Hill: New York, USA, 2001.

[24] A. Drazin and S. Roberts. Exploring design dialogues for ageing in place. *Anthropology in Action*, 16(1):72–88, 2009.

[25] M. R. Endsley. *Designing for situation awareness: An approach to user-centered design*. CRC press, 2016.

[26] M. R. Endsley, C. A. Bolstad, D. G. Jones, and J. M. Riley. Situation awareness oriented design: From user's cognitive requirements to creating effective supporting technologies. In *Proceedings of the Human Factors and Ergonomics Society Annual Meeting*, volume 47, pages 268–272. SAGE Publications, 2003.

[27] C. W. Eriksen. Attentional search of the visual field. *Visual Search*, pages 3–19, 1990.

[28] K. Z. Gajos, J. O. Wobbrock, and D. S. Weld. Automatically generating user interfaces adapted to users' motor and vision capabilities. In *Proceedings of the 20th annual ACM symposium on user interface software and technology*, pages 231–240. ACM, 2007.

[29] GUIDE. Accessed on: 18 May 2014 from http://www.guide-project.eu/.

[30] J. He. Head-up display for pilots and drivers. *Journal of Ergonomics*, 2013, 2013.

[31] C. Ho and C. Spence. *The multisensory driver: Implications for ergonomic car interface design*. Ashgate Publishing, Ltd., 2012.

[32] M. G. Jæger, M. B Skov, N. G. Thomassen, et al. You can touch, but you can't look: Interacting with in-vehicle systems. In *Proceedings of the SIGCHI Conference on Human Factors in Computing Systems*, pages 1139–1148. ACM, 2008.

[33] S. Jenson. The physical web. In *CHI'14 Extended Abstracts on Human Factors in Computing Systems*, pages 15–16. ACM, 2014.

[34] S. Keates and J. Clarkson. *Countering design exclusion—An introduction to inclusive design*. Springer, 2003.

[35] M Kunur, PM Langdon, MD Bradley, J-A Bichard, E Glazer, F Doran, PJ Clarkson, and JJ Loeillet. Reducing Exclusion in Future Cars Using Personas with Visual Narratives and Design Anthropology, In Designing Around People, Pat Langdon, Jonathan Lazar, Ann Heylighen, Hua Dong (Eds.) ISBN: 978-3-319-29496-4, pp. 269–277. Springer, 2016.

[36] P. Langdon, U. Persad, and P. J. Clarkson. Developing a model of cognitive interaction for analytical inclusive design evaluation. *Interacting with Computers*, 22(6):510–529, 2010.

[37] P. Langdon and H. Thimbleby. Inclusion and interaction: Designing interaction for inclusive populations. *Interacting with Computers*, 22(6):439–448, 2010.

[38] J. W. Lasswell. The effects of display location and dimensionality on taxiway navigation. Technical report, ARL-95-5/NASA-95-2, NASA Ames Research Center, Moffett Field, CA, 1995.

[39] D. Navon. Forest before trees: The precedence of global features in visual perception. *Cognitive Psychology*, 9(3):353–383, 1977.

[40] A. F. Newell. Accessible computing—past trends and future suggestions: Commentary on computers and people with disabilities. *ACM Transactions on Accessible Computing (TACCESS)*, 1(2):9, 2008.

[41] A. F. Newell, A. Dickinson, M. J. Smith, and P. Gregor. Designing a portal for older users: A case study of an industrial/academic collaboration. *ACM Transactions on Computer-Human Interaction (TOCHI)*, 13(3):347–375, 2006.

[42] A. Newell. *Unified theories of cognition*. Harvard University Press, 1994.

[43] K. M. Newell, D. E. Vaillancourt, and J. J. Sosnoff. Aging, complexity, and motor performance. *Handbook of the psychology of aging*, pages 163–182, 2006.

[44] MG Ormerod and RA Newton. Moving beyond accessibility: The principles of universal (inclusive) design as a dimension in nd modelling of the built environment. *Architectural Engineering and Design Management*, 1(2):103–110, 2005.

[45] R. Parasuraman, T. B. Sheridan, and C. D. Wickens. Situation awareness, mental workload, and trust in automation: Viable, empirically supported cognitive engineering constructs. *Journal of Cognitive Engineering and Decision Making*, 2(2):140–160, 2008.

[46] U. Persad, P. Langdon, and J. Clarkson. Characterising user capabilities to support inclusive design evaluation. *Universal Access in the Information Society*, 6(2):119–135, 2007.

[47] H. Petrie. Accessibility and usability requirements for icts for disabled and elderly people: A functional classification approach. In *Nicolle, C. and Abascal, J.G. (Eds.), Inclusive Guidelines for Human Computer Interaction*. Taylor and Francis, London. ISBN: 0-748409-48-3, 2001.

[48] N. Poh, A. Ross, W. Lee, and J. Kittler. A user-specific and selective multimodal biometric fusion strategy by ranking subjects. *Pattern Recognition*, 46(12):3341–3357, 2013.

[49] I. Politis, S. Brewster, and F. Pollick. Evaluating multimodal driver displays of varying urgency. In *Proceedings of the 5th International Conference on Automotive User Interfaces and Interactive Vehicular Applications*, pages 92–99. ACM, 2013.

[50] L. J. Prinzel III and M. Risser. Head-up displays and attention capture. *NASA/TM-2004-213000*. Langley Research Center, Hampton, Virginia, 2004.

[51] G. Pullin and A. Newell. Focussing on extra-ordinary users. In *International Conference on Universal Access in Human-Computer Interaction*, pages 253–262. Springer, 2007.

[52] D. D Salvucci and F. J Lee. Simple cognitive modeling in a complex cognitive architecture. In *Proceedings of the SIGCHI conference on human factors in computing systems*, pages 265–272. ACM, 2003.

[53] K. W. Schaie. Methodological issues in aging research: An introduction. *Methodological issues in aging research*, pages 1–11, 1988.

[54] A. Sears, M. Young, and J. Feng. Physical disabilities and computing technologies: An analysis of impairments. *The human-computer interaction handbook: fundamentals, evolving technologies and emerging applications, LEA*, ISBN:0-8058-3838-4, pages 482–503, 2002.

[55] L. Skrypchuk, P. M. Langdon, P. J. Clarkson, and A. Mouzakitis. Creating inclusive automotive interfaces using situation awareness as a design philosophy. In *International Conference on Universal Access in Human-Computer Interaction*, pages 639–649. Springer, 2016.

[56] C. Spence and J. Driver. *Crossmodal space and crossmodal attention*. Oxford University Press, 2004.

[57] C. Spence, F. Pavani, and J. Driver. Spatial constraints on visual-tactile crossmodal distractor congruency effects. *Cognitive, Affective, & Behavioral Neuroscience*, 4(2):148–169, 2004.

[58] NA Stanton and MS Young. Vehicle automation and driving performance. *Ergonomics*, 41(7):1014–1028, 1998.

[59] N. Stanton, P. M Salmon, and L. A Rafferty. *Human factors methods: A practical guide for engineering and design*. Ashgate Publishing, Ltd., 2013.

[60] C. Stephanidis, A. Paramythis, M. Sfyrakis, A. Stergiou, N. Maou, A. Leventis, G. Paparoulis, and C. Karagiannidis. Adaptable and adaptive user interfaces for disabled users in the avanti project. In *International Conference on Intelligence in Services and Networks*, pages 153–166. Springer, 1998.

[61] J. R. Stroop. Studies of interference in serial verbal reactions. *Journal of Experimental Psychology*, 18(6):643, 1935.

[62] R. Swette, K. R. May, T. M. Gable, and B. N. Walker. Comparing three novel multimodal touch interfaces for infotainment menus. In *Proceedings of the 5th International Conference on Automotive User Interfaces and Interactive Vehicular Applications*, pages 100–107. ACM, 2013.

[63] T. E Trimble, R. Bishop, J. F. Morgan, M. Blanco, et al. Human factors evaluation of level 2 and level 3 automated driving concepts: Past research, state of automation technology, and emerging system concepts. 2014.

[64] JBF Van Erp, C Jansen, T Dobbins, and HAHC Van Veen. Vibrotactile waypoint navigation at sea and in the air: Two case studies. In *Proceedings of EuroHaptics*, pages 166–173, 2004.

[65] T. B. Ward, C. M. Foley, and J. Cole. Classifying multidimensional stimuli: Stimulus, task, and observer factors. *Journal of Experimental Psychology: Human Perception and Performance*, 12(2):211, 1986.

[66] C. D. Wickens. Attention and aviation display layout: Research and modeling. Technical report, Final Technical Report AHFDŮ06-21/NASA-05-8, AHFD, NASA AMES Research Centre, CA, 2005.

[67] C. D. Wickens and J. G. Hollands. Attention, time-sharing, and workload. *Engineering psychology and human performance*, pages 439–479, 2000.

[68] C. D. Wickens, J. G. Hollands, S. Banbury, and R. Parasuraman. *Engineering psychology & human performance*. Psychology Press, 2015.

[69] C. D. Wickens and J. Long. Object versus space-based models of visual attention: Implications for the design of head-up displays. *Journal of Experimental Psychology: Applied*, 1(3):179, 1995.

[70] P. KA Wollner, P. M Langdon, and P John Clarkson. Integrating a cognitive modelling framework into the design process of touchscreen user interfaces. In *International Conference of Design, User Experience, and Usability*, pages 473–484. Springer, 2015.

CHAPTER 11

Low Power Wide Area (LPWA) Networks for IoT Applications

Kan Zheng

Intelligent Computing and Communication (IC²) Lab, Key Lab of Universal Wireless Communications, Ministry of Education, Beijing University of Posts & Telecommunications, Beijing, China, 100088, E-mail: zkan@bupt.edu.cn.

Zhe Yang

Intelligent Computing and Communication (IC²) Lab, Key Lab of Universal Wireless Communications, Ministry of Education, Beijing University of Posts & Telecommunications, Beijing, China, 100088.

Xiong Xiong

Intelligent Computing and Communication (IC²) Lab, Key Lab of Universal Wireless Communications, Ministry of Education, Beijing University of Posts & Telecommunications, Beijing, China, 100088.

Wei Xiang

College of Science and Engineering, James Cook University, Cairns, QLD 4878, Australia.

CONTENTS

As an important application scenario of the forthcoming 5th generation (5G) system, the Internet of Things (IoT) has attracted a significant interest in recent years. It is well known that smart IoT devices will have a big impact on our everyday life. Existing wireless transmission techniques (e.g., Bluetooth, Zigbee, etc.) can be applied to IoT applications and provide satisfactory data rates for end-user applications [17]. However, they usually have very limited coverage, which cannot meet the communications requirements of a massive number of smart devices over a large area. Low power wide area (LPWA) networks are specially designed to meet the above requirements due to their long transmission ranges and low energy consumption. Thanks to some key techniques in both the physical layer and medium access control (MAC) layer, LPWA networks are capable of providing long distance communication between servers and devices with low costs.

In this chapter, we mainly focus on low power wide area networks (LPWANs) for IoT applications. First, an overview on the 5G mobile networks is presented. After that, a detailed introduction to typical LPWANs is given, among which Narrow Band Internet of Things (NB-IoT) and IEEE 802.15.4k are emphasized. Based on the IEEE 802.15.4k technique, an air quality monitoring system is designed and implemented. Through a number of PM sensors connected to an access point (AP), air quality data can be collected and sent to the servers of an IoT cloud. Users are able to access the real-time air quality information through either webpages or a mobile APP. Our analysis on the sensed data reveals that the proposed system is reliable in sensing the air quality.

11.1 OVERVIEW ON 5G IoT

As a focus of global research and development efforts, 5G has been paid an increasing attention from both governments and institutions. Following existing wireless communication networks of 4G, 5G is expected to be commercialized around year 2020. It is believed that higher data rates and wider accesses will not only transform our lives, but also unleash enormous economic potential. The evolution from 4G to 5G focuses primarily on four aspects, i.e., data rate, latency, mobility, and capacity, which are illustrated in Figure 11.1.

- Data Rate
  5G aims to provide a much higher data rate than its 4G counterpart. The target peak data rate is 10 Gbps, which represents a 10-fold increase over the existing 4G network (1 Gbps). Given this high peak data rate, users can easily enjoy network services with higher quality, such as downloading high-definition (HD) movies in a few seconds. Several techniques may be employed to realize the

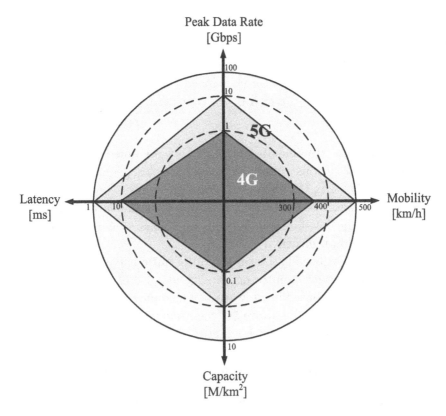

Figure 11.1 Requirements of 5G systems.

target data rate, e.g., network densification, massive MIMO, and millimeter-wave communications;

- Latency
  Latency is another important indicator of the performance of a wireless network. In the 4G era, wireless networks usually suffer from tens of milliseconds, which severely affect the user's experience. 5G networks are believed to be able to reduce the latency to 1 ms. Therefore, more network equipment can be moved to the cloud, which greatly helps save costs and power consumption compared with local deployment;

- Mobility
  With the rapid development of modern transportation technology, users tend to surf the Internet in high-speed environments. Due to the Doppler shift, users in fast-moving vehicles or trains may suffer from poor channel conditions. 5G aims to tackle this challenging issue by allowing the moving speed of user devices up to 500 km/h; and

- Capacity
  Industry analysts have predicted that more than 50 billion smart devices will be connected to the Internet by 2020, posing a serious challenge to the capacity of the existing network. The target capacity of 5G is 1,000,000 simultaneous connections per kilometer, which is ten times greater than that of existing 4G networks.

5G aims to solve the aforementioned challenges stemmed from varying performance requirements in various communication scenarios. Generally, there are four typical scenarios for 5G, i.e., the wide-area coverage scenario, high-capacity hot-spot scenario, low-power massive-connections scenario, and low-latency high-reliability scenario [19], where the low-power massive-connections scenario targets primarily IoT-related services and application.

IoT has been widely utilized in use cases of information acquisition and intelligent control, such as infrastructure monitoring, intelligent agriculture, and smart city, etc. Numerous devices are distributed in wide areas, which may consume large amounts of energy. Therefore, apart from the above 5G requirements, IoT also needs to operate in a cost-effective and energy-efficient manner.

11.2 OVERVIEW ON LOW POWER WIDE AREA NETWORKS (LPWANS)

Compared with the traditional wireless communications techniques, LPWA technology is specifically designed with the objectives of wide coverage and ultra-low energy consumption. This makes LPWAN a promising communications technology for IoT applications. An overview on LPWANs is given in this section. Both the application scenarios and classifications will be presented.

11.2.1 Application Scenarios of LPWANs

With the rapid development of the information and communication technology (ICT) industry, people become more and more dependent on mobile networks and smart devices. Benefiting from its long transmission range and low power consumption, LPWAN has promising application prospects. Typical application scenarios of LPWAN are illustrated in Figure 11.2.

Security As one of the most serious aspects of concern of the contemporary society, security services are developing rapidly in recent years. With the aid of LPWAN technology, emergency information can be timely transmitted to users, which can reduce personal property losses as much as possible;

Health care Health care issues can be alleviated to a certain extent if being combined with LPWA techniques. Through developing user-centric and individually tailored devices, health conditions can be monitored for a long period of time. Furthermore, thanks to the rich medical data, diagnosis can be made more accurately and effectively;

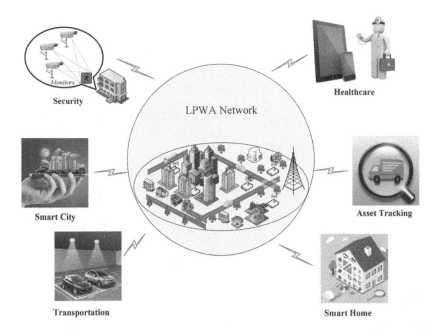

Figure 11.2 Application scenarios of LPWAN.

Asset Tracking Individuals may be concerned with the safety of their valuable assets. LPWA-based asset tracking techniques can provide a real-time visibility about the locations and statuses of their assets, which can be protected from either damage or theft;

Transportation With the development of modern transportation, an increasing number of vehicles are driven on the road, which may bring about severe traffic congestions. Therefore, knowing real-time traffic conditions has become the key to alleviating the traffic congestion problem. LPWAN can help establish direct communications between vehicles and information centers, which can not only help relevant government agencies to improve the efficiency of traffic management, but also to guide drivers to plan optimum routes to their destinations;

Smart Home LPWA techniques can play a vital role in smart home applications. For example, by using a mobile APP, a user can easily control the environment at home from his/her workplace, such as the lighting, heating, ventilation, and air conditioning. Furthermore, based on the analysis of user behaviors, the preferences of a certain user can be extracted. This information can instruct the appliances to operate in a more personalized way without human intervention; and

Smart City Smart city aims to make full use of the urban informatics and technologies to meet city residents' needs and to improve the quality of life. Through LPWA techniques, city officials can easily obtain real-time information about what is happening in the city, which can greatly improve the efficiency of city management.

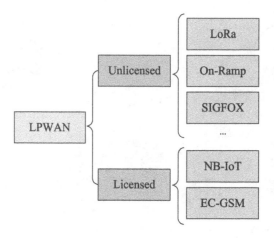

Figure 11.3 Classification of typical LPWAN techniques.

11.2.2   Classification of LPWANs

Typical LPWANs can be roughly classified as the unlicensed networks and licensed networks, which are depicted in Figure 11.3. The unlicensed LPWANs mainly consist of LoRa, On-Ramp, SIGFOX, etc. These networks work in the industrial, scientific, and medical (ISM) bands, and are ready to be or already being deployed. Moreover, these standard protocols are only supported by industry alliances. Apart from the unlicensed ones, the licensed LPWANs include narrow band (NB)-IoT and EC (Extended Coverage)-GSM, which are upgraded from the existing wireless communications techniques [12]. Details about these LPWA techniques are introduced as follow.

- LoRa
  LoRa is specially designed for battery-operated devices in LPWA networks. It employs the star topology in which gateways are used to relay messages between devices and the servers in the backend. Communications between all devices are secure and bi-directional. In order to adapt to varying communications requirements, LoRa offers adaptive data rate (ADR) which is managed by the network server. This helps maximize the battery lifespan.

- IEEE 802.15.4k
  IEEE 802.15.4k is regarded as a promising technology for LPWANs. It was founded and developed by a famous systems provider for LPWA sensor networking and location tracking, i.e., On-Ramp Wireless. One of the features in this standard is the Random Phase Multiple Access (RPMA), which covers wide areas and provides robust communications links. The bandwidth of each channel is 1 MHz, which is larger than other comparative LPWA techniques.

- Sigfox
  Sigfox was developed in 2009, and is regarded as the first LPWAN technique proposed in the IoT market. Compared with the other networks, it has already

provided nationwide low-power connectivity in a number of European countries. In the physical layer, the Ultra Narrow Band (UNB) technique is selected as the modulation scheme, which is effective in improving spectral efficiency. The protocols of the network layer are not open to the public yet. It is claimed that up to a million IoT objects can be handled by a Sigfox gateway. The coverage of Sigfox is 30–50 km in rural areas and 3–10 km in urban areas.

- NB-IoT

  As the only LPWA technique standardized by the 3rd Generation Partnership Project (3GPP), NB-IoT is a narrowband radio technology specially designed for new requirements in IoT scenarios. The aim of this technology is to provide cost-effective connectivity to a massive number of devices with low costs and energy consumption. In particular, indoor coverage is also a focus of this technology. NB-IoT can easily coexist with the GSM and LTE networks. Through use of the existing communications infrastructure, both time and money for deploying this network can be significantly saved.

- EC-GSM

  EC-GSM aims to provide high capacity, long distance, and low power consumption communications among IoT devices and co-exist with the existing wireless networks. This standard is built on the basis of the Enhanced General Packet Radio Service (eGPRS), which is an improved technique allowing for higher data rates than traditional GSM networks. A software upgrade is made on the existing GSM networks. It is supposed that the battery life of IoT devices based upon this standard can be prolonged to up to 10 years. Furthermore, the communications security is also improved with specific techniques such as entity authentication, user identity confidentiality, and mobile equipment identification.

Brief comparisons among these LPWANs are shown in Table 11.1. Among all these LPWANs, two are the focus of this chapter, i.e., NB-IoT and IEEE 802.15.4k.

11.2.2.1 LPWAN Based on NB-IoT

NB-IoT was initiated by 3GPP in September 2015, which was promoted by Nokia, Ericsson, and Intel. In June 2016, the core specifications were completed in Release 13. Till now, it has been supported and improved by many famous communications companies including Alcatel-Lucent, Huawei, and AT&T. The objectives of this technique are to address the communications requirements of indoor coverage, support of massive devices, low cost, low delay, and low power consumption. Requiring only a software upgrade, NB-IoT can be fully integrated into the existing GSM and LTE networks, saving large amounts of time and money for building new infrastructure.

Table 11.1 Comparison among typical LPWANs

	LoRa [6]	IEEE 802.15.4k (OnRamp) [18]	Sigfox [6]	NB-IoT [16]	EC-GSM-IoT [1]
Band	169,433, 915, 868 MHz	2.4 GHz	868, 902 MHz	In-band & Guard-band LTE, standalone	800-900, 1800-1900 MHz (In-band GSM)
Coverage	160 dB (20 dBm); Rural: 14-45 km; Suburban: 15-22 km; Urban: 3-8 km	172 dB (21 dBm); 165 dB (14 dBm); around 16 km	Rural: 40 km; Urban: 3-10 km	164 dB for standalone, FFS others	164 dB (33 dBm); 154 dB (23 dBm)
Max Devices	~100,000 per cell	~384,000 per cell	~100,000 per cell	50,000 per cell	50,000 per cell
Downlink	Wide-band linear frequency modulated pulses	Random phase multiple access (RPMA)	Ultra-narrowband (UNB)	OFDMA, 15 KHz tone spacing, Tail-biting Convolutional code (TBCC), 1Rx	TDMA/FDMA, GMSK and 8PSK (optional), 1 Rx
Uplink				SC-FDMA:, 15 KHz tone spacing, Turbo code	TDMA/FDMA, GMSK and 8PSK (optional)
Bandwidth	125; 250; 500 kHz	1 MHz (40 channels)	200 Hz	180 kHz (36 subcarriers)	200 kHz per channel. Typical system bandwidth of 2.4 MHz
Peak rate (DL/UL)	0.3-50 kbps	0.008-8 kbps	0.1 kbps	DL:~250 kbps; UL: ~250 for multi-tone, ~20 kbps for single tone	For DL and UL (4 timeslots): ~70 kbps (GMSK), ~240 kbps (8PSK)
Battery life	~105 months with 2000 mAh battery	~20 years	~20 years with 3 messages/day	~10 years 5 Wh battery	~10 years 5 Wh battery

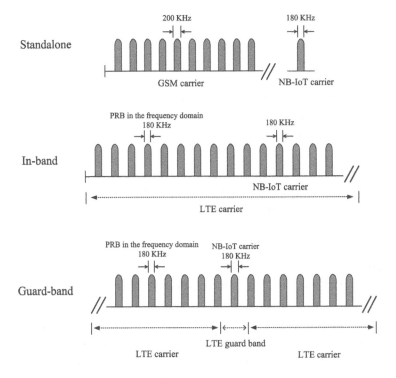

Figure 11.4   Spectrum deployments for NB-IoT.

NB-IoT has widely reused existing techniques in 4G LTE, e.g., downlink orthogonal frequency-division multiple-access (OFDMA), uplink single-carrier frequency-division multiple-access (SC-FDMA), interleaving, the subcarrier spacing, OFDM symbol duration, subframe duration, etc [3]. Therefore, the development time required for NB-IoT devices can be greatly reduced based on existing LTE products. A detailed introduction to NB-IoT will be given focusing on the following aspects, i.e., spectrum deployment, physical channels, and radio protocols.

Spectrum deployment   It is well known that spectrum is one of the most valuable natural resources for the telecommunications industry. Thus nearly all the wireless communication systems regard spectral efficiency as one of the most important performance indicators of their networks. In order to coexist with existing wireless networks (e.g., GSM, WCDMA, and LTE), the spectrum deployment of NB-IoT is designed to be more flexible. Three deployment options are provided for the network operators, i.e., the standalone, in-band, and guard-band, which is depicted in Figure 11.4.

Standalone: Each NB-IoT carrier occupies a 180 kHz bandwidth for both uplink and downlink, which is narrower than a GSM carrier (200 kHz). Therefore, network operators can select one of their GSM carriers to deploy NB-IoT with little impact on their current business thanks to the orthogonality among the carriers. This method is known as standalone deployment.

In-band: In-band spectrum deployment is specially designed for existing LTE techniques. The bandwidth of a NB-IoT carrier is the same as the LTE physical resource block (PRB) in the frequency domain. This provides a feasible way to replace one of the PRBs with the NB-IoT carrier. It has been evaluated that a single NB-IoT carrier can support up to 200,000 devices per cell. With the increasing density of IoT devices, more than one PRB can be replaced with NB-IoT carriers in accordance with the configuration of the network.

Guard-band: For the sake of reducing the interference, a guard-band is inserted between adjacent carriers in LTE. These unused resources can be utilized to deploy NB-IoT carriers. Specific physical layer designs are needed to reduce the resultant interference under this deployment.

Physical channels Physical channels in NB-IoT resemble their LTE counterparts to a large extent. Both the downlink and uplink are introduced in detail as follows.

Downlink: Several physical channels or signals are presented in NB-IoT for downlink transmission, i.e., the Narrowband Primary Synchronization Signal (NPSS), Narrowband Secondary Synchronization Signal (NSSS), Narrowband Physical Broadcast Channel (NPBCH), Narrowband Physical Downlink Control Channel (NPDCCH), and Narrowband Physical Downlink Shared Channel (NPDSCH).

- NPSS and NSSS
  These two signals are responsible for providing time and frequency synchronization information for devices. After the devices receive these signals, they can perform the cell search task including frequency synchronization and cell identity detection;

- NPBCH
  NPBCH carries configuration information about the network and cell. The eNB broadcasts these signals to every user's equipment in the cell. It is located in subframe #0 in every frame;

- NPDCCH
  NPDCCH is designed to transmit the scheduling information in response to random access by the devices. Furthermore, the HARQ acknowledgment is also transmitted through this channel; and

- NPDSCH
  NPDSCH carries data from higher layers including the system information and paging messages.

The transmit block diagram of the NPBCH, NPDCCH, and NPDSCH is depicted in Figure 11.5. The cyclic redundancy check (CRC), forward error correction (FEC), and interleaving are state-of-the-art methods in improving data reliability. Rate matching is responsible for matching the bit number to the allocated resource.

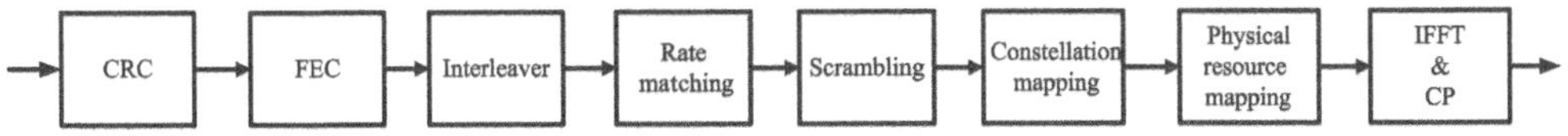

| CRC | → | FEC | → | Interleaver | → | Rate matching | → | Scrambling | → | Constellation mapping | → | Physical resource mapping | → | IFFT & CP | → |

Figure 11.5 Transmit block diagram for the NPBCH, NPDCCH, and NPDSCH.

After that, constellation mapping is invoked based on the modulation scheme, e.g., BPSK, QPSK, and 16QAM. Finally, symbols are mapped to the physical resource before the IFFT.

Uplink: Uplink physical channels mainly consist of the Narrowband Physical Random Access Channel (NPRACH) and Narrowband Physical Uplink Shared Channel (NPUSCH).

- NPRACH
  NPRACH is specially designed for the purpose of random access in NB-IoT. Unlike the PRACH in LTE with 1.08 MHz, the NPRACH occupies four symbol groups, each of which includes one CP and five symbols, which is modulated on to a 3.75 kHz band; and

- NPUSCH
  Two formats are provided for the NPUSCH. Format 1 carries uplink data with a maximum transport block size of 1000 bits. Format 2 is responsible for HARQ acknowledgment.

The transmit block diagram of the NPUSCH is shown in Figure 11.5. Similar to the downlink procedures, FEC and rate matching are also needed for uplink. Then the symbols are scrambled by a certain logical operation. The pilot is produced using the Gold sequence which is generated at the beginning of each burst. Two modulation schemes are utilized for the uplink, i.e., the Gaussian-shaped Minimum Shift Keying (GMSK) and Phase Shift Keying (PSK).

Radio protocols

Random access: Access schemes have a great impact on the performance and capacity of the network. Traditional LTE systems employ the contention-based random

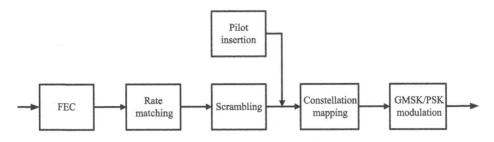

Figure 11.6 Transmit block diagram for the NPUSCH.

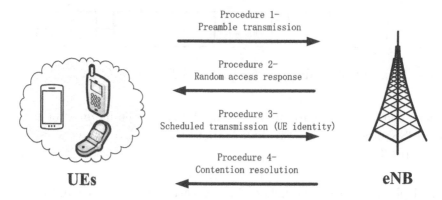

Figure 11.7 Main procedures of random access in NB-IoT.

access method, which usually needs four procedures for a new user equipment to be accessed by the system. As illustrated in Figure 11.7, the four procedures are:

- Procedure 1: Preamble transmission
  When a UE enters a certain cell, it needs to transmit a random access preamble to the eNB to notify its arrival. These messages are sent using the NPRACH;

- Procedure 2: Random access response
  The eNB allocates time-frequency resources to this UE for uplink transmission. Then it transmits the scheduling and timing information back to the UE;

- Procedure 3: Scheduled transmission (UE identity)
  Using the resources obtained in Procedure 2, the UE sends its identity information to the eNB; and

- Procedure 4: Contention resolution
  The eNB deals with possible contentions or collisions among multiple devices and then transmits the contention resolution messages to the UEs.

Based on the traditional random access scheme adopted in LTE networks, some improvements are made in NB-IoT due to the different coverage classes of UEs. The configuration of NPRACH used to send the random access preamble becomes more flexible and adjustable, e.g., the number and offset of the subcarriers in the frequency domain, as well as the starting time and periodicity in the time domain [16].

Device Operation Modes: In order to save energy and prolong the battery lifetime, three operation modes are enabled in NB-IoT devices, i.e., the connected mode, idle mode, and power saving mode, as shown in Figure 11.8.

In the connected mode, a NB-IoT device can communicate normally with the eNB. However, it inevitably consumes a large amount of energy. Once a specified activity has not taken place for a certain period of time determined by the Connected Mode Release Timer, the NB-IoT devices can be placed into either the idle mode or the power saving mode. The choice is based on the configuration of the Active Timer.

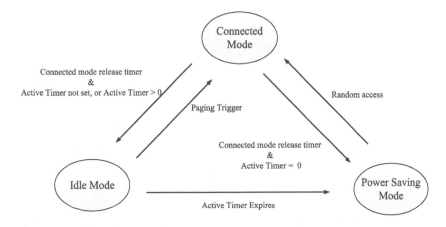

Figure 11.8 Device operation modes in NB-IoT.

Devices can also be accessed by the network even if they are in the idle mode. Instead of monitoring the downlink control channel all the time, devices in the idle mode only need to listen to the paging messages from the network. This mechanism helps save energy because power consumption in the idle mode is much lower than in the connected mode. Upon receiving a paging message, the devices move back to the connected mode. Moreover, they can also move to the power saving mode after the Active Timer expires.

Unlike the above two modes, devices in the power saving mode can no longer be reached by the network. This is a "deep sleep" mode in which even the paging messages are not listened to. When new traffic arrives, devices need to wake up and initiate a random access to the eNB. Then it will move to the connected mode ready for data transmission.

11.2.2.2   LPWAN Based on IEEE 802.15.4k

IEEE 802.15.4k is proposed in [18] to provide a feasible means to handle the problem of long range communications with low energy consumption. Specific techniques in the physical and media access control (MAC) layers are designed and implemented for improving the transmission reliability and minimizing the network infrastructure and maintenance. This specification mainly works in three frequency bands, i.e., 868 MHz, 915 MHz, and 2.4 GHz, which is plotted in Figure 11.9. Different from the other LPWA solutions, this protocol can work in the 2.4 GHz band. However, thanks to a robust physical layer design, it operates well over a long-range wireless link and under the most challenging RF environments. New advanced techniques are proposed to meet these requirements in both the physical and MAC layers [9]. Key techniques in the two layers are introduced in detail as follows.

Physical Layer   As depicted in Figure 11.10, the physical protocol data unit (PPDU) is composed of a synchronization header (SHR) and a physical service data unit (PSDU). The SHR is responsible for retrieving frequency, symbol, and frame

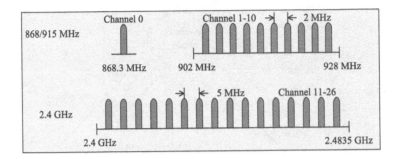

Figure 11.9 Frequency bands for IEEE 802.15.4k.

synchronization information, which includes the preamble and the start of the frame delimiter (SFD). The PSDU is the data field of the PPDU. The PPDU processes in the physical layer are illustrated in Figure 11.10. The major functional blocks employed in the physical layer are introduced in detail below:

- Convolutional FEC and interleaving
  Convolutional FEC encoding and interleaving are conducted at the PSDU. According to the protocol, the convolutional code rate and constraint length are set to be 1/2 and 7, respectively. The encoded data are then interleaved using a pruned bit reversal interleaving algorithm. Data reliability can be significantly improved by using these techniques;

- DSSS
  The direct sequence spread spectrum (DSSS) enables the receiver to detect signals with a very low carrier-to-interference ratio. The process gain depends highly on the type and length of the spreading code. As one of the most commonly utilized spreading codes, the Gold code is chosen by IEEE 802.15.4k, and the spreading factor is up to 2^{15} in order to provide a sufficient link budget. However, as the computational complexity of the DSSS transceiver increases exponentially with the length of the spreading sequence, high efficient digital

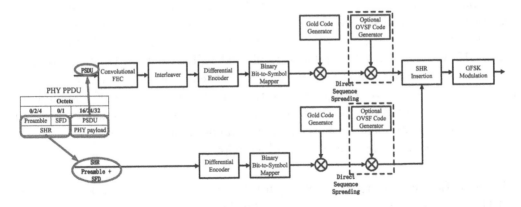

Figure 11.10 Key functional blocks in the physical layer.

signal processing algorithms are need to be designed for this LPWA system; and

- Modulation
  According to IEEE 802.15.4k, the modulation scheme is usually BPSK or O-QPSK. In addition, Gaussian frequency shift keying (GFSK) is an option owing to its good performance in overcoming the effect of the frequency or phase offset. Although performance degradation is unavoidable due to non-coherent demodulation, the DSSS is able to tolerate an acceptable frequency offset at very low SNRs.

MAC Layer LPWANs often need to support massive IoT device accesses with low energy consumption. Therefore, specific techniques in the MAC layer are designed to tackle these issues [9]. These MAC techniques focus primarily on two aspects, channel access and network topology.

- Channel Access
  Channel access plays a vital role in improving the capacity of LPWAN. Along with a rapid increase in the number of devices, the selection of channel access schemes becomes particularly important. Channel access schemes are commonly classified into two broad categories, i.e., the reservation-based and contention-based methods.

  Reservation-based methods are designed to divide radio resources in various dimensions (e.g. time, frequency, etc.), and then distribute these resources to various devices. The reservation of radio resources is relatively fixed, which can effectively reduce possible collisions. However, an increasing number of devices may cause a shortage of radio resources, which is proven to be the bottleneck of the network's capacity.

  In order to improve the capacity of the network with limited radio resources, contention-based methods have been paid an increasing attention for LPWA systems. The contention-based methods take advantage of the fact that the common medium is shared among devices and devices need to compete for access to the medium according to certain rules. The contention-based methods are effective especially when there are massive and unpredictable access activities. Without strict synchronization between the transmitter and receiver, the energy consumed by synchronization signaling is reduced. Devices can enter into the sleep mode to save energy, and be woken up once data transmission takes place. However, due to the competition mechanism, collisions are inevitable when devices compete for the access to the common medium. Several techniques have been designed to tackle this issue, of which the most widely used one is the Carrier Sense Multiple Access with Collision Avoidance (CSMA/CA). In order to reduce channel collisions, this method employs a carrier sensing mechanism which is responsible for sensing occupied carriers and choosing free ones. However, this mechanism is not suitable for some extreme scenarios. When signals from remote devices are severely attenuated attributed to a large path loss, the target device is difficult to detect these signals which may ultimately result in collisions. Furthermore, a CSMA-based system may become inefficient when there are massive simultaneous connections to the LPWAN. As a result, the CSMA/CA mechanism is not always practical for massive access scenarios.

Apart from CSMA-based systems, Aloha-based protocols perform better with the additional physical techniques in LPWANs. For instance, the DSSS technique enables an LPWAN to detect and identify multiple arriving packets simultaneously. Based on specific spreading sequences, data packets from multiple devices can be successfully decoded even if their transmission periods overlap completely. Moreover, some overhead caused by carrier sensing, such as the request to send (RTS) and clear to send (CTS) control messages, can be saved in Aloha-based protocols, making it more energy-efficient for LPWANs.

- Star Topology
  As is known to all, topology is of great importance to the performance of wireless networks. A proper topology can greatly improve the network's efficiency, reliability, scalability, etc. Traditionally, multi-hop relay topologies including the mesh, tree, cluster, and chain are widely utilized in IoT networks, such as the wireless sensor network (WSN) [20]. Based on the multi-hop mechanism, the communications coverage of these networks can be greatly extended. However, these topologies have to face some serious issues. As some devices are responsible for relaying the data from remote ones, a breakdown of one device may have a severe impact on multiple devices. Furthermore, the problems of high transmission latency and large signaling overhead are also nonnegligible for multi-hop networks.

  Thanks to the physical layer techniques in the 802.15.4k specification, the communications range of a single hop from the devices to the AP is greatly extended, which is large enough to cover a wide area. On this occasion, the star topology is superior to its multi-hop counterpart. In the star topology, devices are distributed around the AP and communicate directly with the AP through single-hop links, which is shown in Figure 11.11. As there are no extra relaying tasks for the devices, both energy consumption and signaling overhead can be greatly reduced. In addition, the operation conditions of a certain device are not reliant on the other devices, which makes the entire system more robust to possible failures and easier for scalability.

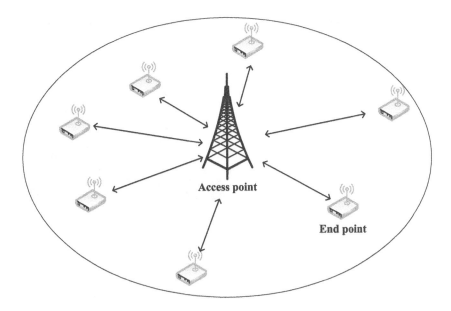

Figure 11.11   Star topology of the LPWAN.

11.3   IMPLEMENTATION OF LPWAN BASED ON IEEE 802.15.4K

Based on the IEEE 802.15.4k specification, we have designed and implemented a prototype LPWAN. In our prototype, the network mainly consists of an AP and several devices. Devices are usually embedded systems which are responsible for data collection, signal processing, and signal transmission. These signals are received by the antenna of the AP and then decoded for further use. The two crucial components are introduced as follows.

11.3.1   Access Point (AP)

The AP is implemented using a software-defined radio (SDR) platform, which is based on the concept that communications modules should be developed in software as much as possible. Thanks to this SDR platform, our prototype LPWA system can be deployed rapidly with low costs [15].

As depicted in Figure 11.12, the AP includes a universal software radio peripheral (USRP) and a general purpose processor (GPP) [14]. The Ettus USRP B210 is selected due to its wide frequency range covering all the operating frequency bands of IEEE 802.15.4k. The GPP is a common computer in this system. A photo of the AP is shown in Figure 11.13.

After receiving signals via the antenna, the USRP is used for essential signal conversion, such as frequency conversion, RF band selection, and AD/DA conversion. Then the signals are transmitted to the GPP through the USB interface for baseband signal processing. All the processing procedures are implemented by software in the C++ language. The codes can be greatly simplified through using specific open-source function libraries. Moreover, the AP uses multi-thread processing to handle concurrence data transmission from multiple devices.

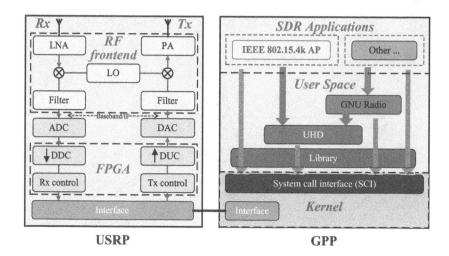

Figure 11.12   Framework of the AP.

11.3.2   Devices

Devices are used to gather useful data and transmit them to the AP through a wireless channel. As shown in Figure 11.14, the LPWA device mainly includes: 1) Controller module; 2) Sensor module; 3) LPWA transmitter module; and 4) Power module.

- Controller module
  The controller module is based on a powerful microcontroller unit (MCU), i.e., STM32F103RC. This MCU is widely used in signal processing due to its powerful hardware, e.g., a 32-bit Cortex-M3 microcontroller operating at the 72 MHz frequency, a Flash memory up to 512 Kbytes, and an extensive range of enhanced I/Os and peripherals connected to two APB buses [2]. Signals processed by the MCU are sent to the RF transceiver through a general-purpose input/output (GPIO) interface.

  Apart from the RUN mode, this controller module also provides several low power consumption modes for energy saving, i.e., the SLEEP, STOP, and STANDBY modes. Once a monitoring node is powered or reset, the controller

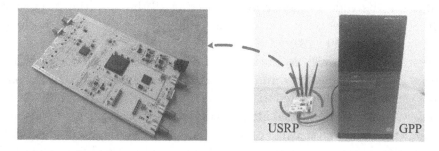

Figure 11.13   Photo of the AP.

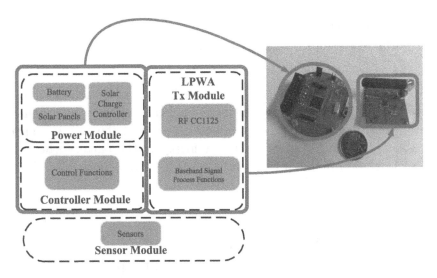

Figure 11.14 Main functional components of the LPWA device.

enters in the RUN mode. Under this circumstance, the monitoring node works at full capacity for data processing and thus consumes a lot of energy. Thus, the monitoring node may enter into the low power modes until being woken up, and then enter into the RUN mode either periodically or evoked by events. In the SLEEP and STOP modes, all the I/O pins maintain the same state as those in the RUN mode, and the contents in the SRAM and register are kept, which consumes some power to a certain extent. Moreover, the STANDBY mode needs the least amount of power but loses the contents stored in the SRAM and register, which is preferred in our design.

Furthermore, computing the spreading sequences has become one of the most energy-consuming procedures with a spreading factor (SF) up to 32,768. Thanks to the flash memory of the MCU, the spreading sequence is only calculated once and then stored in the memory. This information can be kept in any mode. Next time when the node is wakened from the low power modes, the spreading sequence can be easily read from the flash memory.

- Sensor module
  LPWA devices are widely used in wireless sensing applications due to their low cost and long battery life. Data are collected using specific sensors such as the position sensor, temperature sensor, and air quality sensor. Thanks to the variety of the sensors, physical information is converted into electrical signals and then processed by the MCU.

- LPWA transmitter module
  The RF transceiver is based on the Texas Instruments (TI) CC1125 chip, which has ultra-high performance in the narrowband. Signals from the MCU are modulated in this chip and sent to the AP through a wireless channel. This powerful transceiver offers common modulation techniques (e.g., 2-FSK, 2-GFSK, 4-GFSK, etc.) and can work in several frequency bands (e.g., 169 MHz, 433

MHz, 950 MHz, etc.) [23]. In addition, it has very high receiver sensitivity, i.e., −129 dBm at 300 bps, −123 dBm at 1.2 kbps, and −110 dBm at 50 kbps.

- Power module

  The energy of all the components is supplied by the power module. This module is based on a lithium battery which can be recharged via a solar panel. A solar charge controller is used in this module to control the voltage generated by the solar panel and to store electrical energy in the battery safely. With the aid of this controller, the battery is protected from both over-charging and over-discharging.

11.3.3 Experimental Results

For the purpose of validating the performance of the proposed LPWAN, several field tests are carried out in the urban environment. The AP with a wipe antenna is deployed on the roof of a 15-floor building and the devices are distributed around the AP. Through a 10-meter cable, the antenna is connected with the USRP B210 board located in the laboratory. Key parameters of our experiments are stated as follows. The operation frequency of the system is 433 MHz and the symbol rate is 200 k symbols per second. The PPDU of all data packets consists of a four-octet preamble and a 16-octet PSDU, which is spread by the Gold code with an SF of 32,768. The transmit power of the devices is set to be 15 dBm. The CRC is implemented in this system to ensure data integrity. The carrier-to-interference (C/I) is selected as the performance metric of our experiments, which is defined as the ratio of the normalized peak output of the preamble detector to the received signal power. A higher C/I implies more reliable communications. These experiments are designed to validate the LPWA system in various scenarios, e.g., the outdoor scenario, indoor scenario, underground scenario, etc. The multi-user performance of the network is tested as well. Furthermore, experimental results reveal that when the C/I value is greater than −30 dB, the AP can decode the received packets correctly.

Outdoor/indoor In this experiment, the C/I values of 20 spots around the AP are measured, which include both the outdoor and indoor scenarios. The deployment of the LPWA system and the measured C/I values are illustrated in Figure 11.15. Due to the complicated radio propagation characteristics, the C/I values do not always decrease with the increase of the communication distance. However, it is clear that even the furthest spot, which is 3.4 km away from the AP, still has a C/I value larger than −30 dB. Therefore, it can be concluded that this LPWA system provides over 3 km coverage in both the indoor and outdoor scenarios.

Underground The performance of detecting underground devices may be different from the overground ones due to the penetration loss. Hence the underground scenario is tested in this experiment. Five devices are installed in the underground car park of the building, which are 300 meters away from the AP on the roof. The deployment

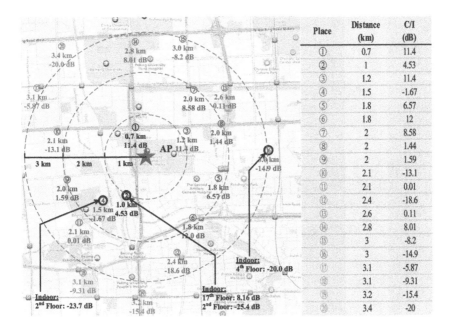

Place	Distance (km)	C/I (dB)
①	0.7	11.4
②	1	4.53
③	1.2	11.4
④	1.5	-1.67
⑤	1.8	6.57
⑥	1.8	12
⑦	2	8.58
⑧	2	1.44
⑨	2	1.59
⑩	2.1	-13.1
⑪	2.1	0.01
⑫	2.4	-18.6
⑬	2.6	0.11
⑭	2.8	8.01
⑮	3	-8.2
⑯	3	-14.9
⑰	3.1	-5.87
⑱	3.1	-9.31
⑲	3.2	-15.4
⑳	3.4	-20

Figure 11.15 Coverage performance of the developed LPWA prototype in indoor/outdoor scenarios.

of this experiment is plotted in Figure 11.16. The C/I values labeled in the figure indicate that the LPWAN is still able to cover reasonable underground areas.

Multi-user performance Five devices are distributed in the scenario illustrated in Figure 11.16. All the devices are set to send data at the same time. Therefore the transmission timing of each device can be regarded as nearly simultaneously. After that, the preamble of each packet is detected in the AP, which is shown in Figure 11.17. It is easily seen that the preambles of all the devices are distinguishable, which buttresses the excellent multi-user performance of our developed LPWAN.

11.4   LPWA-BASED AIR QUALITY MONITORING SYSTEM

Air pollution has recently become a serious problem because of rapid developments in industry and transportation of the modern society. When people are exposed to air pollutants for an extended period of time, they are more likely to suffer from a series of severe diseases [5]. There is therefore increased monitoring of air quality interest in which is considered effective to reveal the trend of the air condition and guide human's daily activities. However, due to the high cost and large volume, traditional air quality monitoring stations are unlikely to be widely deployed. Moreover, complicated calculation is needed offline so as to obtain precise measurements, which usually consumes a large amount of time. These challenges explain the low spatial and temporal resolution of current air quality data [8].

Alongside a rapid development of wireless sensor networks (WSNs), air quality sensors can be widely deployed with much lower costs [4] [13] [7]. However, due to the

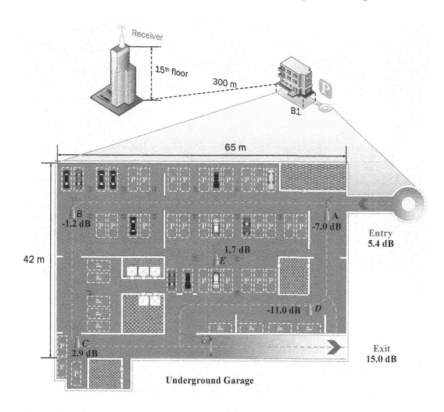

Figure 11.16 Coverage performance of the developed LPWA prototype in the underground scenario.

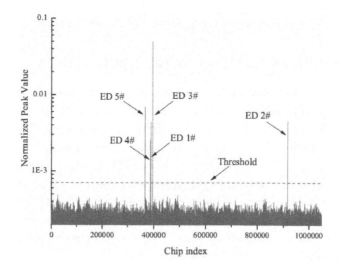

Figure 11.17 Multi-user performance of the developed LPWA prototype.

limited transmission range of traditional wireless techniques (e.g., GPRS, Bluetooth, Zigbee, etc.), air quality sensors can not be deployed too far away from the gateway. Therefore, these methods cannot meet the communications requirements of a massive number of air quality sensors deployed over a large sensing area.

Thanks to the advance of LPWA techniques, ubiquitous connectivity between low-cost sensors and the AP can be established with low energy consumption [11]. Due to the special design in the physical and MAC layers, LPWANs based on IEEE 802.15.4k can provide suburban and urban transmission ranges of around 20 km and 5 km, respectively, which is fairly enough for the purpose of monitoring local air quality [10].

In this section, an LPWA-based air quality monitoring system is designed and implemented. This system focuses on one of the main air pollution sources, i.e., the Particulate Matter (PM) 2.5. Data collected by the sensors are sent to the AP using the IEEE 802.15.4k specification. In order to store and process the sensed data, an IoT cloud with a variety of servers is established. Users can obtain real-time or historical air condition information through the display application conveniently.

11.4.1 System Architecture

As depicted in Figure 11.18, a three-layer hierarchical architecture is proposed in the air quality monitoring system, i.e., the sensing layer, network layer, and application layer.

11.4.1.1 Sensing Layer

As an indispensable component of the system, the sensor layer is mainly responsible for air quality sensing and data reporting. In order to improve the extensity of the air quality data, the monitoring nodes are usually deployed widely across a large geographical area. Thus, large-scale air quality data can be collected by these nodes. Compared to the traditional ultra-red sensors, the PMS5005 sensor uses laser to detect the Particulate Matter 2.5 ($PM_{2.5}$), which can provide high precision without calibration. Therefore, it is selected for our system. In addition, SHT20 is used to sense real-time temperature and humidity information, which can greatly enrich the gathered data. The photo of the $PM_{2.5}$ sensor is shown in Figure 11.19.

Since the air quality usually does not vary too much, the monitoring node can be woken up to sense the air quality and transmit data every ten minutes by default. An adaptive mechanism is also designed and implemented. If the difference between two successive air quality data readings exceeds a given threshold, the sensing cycle becomes shorter automatically. Otherwise, the cycle may be increased.

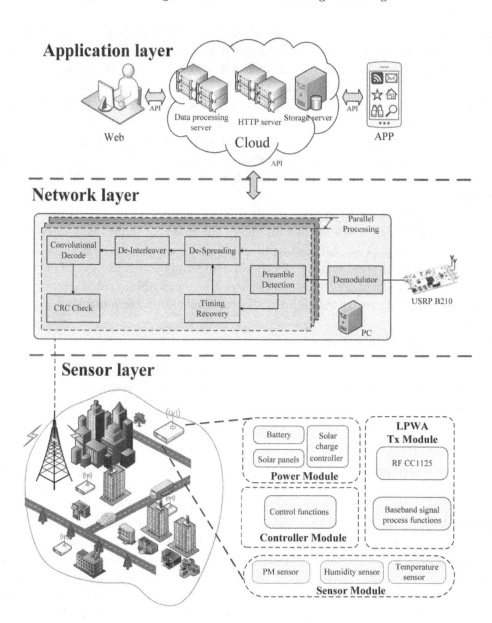

Figure 11.18 Architecture of the LPWA-based air quality monitoring system.

Figure 11.19 Photo of the $PM_{2.5}$ sensor.

11.4.1.2 Network Layer

In the network layer, an LPWAN based on the IEEE 802.15.4k specification is employed to provide ubiquitous connectivity between the monitoring nodes and the AP. A star topology is adopted in this network, which can reduce maintenance and deployment costs of the system. Without having to function as relays in the multi-hop topology, significant energy consumption can be saved by the sensor nodes. Furthermore, a particular DSSS technique is adopted in the physical layer so as to extend the coverage of single-hop communications. Details about the IEEE 802.15.4k-based LPWAN are presented in Section 1.2.3.

11.4.1.3 Application Layer

An IoT cloud is implemented in the application layer with the objective of hosting and processing the sensed data. When the AP receives air quality data reported from the sensor nodes, it first stores the data in the IoT cloud. After that, necessary processing is conducted on these data such as data aggregation and data cleaning. Moreover, client applications are supported by the IoT cloud which can provide a convenient means for users to access the air quality information. Both a web access interface and a smartphone APP is developed in this system.

With the help of the sensor node, the system can obtain rich air quality data. Then, the IoT cloud is in demand to provide a speedy and convenient way to store these data and visualize the data when required.

Server: Three types of servers are used in the IoT cloud with different functions, i.e., the data processing server, storage server, and HTTP server. Thanks to the development of virtualization techniques, these servers can be hosted on visual machines so as to make full use of physical resources, e.g., CPU, memory, etc. The servers deployed in the IoT cloud are introduced in detail as follows.

- Data processing server
  Raw data cannot be directly utilized in the presence of the possible transmission errors or machine failures. Some pre-processing procedures are needed to detect and clean the "dirty data" so as to ensure the integrity and reliability of the dataset. These procedures are carried out in the data processing server;

- Storage server
  Storage server is responsible for storing the air quality data into the database. Traditionally, relational databases are commonly employed in IoT clouds. However, with a dramatic increase in both data volumes and types, the read-write speed of the relational database has become the bottleneck for the performance of the cloud. Consequently, non-relational databases have received growing attention. This system employs Redis as the database. As one of the most promising non-relational databases, Redis stores all key-value data in the memory. As a result, both the flexibility and the I/O speed of the database can be greatly improved;

- HTTP server

  Based on the traditional request-response mechanism, the HTTP server is able to accept user requests and respond to them accordingly. In our proposed system, the HTTP server is developed using Servlet and JSP (JavaServer Pages) [24]. For the sake of managing the lifecycle of servlets, a web container termed Tomcat is deployed in this server. Air quality data can be obtained by a web browser or a mobile APP through the HTTP protocol, which provides a convenient means for user access from varying OS platforms.

Display Applications: Users with various kinds of clients can have access to the air quality information, either through a website or a mobile APP. Figs. 11.20 (a) and 11.20 (b) depict the GUI interfaces of the website and the mobile APP, respectively.

Through the display applications, the air quality information can be shown in real-time including:

- Current Air Quality Indicator (AQI);

- AQI trend of the present day;

- AQI trend in last week/month.

11.4.2 Experimental Results and Analysis

11.4.2.1 Experimental Configurations

In order to verify the reliability of this air quality monitoring system, field tests are carried out in a typical urban environment. The AP is located on the roof of a 15-floor building with a whip antenna which is connected with a USRP B210. Our system is deployed using star topology illustrated in Figure 11.21 (a), which consists of five monitoring nodes distributed over an area of around a radius of 3 km from the center of our university campus. A photo of the monitoring node is shown in Figure 11.21 (b). Key parameters of our experiments are listed in Table 11.2.

As one of the most commonly used air quality indicators, the AQI is selected as the metric of this system. It is calculated based on the concentration of six air pollutants, i.e., $PM_{2.5}$, PM_{10}, carbon monoxide (CO), sulfur dioxide (SO_2), nitrogen dioxide (NO_2), and ozone (O_3) [21]. A number of government agencies or research institutions publish the real-time AQI based upon these six pollutants, and then advise residents on daily activities accordingly. As the AQI increases, the probability for residents to be harmed by the air pollutant becomes higher.

It is noted that different countries have their own AQI calculation methods. According to China's Ministry of Environmental Protection in [21], the AQI is calculated as follows.

$$i_p = \frac{A_{high} - A_{low}}{B_{high} - B_{low}}(\beta_p - B_{low}) + A_{low}, \tag{11.1}$$

where i_p is the Individual Air Quality Index (IAQI) of the pth pollutant, β_p is the concentration of the pth pollutant, B_{high} is the nearest concentration point larger

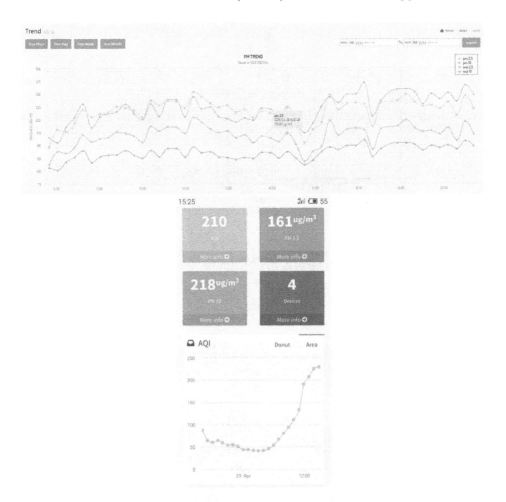

Figure 11.20 Examples of the GUI interface of the air quality monitoring system: (a) Webpage; (b) Mobile APP.

than β_p, B_{low} is the nearest concentration point smaller than β_p, A_{low} is the IAQI point corresponding to B_{low}, and A_{high} is the IAQI point corresponding to B_{high}.

The critical points for six air pollutants are given by China's Ministry of Environmental Protection in [21]. For example, the critical points for $PM_{2.5}$ and PM_{10} are listed in Table 11.3, which are the only two metrics concerned in our air quality monitoring system. Then, the AQI can be given as the highest value of all the measured IAQIs, i.e.,

$$\alpha = \max\{i_1, i_2, ..., i_n\}. \tag{11.2}$$

11.4.2.2 *Results and Analysis*

Data from Mar. 1st to 14th 2016 are chosen for further analysis. The concentration of $PM_{2.5}$ changed dramatically during this period of time. Air quality information including $PM_{2.5}$, PM_{10}, and AQI are depicted in Figure 11.22. As can be observed

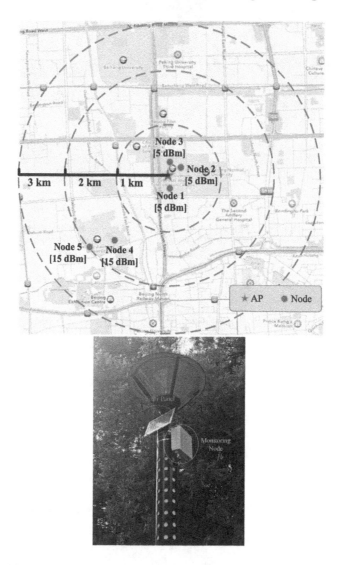

Figure 11.21 Field deployment of the LPWA-based air quality monitoring system: (a) Location of the monitoring nodes in the field trial; (b) Photo of the monitoring node.

from the figure, there is a strong relationship between the concentrations of $PM_{2.5}$ and PM_{10}. In order to verify the reliability of this system, another dataset is also obtained from a $PM_{2.5}$ historical data platform as in [25]. The comparison between these two datasets is shown in Figure 11.23. It can be clearly seen that the trends of the two lines accord well with each other, and the difference may be due to the different measure locations. Therefore, the air quality data gathered from the proposed system can satisfy the accuracy and reliability requirements.

It is well known that air quality is highly related to some meteorological conditions, such as the temperature, humidity, and wind speed. This experiment is conducted in order to reveal the relationship between $PM_{2.5}$ and the wind speed. Wind speed data can be obtained from a well-known meteorological data website [22]. As

Table 11.2 Key parameters of the LPWA-based air quality monitoring system

Parameters	Value
SDR hardware	Ettus Research USRP B210
CPU	Intel Core i7-3632QM 2.2GHz / i7-3770 3.4GHz
Operating system	Ubuntu 12.04 LTS, 64 bit
GNU Radio	Version 3.7.x
IT++ library	Version 4.3.1
AP location	On the 15th floor of the main building in BUPT
Frequency	433 MHz
Symbol rate	200 ksym/s
Devices Tx power	5, 15 dBm
Spreading factor (SF)	32768
Preamble	4 bytes
PSDU	16 bytes

Table 11.3 Concentration points for $PM_{2.5}$ and PM_{10}

IAQI	$PM_{2.5}$ $(\mu m/m^3)$	PM_{10} $(\mu m/m^3)$
0	0	0
50	35	50
100	75	150
150	115	250
200	150	350
300	250	420
400	350	500
500	500	600

plotted in Figure 11.24, during April 3–7, 2016, the wind speed was slower than 4 m/s, which helped gradually increase the $PM_{2.5}$ concentration, and reached the peak on the noon of April 7, 2016. Then, it dropped rapidly and kept low during April 8–11, which was due to the dispersal of air pollutants by strong winds. From the above analysis, it can be concluded that strong winds help greatly with the dispersal of air pollutants.

11.5 CONCLUSION AND OUTLOOK

This chapter focuses mainly on the LPWAN. First, we gave an overview of IoT in the 5G era. Next, a brief introduction to LPWANs was presented. Both the application scenarios and classifications were introduced in detail. Among all the LPWANs, we mainly focused on the NB-IoT and IEEE 802.15.4k. After that, an LPWAN based on IEEE 802.15.4k was designed and implemented. Several field tests were conducted to test the performance of our system. Experimental results show that the system performs well in the indoor, outdoor, and underground scenarios. Furthermore, through applying multi-thread processing, the AP is able to support massive device access.

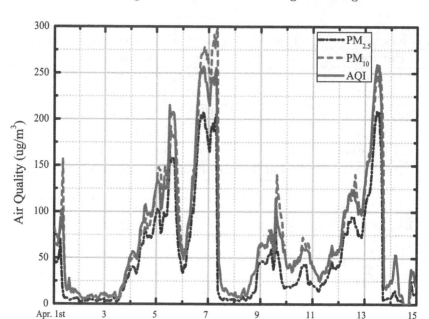

Figure 11.22 Air quality data including $PM_{2.5}$, PM_{10}, and AQI.

In addition, we implemented an air quality monitoring system by using the advanced LPWA techniques. Laser PM sensors were used to collect real-time air quality data and then send the sensed data to the AP for further processing. The reliability of the gathered data was verified compared with a well-known air quality database.

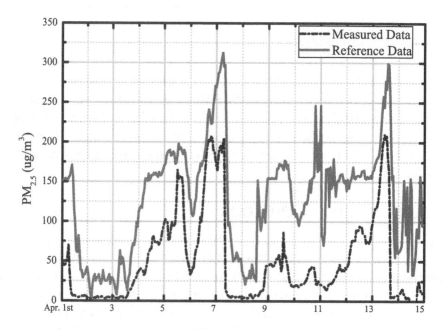

Figure 11.23 Comparison between the measured and reference data on the $PM_{2.5}$ concentration.

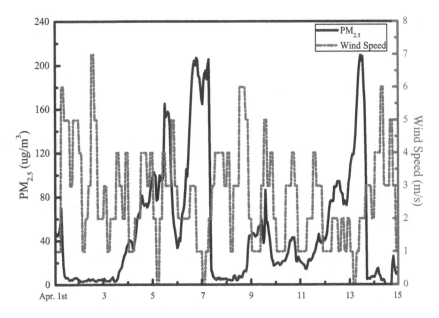

Figure 11.24 Relationship between the $PM_{2.5}$ concentration and the wind speed.

Based on the analysis of the air quality data, some interesting facts were revealed. For example, there exists a strong correlation between the concentrations of $PM_{2.5}$ and PM_{10}. Moreover, air quality is highly related to meteorological conditions, e.g., the wind speed.

Further studies on LPWANs are still needed in certain aspects. For instance, the capacity of the LPWAN can be further improved using methods such as channel diversity and adaptive transmission strategies. Moreover, LPWANs usually have limited data rates, which cannot fully satisfy the requirements of 5G systems. Therefore, other IoT techniques, such as Wi-Fi, Bluetooth, and Zigbee, are likely to cooperate with the LPWA ones in the forthcoming 5G era.

Bibliography

[1] TR 45.820. Cellular system support for ultra-low complexity and low throughput internet of things (ciot). v13.1. *3GPP Standards*, 2016.

[2] Alldatasheet.com. Project Website: http://www.alldatasheet.com/datasheet-pdf/pdf/231959/STMICROELECTRONICS/STM32F103RC.html.

[3] ERICSSON. Nb-iot: A sustainable technology for connecting billions of devices. *ERICSSON Technology Review*, 93(3):2–11, 2012.

[4] C. Peng et al. Design and application of a voc-monitoring system based on a Zigbee wireless sensor network. *IEEE Sensors Journal*, 15(4):2255–2268, 2015.

[5] E. Fotopoulou et al. Linked data analytics in interdisciplinary studies: The health impact of air pollution in urban areas. *IEEE Access*, 4:149–164, 2016.

[6] H. Dhillon et al. Wide-area wireless communication challenges for the internet of things. arxiv:1504.03242. *Networking and Internet Architecture*, 2015.

[7] J. Chen et al. Utility-based asynchronous flow control algorithm for wireless sensor networks. *IEEE Journal on Selected Areas in Communications*, 28(7):1116–1126, 2010.

[8] K. B. Shaban et al. Urban air pollution monitoring system with forecasting models. *IEEE Sensors Journal*, 16(8):2598–2606, 2016.

[9] K. Zheng et al. Challenges of massive access in highly dense lte-advanced networks with machine-to-machine communications. *IEEE Wireless Commun.*, 21(3):12–18, 2014.

[10] K. Zheng et al. Design and implementation of lpwa-based air quality monitoring system. *IEEE Access*, 4:3238–3245, 2016.

[11] L. Lei et al. Delay-optimal dynamic mode selection and resource allocation in device-to- device communications, Part ii: Practical algorithm. *IEEE Transactions on Vehicular Technology*, 65(5):3491–3505, 2016.

[12] M. Centenaro et al. Long-range communications in unlicensed bands: The rising stars in the iot and smart city scenarios, arxiv:1510.00620. *IEEE Wireless Communications*, 2016.

[13] S. He et al. Mobility and intruder prior information improving the barrier coverage of sparse sensor networks. *IEEE Transactions on Mobile Computing*, 13(6):1268–1282, 2014.

[14] X. Xiong et al. Implementation and performance evaluation of lecim for 5G m2m applications with sdr. In *IEEE Globecom Wks.ps (GC Wksps.)*, pages 612–617. IEEE Press, 2014.

[15] X. Xiong et al. Low power wide area machine-to-machine networks: Key techniques and prototype. *IEEE Commun. Mag.*, 53(9):64–71, 2015.

[16] Y. Wang et al. A primer on 3gpp narrowband internet of things (nb-iot). arxiv:1606.04171. *Networking and Internet Architecture*, 2016.

[17] Z. M. Fadlullah et al. Toward intelligent machine-to-machine communications in smart grid. *IEEE Commun. Mag.*, 49(4):60–65, 2011.

[18] IEEE Standard for Local and Critical Infrastructure Monitoring Networks Metropolitan Area Networks Part 15.4: Low-Rate Wireless Personal Area Networks (LR-WPANs) Amendment 5: Physical Layer Specifications for Low Energy. Ieee std 802.15.4k. 2013.

[19] IMT-2020 (5G) Promotion Group. 5G concept. 2015.

[20] Q. Mamun. A qualitative comparison of different logical topologies for wireless sensor networks. *Sensors*, 12(11):14998–13, 2012.

[21] People's Republic of China Ministry of Environmental Protection. People's Republic of China ministry of environmental protection standard: Technical regulation on ambient air quality index. 2012.

[22] Rp5.ru. Project Website: http://rp5.ru/Weather_in_the_world.

[23] Ti.com. Project Website: http://www.ti.com/product/CC1125/datasheet.

[24] Tomcat.apache.org. Project Website: https://tomcat.apache.org/tomcat-4.1-doc/servletapi/.

[25] Young-0.com. Project Website: http://www.young-0.com/airquality.

CHAPTER 12

A Data-centered Fog Platform for Smart Living

Jianhua Li

Swinburne University of Technology

Jiong Jin

Swinburne University of Technology

Dong Yuan

The University of Sydney

Marimuthu Palaniswami

The University of Melbourne

Klaus Moessner

University of Surrey

CONTENTS

Nowadays, smart environments (e.g., smart home, smart city) are built heavily relying on Cloud computing for the coordination and collaboration among smart objects. The Cloud is typically centralized but smart objects are ubiquitously distributed; thus, data transmission latency (i.e., end-to-end delay or response time) between Cloud and smart objects is a critical issue especially to the applications that have strict delay requirements. To address this concern, a new Fog computing paradigm has been recently proposed by the industry. The key idea is to bring the computing power from the remote Cloud closer to the users, which further enables real-time interaction and location-based services. In particular, the local processing capability of Fog computing significantly scales down the data volume towards the Cloud, and it in turn has great impacts on the entire Internet. In this chapter, smart living as one of the primary elements of smart cities has been conceptualized to EHOPES, namely smart Energy, smart Health, smart Office, smart Protection, smart Entertainment and smart Surroundings. And then the data flow analysis has been investigated to disclose a variety of data flow characteristics. Based on these studies, a data-centered Fog platform has been developed to support smart living. Case studies are also conducted to validate and evaluate the proposed platform.

12.1   INTRODUCTION

12.1.1   Smart City

Some ancient people are believed to have lived in a variety of caves. The cave dwellers lived on fishing, hunting and primitive agriculture. People gathered with the developing of productivity to dwell in villages, towns and cities, as shown in Figure 12.1. City dwellers today heavily rely on public services such as water, electricity, gas, telephony and the Internet. However, no matter where people live, the human pursuit of a dreamlike living environment never stops in terms of affordability, security, comfort and automation. In 2008, the abstraction of smart cities was introduced by IBM, as a part of its Smarter Planet initiative. By the start of 2009, the idea had captivated the imagination of lots of nations in Europe, America and Asia. Interestingly, all of a sudden, numerous cities announced their plans to turn into smart cities (to name a few, Melbourne, Beijing, Vienna and Amsterdam). Globally, new cities such as Masdar outside of Abu Dhabi, Paredes in Portugal and Songdo in South Korea label themselves as smart and older cities (such as Silicon Alley in New York City, Silicon Roundabout in London and Akihabara in Tokyo) regenerate themselves as smart.

According to Wikipedia, "a smart city is an urban development vision to integrate multiple information and communication technology (ICT) and Internet of Things (IoT) solutions in a secure fashion to manage a city's assets. The city's assets include, but are not limited to, local departments information systems, schools, libraries, transportation systems, hospitals, power plants, water supply networks, waste

Figure 12.1   From cave dwellers to smart city residences

management, law enforcement, and other community services." "ICT allows city offi-cials to interact with the community and the city infrastructure and to monitor what is happening in the city, how the city is evolving, and how to enable a better qual-ity of life. Through the use of sensors integrated with real-time monitoring systems, data are collected from citizens and devices—then processed and analyzed." In brief, a smart city can be defined as a city in which ICT is merged with its infrastructures and integrated using new hardware and software technologies.

Their target on building a smart city is to improve quality of life by using technol-ogy to improve the efficiency of services to meet the requirements of city dwellers. The city services can be tailored to the needs of individual citizens by using technology to integrate the information systems provided by different service delivery agencies to enable better services. Furthermore, a smart city may reduce crime and respond faster to public threats by processing information in real time. For example, in Chicago, a new public safety system has been in use that allows real time video surveillance and faster, more effective responses to emergencies. In addition, a smart city may monitor and manage an entire water system from rivers to pipes at home. In Galway, Ireland, all the stakeholders from government to family can acquire up-to-date water forecast information. Last but not least, a smart city can help residents to bring down their living costs by reducing consumption. The scope of this study has engaged diverse research fellows in the domains that include city planning, sustainable environment, transportation engineering, public health, economic forecasting and mobility.

12.1.2 Internet of Things

In order to achieve various services required by all kinds of citizens, the Internet has been extended from connecting computers to smart objects in an unthinkable manner, which is called the Internet of Things. The IoT is the network of smart objects, physical devices, vehicles, buildings and things embedded with electronics, software, sensors, actuators and telecommunication connectivity that enable these objects to process data. IoT is defined as "the infrastructure of the information so-ciety" by the Global Standards Initiative on Internet of Things (IoT-GSI) in 2013. IoT is increasing the ubiquity of the Internet by integrating every object for interac-tion via embedded systems, which leads to a highly distributed network of devices communicating with human beings as well as other devices. In brief, IoT is the key physical foundation of smart cites.

The term smart city has been developed as a catchphrase for the way global ICT leaders such as IBM, Cisco, Intel and others, are beginning to generalize their products as they see markets in cities representing the next wave of product inno-vations. Intel predicts there will be 26 things per head at average globally by 2020. Cisco forecasts IoT is a trillion-dollar level market and advocates 90 percent of things have not been connected in its Internet of Everything (IoE) white papers. IBM has invested billions of dollars to improve its networking products family. The reality is that so many intertwined and multi-faceted networks have been designed and im-plemented between people, institutions, places and social networks. We argue that the key insight for understanding the smart city is in understanding the structure of these coupled networks and how this structure evolves. From ICT point of view, data is the chief element in terms of network planning, design, implementation and maintenance.

Internet of Things (IoT) is connecting billions of people and things in an unprecedented manner. Machine-to-Machine (M2M), Machine to People (M2P) and People to People (P2P) network would be interconnected globally, which is significantly changing our society in terms of more comfortable living conditions, unimpeachable healthcare systems, customized relaxation and enjoyable working surroundings. The Internet is considered as the backbone network to link these heterogeneous and pervasive sensor networks. In order to be more affordable, sensors are widely made with limited capability in terms of processing, storage, communication, etc., which lead to a variety of ad hoc networks among those sensors such as Zigbee networks for home appliances and UWB network for home entertainment systems. These ad hoc networks have limited or no capability of supporting TCP/IP while the Internet has been built on a TCP/IP protocol suite thoroughly. Thus today's internetworking of sensor networks can be reviewed as a gateway or proxy based network. Within each ad hoc network, the sensors can be either IP-innocent or limited IP-awareness nodes, i.e., the fully fledged IP feature has not been supported. Those networks can be ZigBee, RFID and BACnet. The losses of transparency cause a number of issues, especially when the interaction is required among different ad hoc networks.

From the perspective of a service provider, we can see that a variety of ad hoc network of things exist on the Internet as a mixed instead of an integrated system. The network protocol used in one may be different from another, which leads to addressing format, data header structure and path selecting being totally in chaos during the communication between senders and receivers. Under this circumstance, the communication among those nodes will have to rely on their gateways. The network performance will be severely degraded because of this bottleneck and imbalanced traffic. Furthermore, this makes the network programming extremely complex and difficult. For example, a node may be required to have multiple addresses and names in order to meet the criteria of different ad hoc networks. A number of routing protocols such as 6LoWPAN, ROLL, CORE, uIP and CoAP have already been in this arena.

From the customer's point of view, this disorder can lead to more issues. First above all, end-to-end transparency loss is a severe issue. As TCP/IP is the predominant protocol on the Internet, any node that cannot support a complete TCP/IP protocol stack (either caused by IP-innocent or header compression) would be considered as a troublemaker, which further contributes to a hard security implementation such as IPSec. Second, the fragmentation and reassembly of data packets consume large amount of memory, which leads to severe packets drop (both user data and control messages) if the buffer has no free space. Third, sensor nodes quite often work in a sleep/wake up/listen to cycle mode in order to save energy, possibly only waking up 10 seconds a day or even 10 seconds a week. The sensor node will listen to see if it has any message once woken up; however, the router typically does not save the message that long. Other known issues are still massive, such as routing packets, limited energy supply and processing capability. In brief, reliability is a hard nut to crack.

12.1.3 Smart Living

In [7], a smart city is well performing in six characteristics, namely smart economy, smart mobility, smart environment, smart people, smart governance and smart living. Aiming to improve the quality of life, smart living guides people to a better, more balanced life with advice and tips by creating homes for families that are functional, energy-efficient, healthy and secure.

In [9], smart living has been further conceptualized as smart energy, smart health-care, smart office, smart protection, smart entertainment and smart surroundings (EHOPES). Smart living networks are developed with a variety of devices and media, which include mobile phones, tablets, personal computers, TVs, wearable devices, interactive message terminals, electronic appliances, etc. The explosion of Internet of Things (IoT) / Internet of Everything (IoE) [6] makes communication and collaboration of intelligent home appliances possible and necessary. A network is the essential component of smart living which is composed of smart objects [8] and a variety of processors. The smart objects include sensors, actuators, controllers and inter-connectors. The processors are used to control, communicate with and monitor the smart objects in the network. Current smart living networks take advantage of Cloud computing [2] (hereinafter referred to as Cloud) and IoT/IoE, thereby offering a number of services such as the smart home [16].

12.1.4 Ad Hoc IoT

The Institute of Electrical and Electronics Engineers (IEEE) group has defined a number of standards for ad hoc networks. Some of the popular wireless ones are 802.15.1 Bluetooth, 802.15.3 UWB, 802.15.4 ZigBee, 802.11 families and their variants.

The ad hoc networks are application-oriented in general. ZigBee is mostly used in home appliance and UWB is favored in entertainment systems. They are supposedly existing in a small area and serve only for one application. As the networks are designed for specific projects, the network components such as hardware and protocols can be fine-tuned to reinforce sensors in terms of processing, storage, communication and power. For example, the sink nodes are introduced to store data on the behalf of sensors; thus the sensor nodes can work at a sleep/wake up mode to extend the life of the battery.

Some of these ad hoc networks may not support predominant Transmission Control Protocol/Internet Protocol (TCP/IP) suites. Hence we call this an IP-innocent sensor network. The others may support a scaled down version of TCP/IP protocol suites. Thereby we name them IP-aware sensor networks. However neither of them supports fully fledged IP features. In the cases that the Internet is employed to connect these remote ad hoc networks, a gateway or proxy is required.

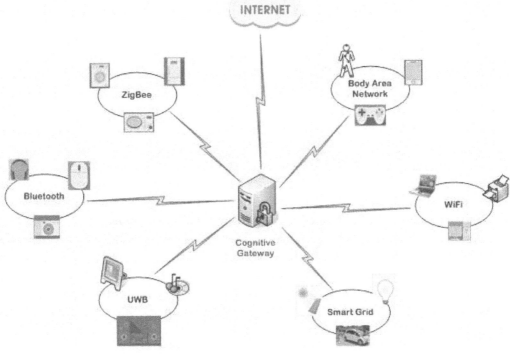

Figure 12.2 Gateway Based IoT

12.1.5 Gateway or Proxy Based IoT

A gateway or proxy typically supports at least two protocol suites, through which data packets can be interpreted between different protocols, as shown in Figure 12.2. As the TCP/IP is the predominant protocol used by the Internet, one of those supported protocols should be TCP/IP. There are two types of interfaces on this gateway, namely internal (the one connected to the ad hoc network) and external (the one connected to public network, such as the Internet) interfaces. By taking advantage of a gateway's Internet capabilities, the data generated by ad hoc network nodes can be delivered to a remote site across the Internet.

Some of the gateways have been enhanced in order to improve their capabilities of local services, such as data processing and internetworking among a number of ad hoc networks. This type of a gateway is called cognitive gateway. It manages the entire local ad-hoc network including application management, network management and network interconnection. As a fact of the concentrated management on access control, security, quality of service (Qos) and radio frequency channel, it serves best in a small office home office (SOHO) network.

However, there is a single point of failure in this type of network architecture. The network downtime not only causes financial and informational loss, but also leads to severe issues. One example is that a patient may lose his life if some emergent health care data cannot be processed immediately. Additionally, the capability of local processing and storage is limited. Moreover, the data security is hard to maintain. Last but not least, multi-tenancy is very poor. Thus Cloud based IoT has been introduced to overcome the above issues.

12.1.6 Cloud Computing Based IoT

According to Berkeley RAD, "Cloud computing refers to both the applications delivered as services over the Internet and the hardware and systems software in the datacentres that provide those services. The services themselves have long been referred to as Software as a Service (SaaS). The datacentre hardware and software is what we will call a Cloud." Evolved from Grid Computing, the Cloud aims to offer on-demand-self-service, broad network access, resource pooling, rapid elasticity, measured service in a "pay-as-you-go" manner. For the reason that the Cloud can be considered as an unlimited capability for processing and storage, it is supposed to be an ideal platform for heterogeneous and pervasive sensor nodes. Some researchers contributed to develop the Cloud base IoT architecture, application, as well as measurement, and a number of service providers (Amazon, Microsoft, etc.) offer service globally on the Internet. For example, a cognitive gateway [16] is employed to link smart objects to the Cloud data center when external interactivity is required. In this chapter we call it a Cloud model.

Although the Cloud is super powerful in terms of storage, processing and big data services, the Cloud based IoT has many limitations. First and foremost, neither a centralized Cloud nor distributed IoT sensor nodes can guarantee QoS, which makes the Cloud based IoT less appealing. In a real time data analytics scenario such as a smart traffic light control system, the traffic data should be collected and processed, then decision making and action taken should be completed within a few seconds. It is nearly impossible for the Cloud to offer such services. Second, it is mandatory to have Internet services to utilize the Cloud, but the Internet connection is less reliable and dependable in many places. Third, a majority of sensor nodes are considered as resource-constrained devices particularly for telecommunications. The trade-off to link ubiquitous sensor nodes to the centralized Cloud is persistent. In order to address the above issues, Fog computing is introduced as a complementary paradigm, as shown in Figure 12.3.

12.1.7 Fog Computing

Fog computing [5] (hereinafter referred to as Fog) is a newly proposed computing paradigm that offers certain local processing capability, which is able to address the above issues. In contrast to the Cloud, Fog has four unique features [4] to greatly support smart living, which are:

- Low latency, i.e., Fog offers millisecond to subsecond level latency, while it is in the minutes level in the Cloud.

- Proximity, i.e., Fog adopts the decentralized model, which is closer to smart objects. Cloud adopts the centralized model.

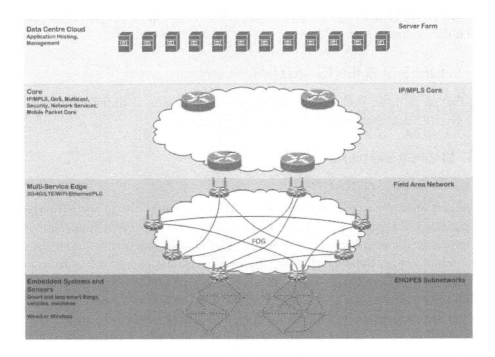

Figure 12.3 From cave dwellers to smart city residence

- Real-time interaction, i.e., Fog computing offers quick, even real-time, interaction. Cloud is perfect at batch processing.

- Multi-tenancy, i.e., both Fog and Cloud support multi-tenancy, but Fog performs better for applications that require low-latency.

Fog is a hierarchical network paradigm composed of two types of Fog nodes in terms of functionality, i.e., Fog Server (FS) and Fog Edge Node (FEN). An FS hosts a variety of management, collaboration, and coordination services between Cloud and FENs. An FEN provides adjacent computing for ubiquitous and heterogeneous smart objects in terms of processing, storage and communication.

In this chapter, smart living coverage to disclose EHOPES in terms of definition, involved devices and dataflow analysis will be investigated. This investigation shows the necessity of local processing for the majority of EHOPES data. Thus Fog is fit to EHOPES. Then the Fog paradigm and its relation to the Cloud is going to be explored. Next we study how Fog can be employed to support EHOPES. Based on the analysis of dataflow, the variety of interplay of Fog nodes (more details in Section 12.3.2) and Cloud will be elaborated. Then a data-centered Fog platform for smart living will be put forward. Following the work above, a case study and the evaluation have been conducted. This case study focuses on the actual latency in the two cases. Case 1: Cloud is employed; case 2, Fog is employed as the platform for EHOPES. A number of EHOPES applications (same amount of data) are conducted on the two platforms. The results demonstrate that Fog can significantly reduce latency in sharp contrast to Cloud. Then finally we conclude this chapter and list a few potential research directions.

12.2 EHOPES ELEMENTS AND DATAFLOW

In this section, we investigate the definition, devices and dataflow of EHOPES. Each EHOPES application brings up a subnetwork.

12.2.1 EHOPES and Dataflow

- Smart Energy

  Smart energy systems are committed to supplying customers with the best standard of service and quality in the energy industry. We define smart energy (electricity, gas and water) [12] in terms of energy generating [10], energy consuming, energy delivery [13] and billing [15]. Smart energy refers to using IoT and networking technologies to dynamically distribute energy in order to maximizes energy as well as minimize their cost, which involves decision-making and action-taking subsystems. In the smart energy network, metering devices pushed the usage data to and retrieved the billing information out from a number of energy providers intermittently. Residents are happy to know their energy consumption at real time from a variety of devices such as personal computers, tablets or mobiles; together they can learn, share, adapt, optimize and reduce their energy usage. A local Fog Server (FS, refer to Section 12.3.3) that helps to minimize energy costs will delight consumers by bringing timely benefits. For example, a program on FS can work out the cost for hot water supply through using different energy sources and take actions to minimize the cost. In a smart energy subnetwork, dataflow mainly exists between energy provider (Cloud) and local decision maker (FS).

- Smart Health care

  Health care is the keeping or improvement of health of resident via the diagnosis, treatment, prevention of injury, disease and other physical and mental impairments. A smart health care system seeks to intervene early in maximizing health and well-being of a population. Smart health care devices refer to wearable Body Area Network (BAN) devices, health care apps, medical robots, etc. They play an important role in terms of daily monitoring, data collection, tele-diagnostic processes and medical services. For example, BAN device provides live feedback on the wearer's health that helps to alert professionals and consumers to potential risks before they become serious. In the above example, massive repetitive data such as heart monitor stream can be filtered in the Fog Edge Node (FEN, refer to Section 12.3.2). A brief periodic report is stored on a FS and in Cloud as backup. Live data stream can be sent to professionals when telediagnostic services are in action. In this case, a majority of the dataflow exists between BAN devices and FEN. A small portion of data is transferred to FS and Cloud.

- Smart Office

  The home is increasingly being used as an office by homeworkers and teleworkers employers, employees and the community can benefit from a number of advantages such as reduced cost, improved motivation, freedom from traveling, and so on. Smart office is about aspects of business processes that drive daily operations in the office on the basis of projects and scheduled work tasks that are regularly performed by office employees. This subnetwork communicates with various project management systems, databases and information systems that control a regular office work [1]. The hardware infrastructure includes laptops, printers, scanners, mobiles, etc. Depending on its business type, dataflow varies significantly among home, Fog and Cloud. For example, a lawyer may conduct a professional legal practice from home using technologies such as word processing and printers, e-mails and billing software. A number of home-based lawyers may partner to become a medium or large syndicate to enforce their competency. In this case, a large portion of dataflow exists between FEN and Cloud.

- Smart Protection

  Security and safety are always the most important factors no matter where you are and home is not an exclusion. Smart protection focuses on physical security in terms of hazard recognition, invasion detection, alarm, surveillance, and protection robots for homes. The elements such as sensors, actuators, cameras, and robots work jointly on a protection project which aims to secure personnel and property from damage or harm (such as espionage, theft, terrorist attacks, etc.). Massive data are processed at the proximity of properties. For example, an FEN stores videos for a certain period. The video can be removed or pushed to an FS if required. In case a hazard is recognized, the FEN with consultancy of FS can inform corresponding robotics to take actions. When external assistance is required, the FEN reports to Cloud with detailed information such as location and required services. Thus a majority of dataflow exist between FEN and FS.

- Smart Entertainment

  City dwellers get pleasure and delight or get rid of pressure and depression from entertainment activities such as music, games and films. Smart entertainment allows people to customize their amusement and relaxation at home with a family cinema on demand, gaming, karaoke and so on. As video streaming is at random and bursting, the dataflow heavily relies on the Cloud at the stage of initialization, which means the latency level may be minutes above in the beginning. FS can host a large amount of data prior to service to scale down the latency. In this subnetwork, the busted dataflow exist between FEN and Cloud.

- Smart Surroundings

  Surroundings are the area around a particular physical point of place, which

Table 12.1 Dataflow analysis and their requirement on Fog edge nodes

EHOPES Service	Dataflow Characteristics	Processing	Storage	Communication
Energy Network	FEN pushes data to Cloud. FS retrieves billing information from Cloud. FS makes decisions and FEN takes actions.	Medium	Small	Medium
Health Care Network	FEN filters repetitive data. Brief report is sent to FS, also to Cloud as backup. Live data stream occasionally exists between FEN and Cloud.	Large	Small	Medium
Office Network	Heavily relying on Cloud, dataflow varies from one business to another.	Large	Small	Large
Protection Network	Dataflow mainly exists between FEN, FS and robotics.	Medium	Medium	Medium
Entertainment Network	During the initialization stage, burst dataflow mainly exists between FEN and Cloud. Afterwards, FS can host a large amount of data (Cloud to FS).	Medium	Large	Large
Surroundings Network	Dataflow mainly exists within this network.	Large	Medium	Small

refers to biophysical, biological and physical factors that affect a city dweller's life. Smart surroundings involves making decisions and taking actions that are in the interests of protecting sustainable living conditions to support human life. Smart surroundings devices include heaters, coolers, air conditioners, lights, windows, doors, cleaners, hot water supplier, waste/recycle rubbish bin, etc. Those devices may work independently or collaboratively with other devices within this subnetwork. This subnetwork involves a number of sensors, actuators, controllers and robotics. A vast majority of the dataflow exist between the above smart objects and FEN.

12.2.2 Summary

Next, we summarize the dataflow characteristics of EHOPES in Table 12.1. We further divide the computing into processing, storage and communication. Accordingly, we investigate the EHOPES network data requirement on FEN and presented in this table.

Through the analysis above, we recognize that the majority of EHOPES data can be processed at the proximity to data source. In contrast to Cloud, Fog brings more benefits in terms of low latency, proximity, real-time response and multi-tenancy with diminished latency. Local processing on incoming dataflow from numerous smart objects not only scales down the latency, thus improving quality of experience, but also notably attenuates the traffic on the Internet. As a result, Fog has an outstanding impact on the entire Internet infrastructure.

12.3 FOG PLATFORM FOR EHOPES

In this section, after outlining the state of the art for Fog computing and its relations with smart objects and Cloud, we explore the required Fog elements in order to support EHOPES applications. The roles of FEN, FS and Foglet are explored.

12.3.1 State of the Art

The term "Fog computing" was initially proposed by the industry. Cisco, HP and IBM collectively contributed to its motivation, paradigm and high-level architecture [14]. Due to its proximity to smart objects, Fog is able to offer appealing features such as mobility support, location-awareness, minimum latency and multi-tenancy. It provides ubiquitous connectivity for heterogeneous smart objects and allows them to directly access, control and manage resources on Fog nodes. Those resources include CPU, memory, network, environment, energy, hypervisors, OSes, service containers, server instance, security, etc. [11]. Fog serves both wired and wireless devices as Cloud does; however, it is much closer to users. In general, Cloud is good at centrally batch processing while Fog is targeted to offer distributed local processing with minimized and predictable latency.

Instead of replacing Cloud, Fog is complementary to Cloud by providing real-time interaction between distributed smart objects and a centralized server farm. On the other hand, Cloud backs up Fog for its unlimited computing power and storage.

Furthermore, Fog is also excellent in resilience and robustness. Fog users (smart objects, apps, people, etc.) do not necessarily rely on the Internet accessibility any more. FS still works even disconnected from the Internet. Table 12.2 reviews user's perspective towards Cloud and Fog. Next, we investigate Fog elements in terms of hardware and middleware.

12.3.2 Fog Edge Node (FEN)

An FEN is adjacent to smart objects, aiming to provide Fog edge computing in terms of processing, storage and communication. As an endpoint of Fog, FEN provides a variety of wired and wireless access methods to empower immediate communication with smart objects. Repetitive data collected from smart objects are filtered. Decision making and action-taking emerge immediately to provide real-time interaction. It has sufficient processing, storage and communication power to run instances of Foglet (refer to Section 12.3.4), through which FEN collaborates with other Fog nodes.

FENs can be mobile phones, set-top boxes, access points, edge routers or switches

Table 12.2 User's perspective to Cloud and Fog

Evaluation Metrics	Cloud	Fog
Distance to the provider	Remote from	Adjacent to
Service Reachability	Relying on Internet access	Relying on local network infrastructure
Variety of Information	Unlimited	Limited from FEN
Latency	Minutes to yearly [6]	Milliseconds to second [6]
Cost	High	Low
Deployment Speed	Slow	Fast
Network Requirement of Device	High, i.e., the Internet access capability is mandatory.	Low, i.e., the Internet access capability is not necessarily required.

(even some smart sensors) located at one-hop proximity of FS. FENs have enough computing power to accommodate immediate operation for smart objects and instances of Foglet, thereby extending the large computing power further to the smart object level.

An FEN is capable of creating, receiving and transmitting information over a dedicated Fog communication channel. It has certain capability of self-configuration, routing, security and QoS. In brief, FEN focuses on local processing of incoming and outgoing IoT dataflows. Generally, FEN varies significantly in terms of its capability of processing, storage and communication. According to EHOPES data characteristics (refer to Table 12.1), different FEN can be deployed for each sub-network.

12.3.3 Fog Server (FS)

Different from FEN that focuses on the interplay among smart objects, another type of Fog node is FS which focuses on the interplay between FEN and Cloud data centers. An FS refers to both underlying hardware and running instances of required software capable of accepting requests and responding to FENs. It hosts predefined applications and stores a large amount of information to support local FENs. It associates with Cloud when required in order to take the advantage offered by Cloud. FS can work both independently and jointly without/with the support of Cloud.

As Fog varies in size, functionality and surroundings, one or more FS can be deployed in one Fog. Some FS provides large storage to host data and applications, some FS provides advanced routing and switching for FENs and others provide services such as configuration, QoS, security and more. Ideally, FS can be remotely accessed from external networks for management and other operations, which include but not limit to application-deployment, data offloading, network configuration, optimization and billing.

To offer one-hop proximity to FEN, the distributions of FS are well organized to provide a seamless coverage of FENs. An individual or combination of permanent, seasonal and temporary FSs may work respectively or collaboratively to support a variety of FENs in an established community. FSs are facilitated with high-speed uplink to the Internet, so once requested, it can quickly pull and push large amount of data to support local FENs. When necessary, FS can organize some of its redundant resources (processing, storage and more) and lease to an FEN for on-demand services or burst traffic (such as telediagnosis and entertainment program watching).

An FS (physically or virtually) can be advanced routers, switches, robotics and servers with large capabilities of processing, storage and communication. It supports state of the art of routing protocols such as segment routing [3], Layer-7 switching and security implementation such as IEEE 802.1x and IPv6 features such as Anycast.

It is worth noticing that as long as a device is able to run Foglet, it can serve as an FEN. For instance, a smart phone can work as an FEN once Foglet is in place in entertainment. This phone may still have resources available for other Fog or non-Fog applications. The device could be run as an FEN, but such operation is not preemptive. Sometimes an FS and an FEN may be interchangeable. One device may be an FEN and an FS simultaneously.

12.3.4 Foglet (Middleware)

Foglet is a reasonably small agent software that can be easily and smoothly employed by Fog nodes. Foglet helps a smart object to enjoy dynamic, dependable and scalable Fog services. These Fog services include network management and hosted applications. Foglet is capable of bearing the orchestration functionality and performance requirements. It can be running on any Fog nodes when required or on-request. It can be used to monitor the health (physical machine and service deployed on it) and control resources (VMs, service instance, etc.), and negotiate to establish, maintain and tear down sessions between Fog nodes and Fog abstraction APIs.

As a middle-ware, Foglet must offer a cross-platform capability and allow smart objects to take advantage of fruitful Fog services without knowing any infrastructure of Fog. An FEN can utilize Foglet to detect Fog resources (such as CPU, memory, bandwidth, real-time throughput, etc.) and proactively select the best path to deliver data units. Fog nodes also use Foglets to collaborate interactively to liaise and organize related resources to offer customizable service based on SLA.

In summary, smart objects are linked to an FEN to form a sub-network. An FEN runs Foglet to collaborate with other FEN and FS. Some FENs may need to talk to Cloud occasionally while the others do not. Figure 12.4 illustrates the interplay between smart objects, FEN, FS and Cloud.

12.4 CASE STUDY AND EVALUATION

Some EHOPES applications are run on Cloud and Fog in order to compare their latency performance for the same amount of smart living data.

12.4.1 The Scenario

We first present data volume in the given scenario followed by the network topology when either Cloud or Fog is employed.

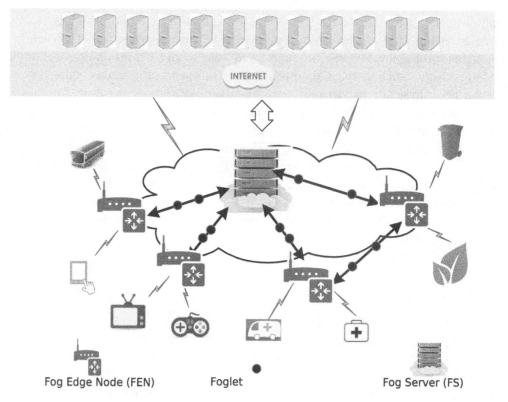

Figure 12.4 Interplay between IoT, Fog and Cloud

At an impairment BAN user's home, Tom is enjoying his smart living services. His BAN sensors generate an average of 8900 bps amount of data by monitoring his vital signs. We assume that his health care apps and medical robots generate a similar amount of data, respectively. The total throughput is about 3375 bytes per second. The security camera resolution is CIF (704x480) level, which generates about 34,290 bps per camera. There are six cameras implemented, altogether generating 25,938 bytes per second. He usually watches TV for four hours each day at home for entertainment. The throughput is about 500 kilobytes per second. Meanwhile, he works as an editor for eight hours each day, five days per week. The average throughput is then about 125 kilobytes per second. The above scenario involves typical EHOPES applications such as smart health care, protection, entertainment and office, whose throughputs are summarized as follows:

Case 1: Cloud computing model

In this case, the data are required to be stored in a centralized Cloud as shown in Figure 12.5. According to Akamai 2014 rankings, the average download data rate is 6.9 Mbps in Australia. Hence, we assume Tom has this speed with the latency between the Cognitive Gateway and Cloud about 250 ms.

Table 12.3 Data volume in the scenario

Application	Throughput (byte/second)
Smart Health care	3375
Smart Office	125000
Smart Protection	25713
Smart Entertainment	500000
Total	654088

Case 2: Fog computing model

In this case, the data are only required to be stored in the local Fog, as shown in Figure 12.6. He has a 1 Gbps link between his FEN and FS. All his FENs share this bandwidth.

12.4.2 The Simulation

The simulation is carried out in the OPNET Modeler 14.5. Cloud and Fog scenario has been set up respectively. For the Cloud model, we use an IP-32 Cloud to simulate the Internet (refer to Figure 12.3), a PPP client to simulate the Cognitive Gateway and a PPP server to simulate the Cloud server that hosts those services required in Section 12.4.1. DS-3 PPP links are used to facilitate the connections. The DS-3 PPP link between Cognitive Gateway and Cloud is fine-tuned to 6.9 Mbps. For the Fog model, we use four Ethernet nodes to act as FENs, and an Ethernet server to act as an FS that hosts required services.

High load e-mail is used to simulate smart health care traffic. To match the traffic volume, we run 42 trials of this application in this simulation. Heavy database query (five trials) is used to simulate smart protection traffic. Image browsing (62 trials) is used to simulate smart entertainment traffic. Heavy load file transferring (297 trials) is used for smart office traffic. The above setting generates the required traffic volume as listed in Table 12.4.

Figure 12.5 Cloud model diagram

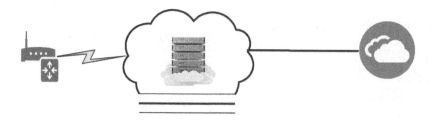

Figure 12.6 Fog model diagram

Table 12.4 Data volume in the simulation

Application	Simulating Application	Average Throughput (byte/second)
Smart Health care	High load e-mail	2800
Smart Office	File transfer	85000
Smart Protection	Database query	28000
Smart Entertainment	Image browsing	530000
Total		645800

12.4.3 Simulation Results

The latency (response time) has been collected for each application. The following are those collected values on a weekly basis.

Figure 12.7 shows instantaneous and average response time for the smart protection network. The blue curve shows this application occurring at that moment in

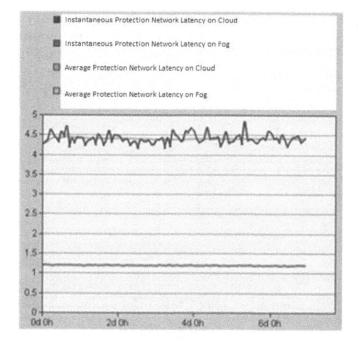

Figure 12.7 Latency curve comparison

Table 12.5 EHOPES latency value on Cloud and Fog

Application	Average Response from Cloud (seconds)	Average Response from Fog (seconds)
Smart Health	2.8	0.8
Smart Protection	4.4	1.2
Smart Entertainment	1.9	0.6
Smart Office	2.8	0.7

Cloud while the red one shows as in Fog. The green curve shows the average response time in Cloud. The cyan curve shows average value in Fog. From the figure, the average delay is 4.4 seconds in Cloud. In sharp contrast, the average delay is about 1.2 seconds on the Fog platform. This result shows that the latency drops 73 percent on average when Fog is employed. Regarding two instantaneous response times, the blue one is a jiggling curve, which implies the latency is unstable in Cloud. While Fog is employed, the latency is relatively stable as shown in the red flat line. Thus the latency on Fog is easier to be predicted.

We can see a significant latency dropping from Cloud to Fog for the same amount of data. The Table 12.5 outlines all the results from the data we have collected.

12.5 CONCLUSION

This chapter investigates Fog computing as a platform for a smart living concept, namely, EHOPES. Because of Fog's proximity to the users, it improves the efficiency and quality of user experience in supporting smart living. As Fog architecture has not been clearly defined, we suggest the required Fog elements such as FEN, FS and Foglet from the IoT user's perspective. Various aspects of FEN and FS in terms of processing, storage and communication are considered for EHOPES. Two use cases are proposed to show the effectiveness of reducing the latency for the same amount of data on Fog compared to Cloud. Although this chapter focuses on Fog platforms for smart living, the framework is ready to be generally applied to other IoT applications wherever Fog is employed. As Fog is merely in its infancy stage, a lot of work is still required to be done, e.g., workload mobility between Cloud and Fog, Fog routing and switching, Fog deployment, Fog security and QoS, interplay between smart object, Fog node and Cloud as well as data storage (pull and push).

Bibliography

[1] O. Akribopoulos, D. Amaxilatis, V. Georgitzikis, M. Logaras, V. Keramidas, K. Kontodimas, E. Lagoudianakis, N. Nikoloutsakos, V. Papoutsakis, I. Prevezanos, et al. Making p-space smart: Integrating iot technologies in a multi-office environment. In *Mobile Wireless Middleware, Operating Systems, and Applications*, pages 31–44. Springer, 2012.

[2] M. Armbrust, A. Fox, R. Griffith, A. D. Joseph, R. Katz, A. Konwinski, G. Lee, D. Patterson, A. Rabkin, I. Stoica, et al. A view of cloud computing. *Communications of the ACM*, 53(4):50–58, 2010.

[3] S. Bidkar, A. Gumaste, and A. Somani. A scalable framework for segment routing in service provider networks: The omnipresent ethernet approach. In *High Performance Switching and Routing (HPSR), 2014 IEEE 15th International Conference on*, pages 76–83. IEEE, 2014.

[4] F. Bonomi, R. Milito, P. Natarajan, and J. Zhu. Fog computing: A platform for internet of things and analytics. In *Big Data and Internet of Things: A Roadmap for Smart Environments*, pages 169–186. Springer, 2014.

[5] F. Bonomi, R. Milito, J. Zhu, and S. Addepalli. Fog computing and its role in the internet of things. In *Proceedings of the first edition of the MCC workshop on mobile cloud computing*, pages 13–16. ACM, 2012.

[6] D. Evans. The internet of things: How the next evolution of the internet is changing everything. *CISCO white paper*, 1:1–11, 2011.

[7] R. Giffinger. European smart cities: The need for a place related understanding, 2011.

[8] G. Kortuem, F. Kawsar, D. Fitton, and V. Sundramoorthy. Smart objects as building blocks for the internet of things. *Internet Computing, IEEE*, 14(1):44–51, 2010.

[9] J. Li, J. Jin, D. Yuan, M. Palaniswami, and K. Moessner. Ehopes: Data-centered fog platform for smart living. In *Telecommunication Networks and Applications Conference (ITNAC), 2015 International*, pages 308–313. IEEE, 2015.

[10] M. Liserre, T. Sauter, and J. Y. Hung. Future energy systems: Integrating renewable energy sources into the smart power grid through industrial electronics. *IEEE Industrial Electronics Magazine*, 1(4):18–37, 2010.

[11] B. AA Nunes, M. Mendonca, X.-N. Nguyen, K. Obraczka, and T. Turletti. A survey of software-defined networking: Past, present, and future of programmable networks. *Communications Surveys & Tutorials, IEEE*, 16(3):1617–1634, 2014.

[12] C. Reinisch, M. J. Kofler, F. Iglesias, and W. Kastner. Thinkhome energy efficiency in future smart homes. *EURASIP Journal on Embedded Systems*, 2011(1):1–18, 2011.

[13] Y. Strengers. *Smart energy technologies in everyday life: Smart Utopia?* Palgrave Macmillan, 2013.

[14] L. M. Vaquero and L. Rodero-Merino. Finding your way in the fog: Towards a comprehensive definition of fog computing. *ACM SIGCOMM Computer Communication Review*, 44(5):27–32, 2014.

[15] P. Wang, JY Huang, Yi Ding, P. C. Loh, and L. Goel. Demand side load management of smart grids using intelligent trading/metering/billing system. In *PowerTech, 2011 IEEE Trondheim*, pages 1–6. IEEE, 2011.

[16] Y. Zhang, R. Yu, S. Xie, W. Yao, Y. Xiao, and M. Guizani. Home m2m networks: Architectures, standards, and qos improvement. *Communications Magazine, IEEE*, 49(4):44–52, 2011.

CHAPTER 13

Resources and Practical Factors in Smart Home and City

Bo Tan

University College London/Coventry University

Lili Tao

University of Bristol

Ni Zhu

University of Bristol

CONTENTS

In PREVIOUS chapters, authors described the technologies that enable the transition from Internet of Things to Smart City. The topics covered communications, networks, data mining, security, and human-machine interface. In this chapter, we focus on practical problems in the smart city: innovative usage of existing resources for challenges in the smart city and practical consideration in deployment.

13.1 INTRODUCTION

As can be interpreted, the smart city is a complex concept which integrates diverse technologies to resolve the new problems in cities. On the one hand, new technologies enable new applications such as intelligent transportation and smart building management. On the other hand, the new applications raise challenges for the current technology. For example, the increasing deployment of autonomous systems such as UAV, self-driving car, and robots need high accurate real-time location information service and context awareness and anomaly detection for decision making. Also, we need better human physical activity information for the purpose of improving security, understanding the human behavior models in the city and health condition monitoring. In this chapter, we discuss how to use the existing resources in the city to resolve these new demands in smart city applications. At the end of this chapter, we will discuss the factors—cost, development and deployment issues needed to implement the smart city in practice.

13.2 NOVEL USAGE OF RADIO RESOURCES

13.2.1 Current Situation and Challenges

In the modern city, licensed and unlicensed radio signals are widely used indoor and outdoor for multiple purposes: from wireless communications, broadcasting to localization, navigation and remote sensing. Here, we generally summarize existing radios' resources and current applications and then describe how the novel signal processing technologies leverage these resources to resolve the challenges in the smart city.

Communication: The most common application of radio signals in urban settings is communications, including the mobile communications over licensed carrier bands and wireless access over ISM bands. Nowadays, these wireless communication radio signals cover urban outdoor areas seamlessly and most indoor areas for bidirectional information exchange. Hence, this pervasive coverage provides a good chance for detecting, localizing and keeping track of the targets of interest within the covered area.

Broadcasting: Broadcasting including AM/FM radio, the analog television signal may be the earliest widely service that carried on radio signals. Recently, DAB and DVBT signals have gradually replaced the transitional analog signal. These broadcasting signals are often emitted in the licensed bands having the comparable high power to cover wider and high altitude areas. These signals will be ideal probes in surveilling low-altitude aircraft and UAVs which may experience an explosion of growth in the near future.

Localization/Navigation: GPS, one of the GNSS systems, is the most used navigation purpose radio signal. The current GNSS system, which also includes GLONASS, Galileo and Beidou, provides convenient navigation service in urban settings. However, coverage of the high density and indoor areas still are an open challenge for the GNSS system.

As can be easily seen, the radio signal provides diverse applications in a modern city. However, new challenges keep on emerging and requiring novel solutions in the coming smart city epoch. Here, we discuss two topics that are involved in many smart city applications that potentially can be resolved with novel RF technologies: accurate short range localization and sensing.

Short Range Localization: The GNSS system almost perfectly resolved the localization and navigation problem in outdoor open air areas, but for the areas with building density and indoor areas, the GNSS signal is not easily accessible. In the meantime, these areas have high demands on smart city services such as traffic information gathering, shopping navigation and security surveillance. In this situation, the short range RF signals, for example, the mobile signal from the GSM/LTE base station and WiFi AP signal emission, are often used in locating the device or assist the GNSS to improve the accuracy.

Passive Sensing: Besides position, motion information is also in demand in the scenarios such as intelligent transpiration, health care and smart building/environment. The real-time detection of target motion information will provide information as to instant traffic/crowd flow and density, activity model of chronic health condition or behavior model in a certain environment. The ubiquitous radio signal existing in modern urban space provides media to interpret target motion.

In the rest of this section, novel applications of existing indoor/outdoor radio signals are introduced to inspire further research on the smart city.

13.2.2 Use of Outdoor Radio Signals

Localization: The mobile signal, either GSM, CDMA or LTE, has been considered as a complement to the GNSS signal in an urban area to provide rough location information for subscribers by using a cell association relation [41]. For the cell association based method, the location service can be provided conveniently without additional signal processing or change in hardware or software. However, the location accuracy of this method depends on several factors: density of base stations, the integrity of the database and users' preference for uploading location information. In addition, there is difficulty in delivering an analytic performance for the cell association based method.

Thus, recently the signal based localization attracts more attention in both research and industry. The type of signal based localization in outdoor scenarios is Positioning Reference Signal (PRS) in the LTE system [12]. PRSs are a downlink signal sent by NBs to allow the UEs to measure time-of-flight (ToF) of signals from different cells. Then, the localization algorithms like Observed Time Difference Of Arrival (OTDOA) are used as shown in Figure 13.1.

Different from data transmission, UEs need to hear PRSs from multiple cells in order to perform OTDOA. The signal sequence, and transmitting scheme on time, frequency and space resources are clearly defined in the LTE Release 9. In addition, a muting mechanism between the LTE cells is introduced in the LTE for PRSs to ensure the UE can hear PRSs from "weak" cells and in the meantime avoid interference from "strong" cells [12]. To ensure a proper performance of OTDOA, not only in

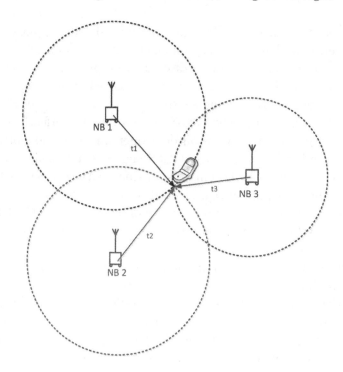

Figure 13.1 Multilateration in OTDOA UE localization. The NBs need to be synchronized in order to use the time delay difference.

LTE OTDOA, another essential requirement is network synchronization, which means PRS occasions in all cells should be aligned in time; usually the cell time alignment is achieved by GPS clock synchronization. The resources for PRS are defined and allocated in the standard. There is still space for the researchers and manufacturers to design the specific waveform for PRS. A typical example can be found like power allocation in [10].

With the timing estimation from PRS, UE can easily obtain its location in the LTE networks. Compared with the GPS signal source, the LTE base stations are easier to access, especially NBs often deployed in very high density in the urban area. Thus, the LTE NBs will be a good complement for the GPS positioning.

Sensing/Tracking: Besides the localization, existing outdoor radio signals can also be used for capturing the instant motion status of targets of interest by using passive detection. As described above, the signal based localization method needs the target to be connected with the network. The target of interest may not always be equipped with the necessary connectivity modules. Thus, passive detection technology is proposed in the research community to resolve the problem, as objects will always reflect the radio signal when the radio signal presets. The passive detection interprets the motion status, such as moving direction, moving speed, even movement trace, by collecting the reflected signals from the objects. In other words, the non-cooperative radio signal source and receivers form a passive bistatic radar. In urban areas, there are adequate radio signal sources to perform passive detection. By extracting the time and frequency differences of the reflected signal from the object and reference signal from the signal source (usually called cross-ambiguity analysis [37]), the receiver can interpret the distance (from the time delay) and speed (from Doppler shifts). In [21], vehicles are captured from space by using the GSM base station signal emission. In actuality, the passive system not only accurately captured speed and moving direction of vehicles, but also discriminated the different types of vehicles from heavy truck to motorbike. In [7], the train is detected when using the GSMR signal along the railway track. Furthermore, to better monitor the increasing amount of UAV in the city, NASA and Verizon have launched a work to use the LTE NB to detect and track the UAV [15]. In future smart transportation, passive sensing data will provide comprehensive data that show the instant motion status.

13.2.3 Use of Indoor Signals

Similar to the outdoor applications, accurate position and instant motion are also important for indoor applications like health care, indoor navigation and robotics. The indoor environment is more severe for radio signal propagation than the outdoor scenario. Thus, actively or passively localizing and tracking the indoor targets and capturing the instant motion status become emerging topics in both research and industrial communities.

Localization: The indoor area is an even more challenging area to receive the GNSS signal due to the lack of the light-of-sight (LoS) propagation path to satellites when compared to the high building density outdoor area. Thus, the indoor radio's signal becomes a good candidate to locating targets in either an active or passive mode. The RF based indoor localization can generally be categorized into two groups: statistics method and signal based method.

The statistics based method, which is also called fingerprint method, uses the statistics relation between receiving signal strength (RSS) and location. Assume there are multiple radio signal sources in the area of interest. Each signal source will result in different RSS in a specific location, e.g., signal source S_i will generate R_i^A on location A. R_i^A is often presented as random Gaussian distribution (μ_i^A, δ_i^A). μ_i^A is determined by statics environment facts such as the distance between signal source and receiver, as well as the obstacles during the signal propagation like walls in the building. δ_i^A is often impacted by the dynamic situation of the environment, for example, moving objects may generate a shadow effect to the receiver. RSS measurements of one specific location from different signal sources can be considered as a unique fingerprint of this location. A typical fingerprint indoor localization system is shown in [18]. Then, the fingerprint information can be stored in the database as reference to locate the receiver who is taking the new RSS measurements. The statistics indoor localization methods just require the based signal strength measurement function, which is the basic function of most COTS radio receivers. However, this method has some obvious drawbacks of practical deployment. First, it needs pre-measurement of RSS information on each potential location in the area of interest. Second, it requires a certain amount of signal sources in order to achieve accuracy level. Last, the method is vulnerable to changes in the environment, for example, change of the layout of the objects in the concerned space and change of the signal sources.

To overcome the drawbacks of the fingerprint methods, people started to investigate the signal based method to locate the indoor targets. Different from PRS in LTE, indoor localization often uses the ISM band signals for timing or direction estimation that is used for localization. In [11], the authors use the phase lock loop (PLL) in a radio receiver to estimate the fractional time delay for better precise distance estimation between transmitter and receiver. The estimated distances of different transmitters are used for locating the receiver by using the TDOA or OTDOA in [12]. The fractional delay method in [11] claims nanosecond timing estimation by using 802.11b signal, however, it requires network synchronization to guarantee the localization accuracy, which means all transmitters in need to be synchronized. Besides the timing measurement, the angle of arrival (AoA) is another parameter often used for indoor localization. [46] is a typical AoA based localization system using the MUSIC algorithm [34]. The AoA based method exempts the requirement of high level network synchronization, but it requires the multiple antennas to form up a phased array to enable the angle estimation function. There is also the COTS system [32], using AoA for high accurate indoor localization. Recently, more and more researchers have started to use the Intel 5300 [16] network interface card (NIC) to perform the AoA estimation as the NIC outputs the channel status information (CSI) that can be used for performing AoA estimation. As a subspace method, the MUSIC algorithm can also be applied for ToF estimation. The concept has been proved in [47] for XX level range estimation. Furthermore, the work in [19] performs joint angle and time delay estimation by using CSI output from Intel 5300 NIC for XX level indoor localization accuracy, which meets the requirements of many different types of smart home applications.

Sensing: Similar to the outdoor scenarios, instant motion also has wide application potential for the indoor scenarios. In future smart home or health care applications, physical activity information of residents will be valuable data for analyzing chronic health conditions and activity levels; a behavior model can even predict dangerous activities like falling. Currently people have tried a different type of sensor to collect this activity information. For example, accelerometers are widely used on the smart band or smart watch to monitor the user's daily activity. The optical systems like MS Kinect [45] are used for capturing human body gestures for both gaming and health care purposes. However, these body gestures cannot provide seamless activity monitoring due to issues of privacy or use. Thus, indoor radio signal starts to be another emerging candidate to provide pervasive monitoring of residents' activity. The researchers in [39], [1] and [31] have proved that with the common WiFi signal, the design sensing system can capture the human movement with LoS and through-wall condition, which is suitable for whole home surveillance. In [38], the authors proved they can use detailed Doppler information to recognize human gestures by using the COTS WiFi AP emission. Furthermore, the work in [22] and [6] captured the accurate human respiration rate by using indoor energy harvesting signals and WiFi AP emission, respectively. In particular the respiration rate was captured with COTS WiFi AP signal in a through-wall situation in a layout. The novel passive/active sensing technology turns the short range indoor radio signal into media to probe a

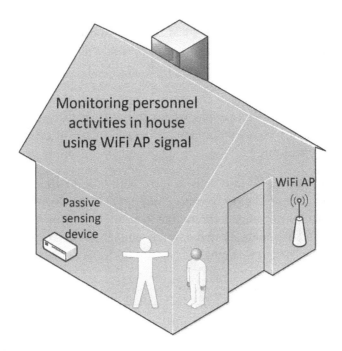

Figure 13.2 Deployment of a passive sensing device in a house to monitor human daily activities for security and health care purposes.

large range of residents' activities that are valuable for the health care and smart home applications as shown in Figure 13.2.

13.2.4 The Trend

From the above sections, we can see diverse novel applications of radio signals in urban areas beyond communications. The applications mainly focus on accurate localization and instant motion sensing for potential outdoor and indoor applications. In addition to the applications mentioned above, here we point out some future research direction on innovative applications of the urban area radio signals:

- **Ultra short range:** A recent experiment [3] has shown that tiny human limb movements, even hand waving and finger tapping, can generate signal distortion that can be captured by radio receivers in a very short distance. This short range information has the potential to be used as the new human machine interaction media for contactless control or gaming.

- **Target recognition:** In experiments, researchers have found that different motion will result in different sensing results when using Doppler as the metric. These differences can be used as a signature to identify different motion status and meanwhile can be potentially used to identify individuals as initially shown in [13]. The individually identified data will be more valuable from the application side.

- **Robotics:** Using the accurate location information for navigating the autonomous system, especially in building robots will also be an emerging area to explore applications of the indoor radio signals.

- **Security:** As the radio signal can work in any light condition with panoramic coverage and through-wall propagation capability, it will also be a good media for security and surveillance applications in a hostage situation in urban areas, especially in a building.

In summary, the urban area radio signals have wide applications in the context of the smart city and a smart home beyond communications. Novel signal processing technologies need to be developed to combat the multipath propagation and interference elimination in the complex urban environment for the purpose of leveraging the benefit of these existing resources.

13.3 VIDEO RESOURCES

13.3.1 Introduction

With the enrichment of the physical space and infrastructures of cities, along with the increased capability of smart devices and agents across different scales, the idea of a smart city has moved towards reality. Among a wide variety of sensing technologies, visual sensors have the potential to address several limitations and may play an important role in some contexts. Specifically, they don't require the user to wear them, and they are able to detect multiple events simultaneously. In fact, due to the low-complexity approaches of the state-of-the-art video analytic solutions, they can be implemented on low-cost devices, and they are available on the market at a low price. Such solutions allow the computer vision approaches to deploy for real-time applications.

Intelligent visual monitoring has received a great deal of attention in the past decade, and has already contributed in many ways to the development of smart cities in both indoor and outdoor environments. The video component of the smart cities is tasked with developing real-time multicamera systems for human activity and behavior analysis, crowded scene understanding, environmental monitoring, traffic control, etc. Apparently, a single camera is unable to achieve these goals. Hence, multi-camera architectures are widely exploited. Two main architecture schemes have been presented so far; they are centralized and distributed networks [50]. Various types of camera sensors are considered to be used in different scenarios. Nowadays, standard video cameras are broadly used for surveillance purposes due to their low-cost and ability to integrate with other sensing modalities [2], but RGB data are highly sensitive to viewpoint variations, human appearance and lighting conditions, and thus stand very little chance of capturing useful information in many scenarios, whereas recent depth sensors have helped to overcome some of these limitations. Some commercial devices such as the Microsoft Kinect and Asus Xotion are available, for which the depth is computed from structured light. However, to track and detect pedestrians in an outdoor environment at night, thermal cameras can be used [44],

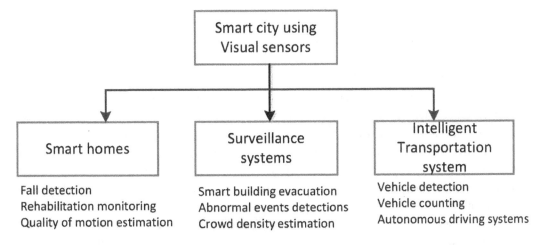

Figure 13.3 Overview of visual sensor-based applications featured in this part

as thermal cameras work on the basis of temperature instead of on reflected light. Thus thermal based vision techniques are less susceptible to lighting.

In the following, we describe applications and the latest advances in intelligent visual monitoring for smart cities and provide our insights and perspectives on future directions. An overview of the applications included in this part is presented in Figure 13.3.

13.3.2 Applications and Current Systems

Smart Homes: Developing a reliable home monitoring system has drawn much attention in recent years due to the growing demands for integrated health care. Existing approaches to current home monitoring systems often include custom-fit environmental, physiological and vision sensors, such as in [50]. Such systems can enable several types of applications, to increase personal safety for elderly patients and to facilitate clinicians to diagnose and monitor patients. This new patient-clinician interactive mode improves the reliability and effectiveness of diagnosis and to some extent significantly shortens the travel time and hospital stay for patients and reduces the workload for clinicians [5].

Fall detection is one of the major challenges in health care for the elderly, with video based technologies offering many advantages over popular wearable alarms because they don't require user action and they are always active. RGB-depth (RGB-D) devices have successfully outperformed other sensing technologies for fall detection [25]. In addition, musculoskeletal problems and recovery progress can be reflected by assessing the quality of motion, which is increasingly in demand by clinicians in health care and rehabilitation monitoring of patients. Researchers have recently proposed a general method for online estimation of the quality of movement on stairs, walking on a flat surface and transitions between sitting and standing [40]. Visual sensors also can be used together with other sensors to complement each other in smart home systems, such as the work in [24] that presented a system using a combination of non-invasive sensors to assess and report sleep patterns: a contact-based pressure mattress and a non-contact 3D image acquisition device.

Surveillance systems: With the rapid growing of the Internet and storage capacity, video monitoring for security purposes has become a popular application. As network video technology has improved, the cost of installing a surveillance system has dropped exponentially, leading to a significant increase in the use of security cameras. An infrastructure for smart surveillance systems already exists with the availability of conventional CCTV systems for surveillance. The computer vision field is nowadays mature enough to demonstrate its possibilities beyond the convention which run by human operators who are prone to distraction.

For indoor systems, such as in smart buildings, cameras can provide a more efficient solution for building evacuation than the conventional evacuation systems which assume that people are aware of the routes to leave the building, while smart evacuation systems are able to direct people through the quickest routes to the exits based on the real-time congestion information obtained and analyzed from computer vision techniques [14].

In the instance of gatherings in outdoor environments, unusual or abnormal events detection is one of the most important issues to be addressed in the development of the intelligent surveillance systems. To encode such high level understanding, one aims at using crowd motion characteristics including the motion directions and the particle energy. The former one is mostly applied to a scene with a small number of people, in which the motion direction can be measured by tracking individuals. Realizing the movements of pedestrians are not completely random, as they tend to move with the intention of reaching their destination and to avoid other pedestrians and obstacles, it is possible to detect abnormal or suspicious individual behaviour when the observed information deviates from the expected model, or even predict an object's motion [27]. An extensive survey in [30] reviews methods for visually recognizing human interactions and predicting rendezvous regions in both observable and unobservable areas. The particle energy can be determined by estimating crowd density and computing optical flow derived from two consecutive frames [26].

Intelligent Transportation Systems (ITS): With the aim of creating an effective solution in next-generation pervasive ITS, interest towards a new field of research and application has been triggered, including vehicle detection and counting, intelligent traffic management and automatic vehicle driving. Several requirements must be addressed such as computational performance, cost, size, power consumption, etc., to increase road traffic safety.

Vehicle detection and counting is one of the most fundamental tasks. A real-time vehicle counting system was recently presented in [33], describing a prototype based online sensor camera as a low-complexity and low-cost solution. A reconfigurable embedded vision the system was described in [42] and was implemented by a system on chip composed of a programmable logic that supports parallel processing and a microprocessor suited for serial decision making, which meets the requirement of fast pixel-level image analysis. More works attempt to improve the performance from the algorithm side, based on either detection [9] or clustering [49] or regression [23].

In autonomous driving systems, three of the most crucial elements are sensing, mapping and planning [36], of which they must be simultaneously handled as they influence each other. Apart from visual sensors, radars and laser scanners are also very popular choices. Compared with the above technologies, the camera has much higher resolution than the others. The state-of-the-art camera has 1.3–1.7 MegaPixels running at 36 frames per second (fps), which is equivalent to 41–67 M per second while in the meantime, LiDAR technology can only run about 60–300 K per second. Most advanced future radar (e.g., MIMO) would go up to 300 K points per second. High resolution enables the detailed description of complex scenes. Furthermore, the camera is the only sensor that provides not only shapes but appearance. In fact, most complex situations are defined based on appearance more than on shapes. For example, road marking, traffic signs and traffic lights are based on textures. *Mapping* is a more challenging but not well-defined problem. A very detailed map is needed for gathering all the information on the road. The task includes detecting lane boundaries of all drivable paths even when the lanes are not existent, attaching semantic meaning to each path and detecting the key action points (merge, split, etc.). *Planning* deals with deciding on what immediate actions to take so as to optimize a long term objective. It is essential to learn other vehicles' behaviors and to negotiate in a multi-agent environment. A most recent work [35] gives an example of when a car tries to merge in a roundabout, of handling the policy learning problem with the notions of multi-agents games.

13.3.3 Future Trends

While visual-based applications for smart cities have been an active research area for quite some time the video-based technology provides an optimal solution for a wide field of smart city applications in IoT, open challenges still remain. Most current systems are proposed based on individual projects, while there is still much potential in spreading the technology to a large number of cameras which are already being used for current video surveillance but are monitored by humans. In the short term, these cameras could be used for the purpose of developing automatic video surveillance systems.

Although target-oriented deployment is preferable at the moment, the observations/decisions made by one system could potentially benefit the others. In the long term, the ultimate goal will be to detect complex events based on longer temporal and large spatial scales. Addressing the limitations of current systems demands more robust and intelligent solutions. New areas still need to be considered to keep

13.4 PRACTICAL CONSIDERATIONS

13.4.1 Pervasive Sensing

Despite the fact that there are many designs and research on frameworks or architectures of overall smart city platforms [17, 43], there is not (or not yet) a fine and solid definition for *Smart Cities*, mainly due to many reasons, such as:

- employing heterogeneous IoT technologies;

- driven by diverse service requirements;

- embodied in the diversiform and combined applications;

- under existing city policies and agreements compulsorily

Nevertheless, in such a comprehensive topic, there is indeed one thing which can manifest: one of the cornerstones of complex smart city platform is the capability of pervasive sensing âĂŤ collecting information from the target objects anywhere, anytime in the urban areas. It requests heterogeneous network connectivity where previous chapters have had enough discussion, as well as rich sensing source [29]. A landscape of potentially applied sensing source across applications is shown in Table 13.4.1.

Application Fields	Sensing Source
Structural Condition	Crack detection, crack propagation, accelerometer, water detection, ice detection, ultrasound
Environment	Noise level detection, temperature, humidity, luminosity, solar radiation, ultraviolet radiation, PM10/2.5/1, air pollution detection (optional from CO, CO_2, NO_2, O_2, O_3, NH_3 and other gases)
Public traffic	Magnetic field, GPS, distance detection by infrared
Energy consumption	Electricity metering, water metering, gas metering
Parking	Magnetic field, infrared, RFID
Atmosphere and weather	Temperature, humidity, luminosity, atmospheric pressure, anemometer, pluviometer, wind vane
Surveillance and security	Camera

There are certainly more sensing sources existing in addition to this tables and more dedicated ones for specific applications. It has to be remembered that it is artificial by any means to classify those sensing sources associated to the applications. Eventually, the boundary between them become blurred in practical cases. Although there is rich sensing source available, practical considerations on design and deployment should be raised for allowing pervasive sensing implementation. Gathering so-called "big data" for smart cities will never be easy in reality in terms of several aspects not limited to:

1. **Cost vs. Sensing accuracy**
   Because of a large number of sensors and sensor nodes (including access points) covering the urban region, the affordable solution is desired where rich and a large number of data are collected for further data manipulations and services. There is a massive body of off-the-shelf sensor selections available in the market, which have different costs based on their sensing accuracy, quality, properties, etc. Moreover, more data points and larger deployment in the areas increase cost.

2. **Energy source vs. power consumption**
   After sensors, sensor nodes, and sensor networks with the Internet connection have been deployed, the high expectation is upon long-term (even as longer as possible) operation of smart city systems. In this scenario, sensor nodes which are mounted by sensors, microcontrollers with their analogue/digital peripherals; communication modules request extra low power consumption. Different energy sources ranging from green energy by energy harvesting technologies to batteries can be applied, adoption of which depends on the requirement of entire hardware modules and the service. A. Zanella, et el. [48] give a fine example of the Padova Smart City project and summarize available energy sources in the individual applications.

3. **Deployment challenge**
   Hardware modules including sensors are unavoidably deployed in the outdoor (even harsh) environments. Protection solutions normally encapsulate all components in IP-standard (i.e., IP67) enclosures against corrosion to the circuitry. However, many of the sensors are required to be exposed to the air, open to contact with measurement objects in order to gain the best sensing performance. Huge engineering efforts have to be invested in practical deployment.

13.4.2 Smart Cities in Reality

Comprehensive surveys were conducted from different perspectives to dissect smart cities [48]. An ideal smart city would be a generic platform, given a more conceptual characterization—a dynamic platform of hybrid systems. It should be of some essential functionalists within a single system, for examples, compatibility, reconfigurability, scalability and interoperability. Many endeavors have been invested worldwide in order to achieve a smart city paradigm [28], but design, deployment, operation and maintenance of such smart city platforms in practice are encountering enormous challenges. As also discussed in Section 13.4.1, the characteristics of smart cities determine that they should fulfill several essential requirements. Therein, the first consideration is the usage of heterogeneous IoT technologies to integrate hybrid devices or/and subsystems provided by different vendors, especially in the situations when they provide diverse services for the diversiform and even for combined applications. Easy integration and cross-platform compatibility in a Plug-and-Play manner are highly expected for effortless deployment in practice. However, complexity of integration and compatibility across the entire smart city system falls into many aspects, which are mainly involved in:

- adaptation of different hardware modules and their I/O ports for various sensors;

- composability of heterogeneous wired/wireless communication protocols;

- acceptancy of different data/message forwarding schemes via the backbone network;

- recognition of different data format;

- operation across assorted data storage scenarios;

- cross-platform service in the application layer.

A fine example addressing one of the aspects can be IFTTT, abbreviated as "If This Then That," which is a web-based application allowing information exchange amongst different web service vendors. Recently cross-platform middleware solutions for multivendor standardizations and consortium are merging. The concept of open APIs (application programming interfaces) is being accepted gradually. Coronado and Iglesias in their paper [8] provided a survey comparison of current task automation services platforms. Although pervasive sensing data can be acquired from sensors and be transmitted all the way through the networks to the data center, there are other must have required implements in order that the data can be further manipulated. For example, associated with the sensing data, descriptive information is needed to label them, such as timestamps while collecting the raw data, message ID, identification and location of the data point, network IP, and metadata. Configuring each data point in such a large deployment is an enormous workload and cost. Additionally, smart cities are dynamic systems, which means systems should allow leave-end devices connecting and disconnecting to the system at all times. In that case, the data points may even need to be reconfigured after deployment. Thus, in regards to the first consideration discussed above, reconfigurability, or more precisely, reconfigurability should be considered in smart city platform design. Many research efforts have been invested to address this, on topics such as reconfigurable networking, software-defined networks, network visualization, etc. [20, 51].

There are many other practical considerations to smart cities' design and deployment, for instance, system, scalability allowing additional devices, and systems and services integrated into the smart city infrastructures dynamically. Maintenance of large scale networks including wireless sensor networks is a challenge in reality and usually costly. Additionally, when the router node fails and drops off from the network, the rest of the network should self-heal or self-organize to avoid the failure of the entire network [4]. Data ownership, security and privacy bonding with their services and user experience also face further issues as discussed in previous topics.

GLOSSARY

360 Degree Review: Performance review that includes feedback from superiors, peers, subordinates and clients.

Bibliography

[1] F. Adib and D. Katabi. See through walls with wifi! *SIGCOMM Comput. Commun. Rev.*, 43(4):75–86, August 2013.

[2] J. K. Aggarwal and M. S. Ryoo. Human activity analysis: A review. *ACM Computing Surveys (CSUR)*, 43(3):16, 2011.

[3] K. Ali, A. X. Liu, W. Wang, and M. Shahzad. Keystroke recognition using wifi signals. In *Proceedings of the 21st Annual International Conference on Mobile Computing and Networking*, MobiCom '15, pages 90–102, New York, NY, USA, 2015. ACM.

[4] A. P. Athreya and P. Tague. Network self-organization in the internet of things. In *2013 IEEE International Conference on Sensing, Communications and Networking (SECON)*, pages 25–33, June 2013.

[5] B. R. Bloem, Y. A. M. Grimbergen, M. Cramer, M. Willemsen, and A. H. Zwinderman. Prospective assessment of falls in Parkinson's disease. *Journal of Neurology*, 248(11):950–958, 2001.

[6] Q. Chen, K. Chetty, K. Woodbridge, and B. Tan. Signs of life detection using wireless passive radar. In *2016 IEEE Radar Conference (RadarConf)*, pages 1–5, May 2016.

[7] K. Chetty, Q. Chen, and K. Woodbridge. Train monitoring using GSM-r based passive radar. In *2016 IEEE Radar Conference (RadarConf)*, pages 1–4, May 2016.

[8] M. Coronado and C. A. Iglesias. Task automation services: Automation for the masses. *IEEE Internet Computing*, 20(1):52–58, Jan 2016.

[9] P. Dollar, C. Wojek, B. Schiele, and P. Perona. Pedestrian detection: An evaluation of the state of the art. *IEEE Transactions on Pattern Analysis and Machine Intelligence*, 34(4):743–761, 2012.

[10] M. Driusso, M. Comisso, F. Babich, and C. Marshall. Performance analysis of time of arrival estimation on OFDM signals. *IEEE Signal Processing Letters*, 22(7):983–987, July 2015.

[11] R. Exel. Receiver design for time-based ranging with IEEE 802.11b signals. In *International Journal of Navigation and Observation*, page vol. 2012, March 2012.

[12] S. Fischer. Observed time difference of arrival (OTDOA) positioning in 3GPP LTE.

[13] O. R. Fogle and B. D. Rigling. Micro-range/micro-Doppler decomposition of human radar signatures. *IEEE Transactions on Aerospace and Electronic Systems*, 48(4):3058–3072, October 2012.

[14] G. P. Hancke, G. P. H. Jr., et al. The role of advanced sensing in smart cities. *Sensors*, 13(1):393–425, 2012.

[15] M. Harris. NASA and Verizon plan to monitor US drone network from phone towers, 2015.

[16] Intel. Intel ultimate n wifi link 5300.

[17] J. Jin, J. Gubbi, S. Marusic, and M. Palaniswami. An information framework for creating a smart city through internet of things. *IEEE Internet of Things Journal*, 1(2):112–121, April 2014.

[18] K. Kaemarungsi and P. Krishnamurthy. Modeling of indoor positioning systems based on location fingerprinting. In *INFOCOM 2004. Twenty-third Annual Joint Conference of the IEEE Computer and Communications Societies*, volume 2, pages 1012–1022, March 2004.

[19] M. Kotaru, K. Joshi, D. Bharadia, and S. Katti. Spotfi: Decimeter level localization using wifi. *SIGCOMM Comput. Commun. Rev.*, 45(4):269–282, August 2015.

[20] D. Kreutz, F. M. V. Ramos, P. E. VerÃŋssimo, C. E. Rothenberg, S. Azodolmolky, and S. Uhlig. Software-defined networking: A comprehensive survey. *Proceedings of the IEEE*, 103(1):14–76, Jan 2015.

[21] P. Krysik, P. Samczynski, M. Malanowski, L. Maslikowski, and K. S. Kulpa. Velocity measurement and traffic monitoring using a GSM passive radar demonstrator. *IEEE Aerospace and Electronic Systems Magazine*, 27(10):43–51, Oct 2012.

[22] W. Li, B. Tan, and R. J. Piechocki. Non-contact breathing detection using passive radar. In *2016 IEEE International Conference on Communications (ICC)*, pages 1–6, May 2016.

[23] X. Liu, Z. Wang, J. Feng, and H. Xi. Highway vehicle counting in compressed domain. In *Proceedings of the IEEE Conference on Computer Vision and Pattern Recognition*, pages 3016–3024, 2016.

[24] V. Metsis, D. Kosmopoulos, V. Athitsos, and F. Makedon. Non-invasive analysis of sleep patterns via multimodal sensor input. *Personal and Ubiquitous Computing*, 18(1):19–26, 2014.

[25] M. Mubashir, L. Shao, and L. Seed. A survey on fall detection: Principles and approaches. *Neurocomputing*, 100:144–152, 2013.

[26] Y. Nam and S. Hong. Real-time abnormal situation detection based on particle advection in crowded scenes. *Journal of Real-Time Image Processing*, 10(4):771–784, 2015.

[27] S. Pellegrini, A. Ess, and L. V. Gool. Predicting pedestrian trajectories. In *Visual Analysis of Humans*, pages 473–491. Springer, 2011.

[28] S. Pellicer, G. Santa, A. L. Bleda, R. Maestre, A. J. Jara, and A. G. Skarmeta. A global perspective of smart cities: A survey. In *Innovative Mobile and Internet Services in Ubiquitous Computing (IMIS), 2013 Seventh International Conference on*, pages 439–444, July 2013.

[29] C. Perera, A. Zaslavsky, P. Christen, and D. Georgakopoulos. Sensing as a service model for smart cities supported by internet of things. *Trans. Emerg. Telecommun. Technol.*, 25(1):81–93, January 2014.

[30] F. Poiesi and A. Cavallaro. Predicting and recognizing human interactions in public spaces. *Journal of Real-Time Image Processing*, 10(4):785–803, 2015.

[31] Q. Pu, S. Gupta, S. Gollakota, and S. Patel. Whole-home gesture recognition using wireless signals. In *Proceedings of the 19th Annual International Conference on Mobile Computing & Networking*, MobiCom '13, pages 27–38, New York, NY, USA, 2013. ACM.

[32] Quuppa. Unique technology, 2015.

[33] C. Salvadori, M. Petracca, S. Bocchino, R. Pelliccia, and P. Pagano. A low-cost vehicle counter for next-generation ITs. *Journal of Real-Time Image Processing*, 10(4):741–757, 2015.

[34] R. Schmidt. Multiple emitter location and signal parameter estimation. *IEEE Transactions on Antennas and Propagation*, 34(3):276–280, Mar 1986.

[35] S. Shalev-Shwartz, N. Ben-Zrihem, A. Cohen, and A. Shashua. Long-term planning by short-term prediction. *arXiv preprint arXiv:1602.01580*, 2016.

[36] A. Shashua. The three pillars of autonomous driving. *20th International Congress on Advances in Automotive Electronics*, 2016.

[37] S. Stein. Algorithms for ambiguity function processing. *IEEE Transactions on Acoustics, Speech, and Signal Processing*, 29(3):588–599, Jun 1981.

[38] B. Tan, A. Burrows, R. Piechocki, I. Craddock, Q. Chen, K. Woodbridge, and K. Chetty. Wi-fi based passive human motion sensing for in-home healthcare applications. In *Internet of Things (WF-IoT), 2015 IEEE 2nd World Forum on*, pages 609–614, Dec 2015.

[39] B. Tan, K. Woodbridge, and K. Chetty. A real-time high resolution passive wifi Doppler-radar and its applications. In *2014 International Radar Conference*, pages 1–6, Oct 2014.

[40] L. Tao, A. Paiement, D. Damen, M. Mirmehdi, S. Hannuna, M. Camplani, T. Burghardt, and I. Craddock. A comparative study of pose representation and dynamics modelling for online motion quality assessment. *Computer Vision and Image Understanding*, 148:136–152, 2016.

[41] u-blox. CellLocate - enhance GNSS positioning indoors.

[42] G. Velez, A. Cortés, M. Nieto, I. Vélez, and O. Otaegui. A reconfigurable embedded vision system for advanced driver assistance. *Journal of Real-Time Image Processing*, 10(4):725–739, 2015.

[43] P. Vlacheas, R. Giaffreda, V. Stavroulaki, D. Kelaidonis, V. Foteinos, G. Poulios, P. Demestichas, A. Somov, A. R. Biswas, and K. Moessner. Enabling smart cities through a cognitive management framework for the internet of things. *IEEE Communications Magazine*, 51(6):102–111, June 2013.

[44] Jiang-tao Wang, De-bao Chen, Hai-yan Chen, and Jing-yu Yang. On pedestrian detection and tracking in infrared videos. *Pattern Recognition Letters*, 33(6):775–785, 2012.

[45] Wikipedia. Kinect, 2016.

[46] J. Xiong and K. Jamieson. Arraytrack: A fine-grained indoor location system. In *Proceedings of the 10th USENIX Conference on Networked Systems Design and Implementation*, NSDi'13, pages 71–84, Berkeley, CA, USA, 2013. USENIX Association.

[47] J. Xiong, K. Sundaresan, and K. Jamieson. Tonetrack: Leveraging frequency-agile radios for time-based indoor wireless localization. In *Proceedings of the 21st Annual International Conference on Mobile Computing and Networking*, MobiCom '15, pages 537–549, New York, NY, USA, 2015. ACM.

[48] A. Zanella, N. Bui, A. Castellani, L. Vangelista, and M. Zorzi. Internet of things for smart cities. *IEEE Internet of Things Journal*, 1(1):22–32, Feb 2014.

[49] R. Zhao and X. Wang. Counting vehicles from semantic regions. *IEEE Transactions on Intelligent Transportation Systems*, 14(2):1016–1022, 2013.

[50] N. Zhu, T. Diethe, M. Camplani, L. Tao, A. Burrows, N. Twomey, D. Kaleshi, M. Mirmehdi, P. Flach, and I. Craddock. Bridging ehealth and the internet of things: The SPHERE project. *IEEE Intelligent Systems*, 2015.

[51] N. Zilberman, P. M. Watts, C. Rotsos, and A. W. Moore. Reconfigurable network systems and software-defined networking. *Proceedings of the IEEE*, 103(7):1102–1124, July 2015.

Index